Pay
21/08

Building and Surveying Series

Acc
Adv
App
Buil
Buil
Buil
Buil
Buil
Buil
Buil
Civi   rray
Con
    a
Con
Con
Con
Con
    a
Eco
Env
Fac
Gre
    St
Intr
    E.
Intr
JCT
    R
Mea
Prin
    V

(co

List continued from previous page

Building and Surveying Series
Series Standing Order
ISBN 0–333–71692–2 hardcover
ISBN 0–333–69333–7 paperback
(outside North America only)

You can receive future titles in this series as they are published by placing a
standing order. Please contact your bookseller or, in the case of difficulty, write
to us at the address below with your name and address, the title of the series
and the ISBN quoted above.

Customer Services Department, Macmillan Distribution Ltd
Houndmills, Basingstoke, Hampshire RG21 6XS, England

BUILDING MAINTENANCE

IVOR H. SEELEY
B.Sc. (Est. Man.), M.A., Ph.D., F.R.I.C.S., C.Eng.,
F.I.C.E., F.C.I.O.B., M.I.H.

Emeritus Professor
The Nottingham Trent University

Second Edition

palgrave

First edition 1976
Reprinted six times
Second edition 1987

Published by
PALGRAVE
Houndmills, Basingstoke, Hampshire RG21 6XS and
175 Fifth Avenue, New York, N.Y. 10010
Companies and representatives throughout the world

PALGRAVE is the new global academic imprint of
St. Martin's Press LLC Scholarly and Reference Division and
Palgrave Publishers Ltd (formerly Macmillan Press Ltd).

ISBN 0–333–45700–5 hardcover
ISBN 0–333–45701–3 paperback

This book is printed on paper suitable for recycling and
made from fully managed and sustained forest sources.

A catalogue record for this book is available
from the British Library.

Transferred to digital printing 2003

Printed and bound in Great Britain by
Antony Rowe Ltd, Chippenham and Eastbourne

This book is dedicated to my elder daughter
LINDA
for her kind and sympathetic help and encouragement
with my book-writing activities over many years

"Whosoever cometh to me, and heareth my sayings, and doeth them, I will shew to you whom he is like.

He is like a man which built an house, and digged deep, and laid the foundation on a rock; and when the flood arose, the stream beat vehemently upon that house, and could not shake it: for it was founded upon a rock.

But he that heareth, and doeth not, is like a man that without a foundation built an house upon the earth; against which the stream did beat vehemently, and immediately it fell; and the ruin of that house was great."

Luke 6. 47–49

CONTENTS

maintenance feedback; maintenance manuals; costs in use/life cycle costing; effect of taxation and insurance.

Choice between direct and contract labour; building maintenance departmental structures and arrangements; maintenance depots; programming of maintenance work; organisation of maintenance work; programming and progressing maintenance work; maintenance management of condominiums; training for maintenance; maintenance incentive schemes.

Clerk of Works; site meetings; setting out; supervision of building work; records.

LIST OF FIGURES

LIST OF TABLES

LIST OF PLATES

PREFACE

Building Maintenance has too often been regarded as the 'Cinderella' of the building industry. Yet in the mid 1980s Britain was spending about £10 billion per annum on the maintenance of buildings and over 50 per cent of the building labour force was engaged on this class of work.

The maintenance of the built environment affects everyone continually, for it is on the state of our homes, offices and factories that we depend not only for our comfort, but for our economic survival. The building stock in the United Kingdom had a replacement value of about £250 billion in 1987 and this alone indicates the importance of effective upkeep. There is still a pressing need for the improvement of large numbers of older but substantial dwellings which lack some of the basic amenities.

Maintenance starts the day the builder leaves the site. Design, materials, workmanship, function, use and their interrelationships, will determine the amount of maintenance required during the lifetime of the building. Furthermore, the client's economic interests may work against the elimination of high maintenance costs in the building design. Case studies undertaken by the DOE showed that about one-third of the maintenance work on the buildings investigated could have been avoided if sufficient care had been taken at the design stage and during construction. The design faults resulted either from failure to appreciate how various constructional details would perform in use, or because certain parts of the building that failed through normal wear and tear could not be replaced without extensive repairs to adjacent parts. A spokesman for the Building Research Establishment has also commented on the frequent failure by designers to make use of authoritative design guides such as British Standards and Codes of Practice, and of the tendency to adopt a careless attitude to detail design.

The building fabric has to satisfy different user needs and occupational factors. The designer should identify what performance is required from the fabric in terms of weathertightness, noise reduction, durability, resistance to heat loss and other relevant criteria, in addition to comfort and visual requirements. Many of the design faults which result in high maintenance expenditure could conceivably be avoided if a maintenance manager, or someone with similar technical knowledge, joined the design team. Not many architects or builders revisit their jobs after the expiry of the defects liability period and few have a continuing responsibility for maintenance.

There is rarely an obvious end-product in building maintenance, and the effect of neglected industrial buildings, for instance, will seldom be as serious for the owner as the disruption caused by a breakdown in production following neglect of plant maintenance. For this reason building maintenance is often considered as one of the first items for budget cuts when retrenchment becomes necessary. Indeed some building owners regard maintenance costs as part of the debit side of the balance sheet and an erosion of legitimate profits, because they fail to appreciate its true value. It is, however, only common sense to ensure that when a decision is made on maintenance work, account is taken of the aggravation of the defect which is likely to occur if work is delayed, with consequent increase in cost.

The satisfactory maintenance of a dwelling makes it fit to live in, but the occupant may have a very narrow view of what this entails. If the household equipment works, the internal decorations are cheerful and the external appearance respectable, he will probably be satisfied. The maintenance manager must take a deeper view—loose roof tiles, unsound timber, defective damp-proof courses or powdering mortar are symptoms of a state of disrepair which cannot be cured by superficial measures. Unless basic repairs are carried out in time, the property will become damp or its structure will deteriorate so that normal jobbing repairs and repainting are no longer sufficient to restore even an appearance of well-being.

Building maintenance is assuming increasing importance which is shown by the holding of several large national conferences on this subject, the mounting of diploma and postgraduate courses in maintenance management, and the considerable volume of maintenance research by government departments, universities and polytechnics, apart from the valuable work undertaken by the leading professional bodies connected with the building industry. Nevertheless, a DOE Committee on Building Maintenance drew attention to the need to give increased emphasis in appropriate degree and professional examination curricula to building maintenance. The Committee believed that the main subject areas that need to be covered are the relationship between the design and performance of buildings and services; the organisation and control of maintenance work; economics and finance; law and liability; and technology. These subject areas form the core of this book which it is hoped will be of value to students and practising surveyors, builders, architects, estate, housing and maintenance managers, and environmental control officers alike.

Effective building maintenance requires the correct diagnosis of defects, and implementation of the correct remedial measures, all based on sound technical knowledge, otherwise there can be additional waste of materials, labour and money since the work will in all probability have to be done again. We need more uniformity in the method of recording maintenance data and greater feedback of information on the performance of materials and running costs of buildings in particular. The increased use of maintenance manuals and more regular maintenance inspections and schemes of planned maintenance will assist in producing more efficient maintenance. Effective maintenance control requires the formulation of sound plans, recording of performance, comparison of performance with the plan and the taking of corrective action where appropriate. Finally there is the need to make better use of available resources.

The second edition has been updated and extended to cover the latest techniques, procedures and research, both in the UK and overseas. In addition, 25 plates have been incorporated to illustrate a representative selection of major defects found in buildings.

Nottingham, Autumn 1987   IVOR H. SEELEY

ACKNOWLEDGEMENTS

The author acknowledges with gratitude the willing co-operation and assistance received from many organisations and individuals, so many that it is not possible to mention them all individually.

Crown copyright material is reproduced from BRE Digests and other publications by permission of the Director of the Building Research Establishment; in this connection it should be mentioned that copies of the digests quoted are obtainable from the Building Research Establishment, Bucknalls Lane, Garston, Watford, Herts WD2 7JR, and Building Centres.

Building Maintenance Information Ltd, (formerly BMCIS) kindly gave permission for the inclusion of the occupancy cost analyses (appendixes 2 and 3) and the energy cost analysis (appendix 4). Tables 13.1, 13.2 and 13.6 are based on forms prepared by Bath City Council, and tables 13.3 and 13.7 on forms prepared by Nottingham Community Housing Association Ltd.

Drawings of improvement schemes are based on the following published articles

J. H. Cheetham, *Building Trades Journal*, 9, 23 and 30 September 1966
J. A. Foreman, *Building Trades Journal*, 3 May 1968
The Louis de Soissons Partnership, *Architects' Journal*, 30 January 1974

The author is indebted to Cluttons, Chartered Surveyors, 5 Great College Street, London SW1P 3SD, and to Eric Stevens FRICS in particular, for supplying plates 1 to 6 inclusive, 8 to 19 inclusive and 22; and to the Paintmakers Association of Great Britain Ltd, Alembic House, 93 Albert Embankment, London SE1 7TY, for kind permission to use plates 7, 20, 21, 23, 24 and 25.

Grateful thanks are due to the publisher for consent to quote from *Building Technology* and *Building Surveys, Reports and Dilapidations*, to the Ellis School of Building, Worcester and NALGO Education Department to use some of the concepts contained in course material prepared by the author in years past, and to Nottingham City Council and Nottingham Community Housing Association Ltd for assistance with case studies.

Much friendly and helpful advice was received from Peter Murby, Editorial Director of Macmillan Education Ltd, during the preparation and production of the second edition, for which I am most grateful. I am also indebted to my wife for her ever-continuing support.

1 NATURE AND IMPORTANCE OF BUILDING MAINTENANCE

Building maintenance has until recently been a neglected field of technology, being regarded by many as a 'Cinderella' activity. It possesses little glamour, is unlikely to attract very much attention and is frequently regarded as unproductive, although many of the managerial and technical problems are more demanding of ingenuity and skill than those of new works. A Government Committee on Building Maintenance[1] described how this class of work is accorded little or no merit and that while it remains a neglected backwater, the morale of those involved in its management and execution must suffer and productivity will remain low.

Property owners all too frequently endeavour to keep maintenance expenditure to a minimum, ignoring or misunderstanding the adverse long-term effects of such a policy. Neglect of maintenance has accumulative results with rapidly increasing deterioration of the fabric and finishes of a building accompanied by harmful effects on the contents and occupants. Buildings are too valuable assets to be neglected in this way. In excess of one-third of the total output of the construction industry is devoted to this activity, inadequate though it is to keep the nation's buildings in a satisfactory condition.

Concept of Building Maintenance

It is highly desirable but hardly feasible to produce buildings that are maintenance-free, although much can be done at the design stage to reduce the amount of subsequent maintenance work. All elements of buildings deteriorate at a greater or lesser rate dependent on materials and methods of construction, environmental conditions and the use of the building.[2]

Definition of Maintenance

BS 3811[3] define 'maintenance' as: "The combination of all technical and associated administrative actions intended to retain an item in, or restore it to, a state in which it can perform its required function." The requirements for maintenance must not be less than those necessary to meet the relevant statutory requirements, and 'maintained' is defined in the Factories

1

Act 1961 as: "maintained in an efficient state, in efficient working order and in good repair."

The Committee on Building Maintenance[2] defined 'acceptable standard', as quoted in the first edition of BS 3811, as "one which sustains the utility and value of the facility" and this is found to include some degree of improvement over the life of a building as acceptable comfort and amenity standards rise. Cleaning will also constitute part of building maintenance activities. BS 3811 subdivides maintenance into 'planned' and 'unplanned' maintenance, as illustrated in figure 1.1.

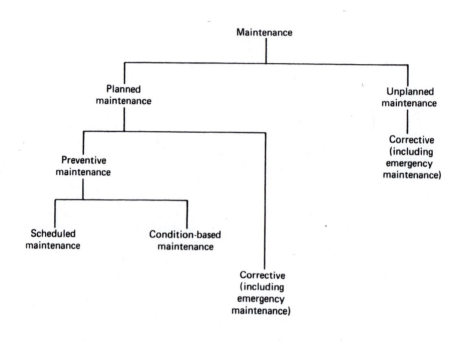

Figure 1.1 Types of maintenance (source: *BS 3811: 1984*[3])

BS 3811[3] categorises building maintenance by means of the following terms and definitions.

(1) *Planned maintenance*: "The maintenance organised and carried out with forethought, control and the use of records to a predetermined plan."

(2) *Unplanned maintenance*: "The maintenance carried out to no predetermined plan."

(3) *Preventive maintenance*: "The maintenance carried out at predetermined intervals or corresponding to prescribed criteria and intended to reduce the probability of failure or the performance degradation of an item."

(4) *Corrective maintenance*: "The maintenance carried out after a failure has occurred and intended to restore an item to a state in which it can perform its required function."

(5) *Emergency maintenance*: "The maintenance which it is necessary to put in hand immediately to avoid serious consequences." This is sometimes referred to as day-to-day maintenance, resulting from such incidents as gas leaks and gale damage.

(6) *Condition-based maintenance*: "The preventive maintenance initiated as a result of knowledge of the condition of an item from routine or continuous monitoring."

(7) *Scheduled maintenance*: "The preventive maintenance carried out to a predetermined interval of time, number of operations, mileage, etc."

Another approach to maintenance classification has been adopted by Speight[4], who subdivided maintenance into three broad categories:

(1) Major repair or restoration: such as re-roofing or rebuilding defective walls and often incorporating an element of improvement.

(2) Periodic maintenance: a typical example being annual contracts for decorations and the like.

(3) Routine or day-to-day maintenance: which is largely of the preventive type, such as checking rainwater gutters and servicing mechanical and electrical installations.

Maintenance work has also been categorised as 'predictable' and 'avoidable'.[4] Predictable maintenance is regular periodic work that may be necessary to retain the performance characteristic of a product, as well as that required to replace or repair the product after it has achieved a useful life span. Avoidable maintenance is the work required to rectify failures caused by poor design, incorrect installation or the use of faulty materials.

With building services, minimal neglect can result in potential danger. 'Appropriate condition' could be interpreted as the maintenance of buildings in a state which allows them to be used for the purpose for which they were provided for the minimum capital expenditure. The main problem is to determine the standards to be applied in a particular situation, and these are more readily assessed for services and finishings than for the fabric. The appropriate condition will be influenced by many factors, including the function of the building, its public image, or even national prestige.[4]

Building maintenance is characterised not only by the diversity of activities but also of the interests involved. A prime aim should be to obtain good value for the money spent on maintenance but there are conflicting views on this— public/private; long/short term; and landlord/tenant.

A system which is based on planned inspections and maintenance will have higher overhead costs than one that is not, but the planning should lead to lower maintenance expenditure. A fully planned system is not always the most appropriate and care is needed in devising the best system for the particular estate. Figure 1.2 shows the cost relationship of planned and unplanned systems.

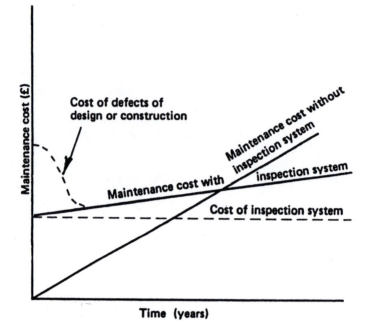

Figure 1.2 Cost relationship between planned and unplanned systems (source: *Managing Building Maintenance*[5])

Significance of Building Maintenance

Repair and maintenance work, including house improvements, increased its proportion of the total construction workload from about 25 per cent in the 1950s to over 50 per cent in the mid 1980s, and totalled about £10 billion per annum. Hence in 1987 over one-half of the total construction work force produced about one-third of the output in terms of the value of work done.

The labour-intensive nature of maintenance work is to some extent inevitable; many maintenance tasks can only be performed manually and demand is seldom other than dispersed and unco-ordinated. Firms undertaking maintenance are rarely large and often have only two or three employees. They are generally under-capitalised and are less subject to market forces than those on new works. These smaller firms have neither the means nor the incentive to invest in new and more efficient techniques and materials. For instance, many small contractors have shown considerable reluctance to invest in labour-saving smallpower tools.

The Committee on Building Maintenance[2], as long ago as 1972, asserted that building maintenance was of great significance to the economy not only because of the scale of expenditure involved, but also because it was important to ensure that the nation's stock of buildings, both as a factor of production and of accommodation, was used as effectively as possible. The committee saw no early prospect of restoring, let alone keeping, a majority of buildings to an acceptable standard. It emphasised that more rather than less maintenance work was necessary if the value and amenity of the nation's building stock was to be kept at current levels.

It was calculated in 1972 that arrears of housing maintenance in Great Britain could amount to eight or nine times the volume of work actually carried out each year, and there was ample evidence of neglect in other classes of building.[2] Since then there has been a serious deterioration in the standard of maintenance, particularly in the public sector, mainly because of cuts in central government funding. If further serious deterioration is permitted, future generations could be faced with a major capital burden.

The standard of maintenance achieved has an important influence on the quality of the built environment and there seems little doubt that society will continue to expect higher standards in new and existing buildings. Consequently, for many years to come, maintenance will remain a significant and important part of the work of the construction industry.

Magnitude of Building Maintenance Problems

Housing

The 1981 English Housing Condition Survey found that 1.1 million dwellings failed to meet the minimum standard of fitness for human habitation, and the equivalent Welsh survey identified a further 87 000 unfit dwellings. Another 1.3 million in England and a quarter of a million in Wales were found to be unsatisfactory in some other way; they either lacked one or more basic amenities or needed extensive repairs. The number of dwellings needing extensive repairs had increased by more than 17 per cent since the previous survey in 1971.

A RICS report in 1986[6] emphasised that most of the dwellings classed as unfit were built before 1919 and were predominantly privately owned. However, substantial numbers of local authority dwellings built since World War II, mainly high-rise system-built blocks of flats, are either structurally unsound or uneconomic to put into good order, although they are not classed as unfit.

In 1985, the government's Urban Housing Renewal Unit[7] showed that £18.8 billion needed to be spent on repairing council houses in England alone, amounting to about £5000 per house. It can reasonably be estimated that the total backlog in housing repairs and maintenance in both the public and private sectors in the United Kingdom in 1986 was in the order of £45 billion.

A RIBA publication in 1985[8] gave some typical examples of serious local authority housing maintenance problems. For instance, the London Borough of Newham needed to spend £5 million for each of five years in carrying out extensive repairs to tower blocks because of the deteriorating concrete fabric and fractured brickwork. In Leicester, chemical changes in concrete housing caused corrosion of reinforcing bars and the necessary replacement cost was assessed at £35 million and in the meantime the buildings could collapse. According to a study by the Office of Population

Censuses and Surveys in 1982, in Liverpool over half the population was living in adverse conditions. In Middlesbrough, dampness penetration and condensation was widespread, particularly in flat-roofed blocks.

Leeds City Council[9] had a housing stock of 94 000 dwellings in 1984 with a high proportion of system-built houses containing features which made them actually or potentially expensive to maintain. The essential housing repair and improvement programme was estimated at £460 million. Condensation problems were arising with greater frequency and it was estimated that as many as 20 000 properties could be affected. At no time since local government reorganisation in 1974 had the authority faced such a daunting array of problems with its housing stock, nor had the need for capital investment in repairs and renewals ever been so high. The City of Glasgow City Council estimated that it needed £77 million to keep its existing housing stock in a wind and watertight condition.

A RIBA survey[10] identified a considerable repair and maintenance backlog on local authority housing in 1985. Furthermore, unless additional resources are found for this work the backlog will continue to increase. Divergences between local authorities' housing investment programme bids and actual allocations have meant that authorities have had to undertake a higher proportion of responsive and emergency repair and maintenance work than is desirable. This represents an inefficient allocation of resources and in the long term adds to the burden of repair that will fall on ratepayers and taxpayers. In the meantime, much stock remains in a poor state of repair to the detriment of many thousands of householders' quality of life. An Audit Commission report in 1985[11] concluded that current levels of local authority capital spending were below the level necessary to maintain, let alone improve, the state of existing local authority housing stock, schools and roads.

Hospitals

The Davies Report of 1983 commissioned by the Department of Health and Social Security estimated that in England alone there was a backlog of maintenance work of £2 billion. A report by the National Economic Development Office in 1985[12] found examples of the short-term patching of defective flat roofs and of hardware not being replaced. Painting, a matter of hygiene, was often regarded as a luxury. There was a generally expressed view among health authority officials that to cope with the restricted maintenance expenditure allocations, measures were being taken which were neither cost-effective and/or were building up further severe difficulties in the years ahead.

The maintenance of new hospitals was also being neglected. Because of excessively tight cost limits and the experimental methods by which many were constructed, hospitals built in the 1960s and 1970s can cost up to three or four times as much to maintain as older hospitals.[8]

Schools

In 1986 many of Britain's school buildings were in an appalling state of repair and through continuing neglect they were getting worse. This decline should be checked, not only to protect a national asset from premature disintegration but also for the sake of the pupils. There is plenty of evidence to suggest that shabby buildings reduce morale, lead to lower quality of work and encourage antisocial behaviour. Many of the problems are rooted in the educational building boom of the 1960s when decades of common sense in materials and detailing were discarded in favour of non-durable and inadequately researched materials, poor and sometimes 'unbuildable' detailing, and lax supervision of construction.[13]

Moreover, the mistakes of the 1960s have been compounded by two decades of neglect. Cuts in spending, or at the very least a failure to increase spending to match the expanding scale of the problem, have led to prolonged delays in essential maintenance with disastrous results. As in all types of maintenance, delay costs money. The failure to patch a leaking asphalt roof immediately may mean a new roof is required a year later; erecting shoring to support a weakened wall can cost many times as much as the original repair.

Her Majesty's Inspectors of Schools summed up the situation effectively in a report published in 1985 as shown in the following extracts. "Much of the nation's school building stock is now below an acceptable standard. In some schools the conditions in which teaching and learning take place adversely affect the quality of the pupils' work. In many more the decorative state of the accommodation does little to create the decent and civilised environment usually associated with education. Without urgent attention to these problems the cost of putting things right may become prohibitive." In 1986 they reported a worsening of the already poor position.

In 1984 the Audit Commission stated that "Authorities have responded to financial pressures by reducing expenditure on long term maintenance to the point where the state of many school buildings gives legitimate cause for concern."

A RIBA report[8] described how in the mid 1980s, despite falling school rolls, an enormous backlog of repair and refurbishment had to be overcome if existing schools were to be made suitable for new teaching methods and community uses. Though still serviceable, much school accommodation is unsuited to modern use. In 1986, 1.5 million primary school pupils were being taught in pre-war buildings. A million pupils lacked access to basic facilities such as sinks and electric power points. Over a quarter of a million pupils were taught in temporary accommodation.

Surveys by the Society of Chief Architects in Local Authorities (SCALA) showed that maintenance of school buildings had fallen well short of the level necessary to stabilise deterioration, let alone carry out desirable improvements. Resources were having to be absorbed to deal with such health and safety hazards as asbestos and the likelihood of increases in major failures of plant and services, causing emergency closures with all the resultant social disturbances.

Much needs to be done in reducing energy costs through energy conservation, particularly in older buildings designed when energy was relatively cheap. The Audit Commission estimated that in 1984 heating and lighting a typical secondary school cost around £30 000 per year; or perhaps as much as £150 million for all secondary schools in England and Wales. Considerable savings (10 per cent or more) could be made through the introduction of energy-saving measures such as thermal insulation.

Urban Regeneration

The country's worst building stock is in the inner cities where unemployment, deprivation and its associated problems are concentrated. A RIBA report[8] describes how government cuts in construction bear with disproportionate harshness on inner city communities, who are in the greatest need of environmental improvement, and also of the employment that more construction work would bring.

HRH The Prince of Wales stated in 1985: "the desperate plight of the inner city areas is well known, with the cycle of economic decline leading to physical deterioration and countless social problems. It is only when you visit these areas . . . that you begin to wonder how it is possible that people are able to live in such inhuman conditions" — a sobering thought with its harsh realities.

Liability for Defects in Buildings

Liability for defects in buildings arises in various ways. It may emanate from the initial building contract between the contractor and the building owner, the scheme administered by the National House-Building Council (NHBC scheme), or statutory requirements. Each is now considered.

1   Liability under Building Contracts

A contractor carrying out building work, whether it be new work or repairs or replacements, is normally under contract to undertake the work in a good and workmanlike manner using suitable materials. Common law rights permit the building owner to claim against the contractor up to 6 years from completion of the work or 12 years in the case of a contract executed under seal. The Standard Forms of Building Contract[14] restrict these common law rights and a defects liability period of six months is common. The majority of building defects are unlikely to become apparent in so short a period. The extension of the liability period would result in higher tenders and the building owner would suffer were the contractor to become insolvent. Persons designing building work also have a duty of care to their clients in carrying out their professional tasks, as they may be held liable for damages where any negligent act or omission on their part prejudices the client's interests.

2 NHBC Scheme

A substantial improvement in the standard of private housing has been achieved by voluntary collective action with government support through the National House-Building Council. This scheme applies to the majority of houses built for private sale or letting, and requires participating house-builders to build to certain minimum standards of design and workmanship subject to inspection by NHBC inspectors, and to undertake to make good any defect during the first two years after the agreement with the purchaser. This latter undertaking is guaranteed by the Council itself should a builder fail to honour his obligations. The Council also insures against any major structural defects up to the end of the tenth year.

The NHBC Scheme is to be welcomed as, unless designers and contractors are held responsible for the performance of their buildings for a sufficient period, there is no effective sanction to influence them to take account of the maintenance implications of their designs or to avoid the errors of judgement and quality control, both in design and in construction, which can lead to unnecessarily high maintenance costs early in a building's life.

3 Statutory Obligations

One of the most important legislative measures concerning building maintenance is the Defective Premises Act 1973 which came into force on 1 January 1974. This Act placed additional responsibilities on contractors who build, improve or repair dwellings, and provided an extension in law of the practical steps formulated by the NHBC. It imposes a statutory obligation on all who are involved in the provision of building work—contractors, sub-contractors, suppliers of materials and the design team. They must all do their work properly and effectively and ensure that the dwelling will be fit for human habitation. Furthermore, a subsequent purchaser of the building who was not a party to any contract with the original contractor or sub-contractors is able to sue them.

The Act provides that any person taking on work for or in connection with the provision of a dwelling (including repairs, maintenance and improvements) owes a duty to see that the work he undertakes is done in (a) workmanlike, or as may be applicable, in a professional manner, with (b) proper materials, and (c) so that, as regards their responsibilities in the work, the dwelling will be fit for habitation when completed. Furthermore, the provisions of the Act extend beyond the parties to the building agreement to embrace any person who acquires an interest in the dwelling such as a subsequent purchaser.

Professional men are to do their work in a professional manner, that is, with all due care and skill. Should there by a defect in their instructions, the Act gives the owner or purchaser a right of action in negligence for breach of statutory duty. Sub-contractors are under the same duties as the main contractor in so far as they take on work or provide materials or services for a dwelling. Suppliers of purpose-built components incorporated in a dwelling also owe a statutory duty to the owner or purchaser, but suppliers of materials and mass produced components are not included in the general liability.

Periods within which a breach of the new statutory duty may be claimed are 12 years in the case of sealed contracts, 6 years in other cases and 3 years where the claim is for personal injuries arising from the defect. The previous common law implications were sometimes restricted by the use of exclusion clauses but in future there can be no exclusion of the statutory obligations. Some relief is however offered where a dwelling has been provided or sold under an approved scheme such as that operated by the NHBC.[15]

A contractor who built on his own land and then sold the completed building was until recently believed to be under no common law liability for negligence (*Dutton* v. *Bognor Regis Building Company*, 1972). This distinction is now removed and all contractors are liable for negligence in accordance with the principles established in *Donoghue* v. *Stephenson*, 1932 (the snail in the ginger beer bottle). A landlord with an obligation to repair premises also has to ensure that no one will suffer injury or damage through the landlord's neglect to maintain the building satisfactorily.[15]

A considerable amount of building maintenance is inescapable because of legal requirements. The Factories Act and the Offices, Shops and Railway Premises Act impose maintenance obligations upon owners and occupiers of these classes of buildings. Statutory undertakings, such as Water Authorities, make their own regulations which often include maintenance clauses designed to prevent danger or wastage. The object of the law in requiring maintenance is not usually to preserve amenities or to safeguard investment, but to protect persons from risk. Public Health Acts contain provisions for the compulsory repair of dilapidated property, while Housing Acts include provisions for the compulsory repair of houses unfit for human habitation and also for financial assistance towards the cost of improvements and conversions to dwellings. Occupiers of commercial and industrial properties are often required to assume liability for structural maintenance and repairs under leases, and even with residential properties, owners and tenants may make such arrangements for maintenance as they see fit, subject to the provisions of the Housing Acts. Employers are required to ensure the safety of their employees at work by maintaining safe plant, systems of work and premises, and by ensuring adequate instruction, training and supervision, under the Health and Safety at Work Act 1974.

Maintenance Needs

A prime aim of maintenance is to preserve a building in its initial state, as far as practicable, so that it effectively serves its purpose. Some of the main purposes of maintaining buildings are:

(1) retaining value of investment;
(2) maintaining the building in a condition in which it continues to fulfil its function; and
(3) presenting a good appearance.

The amount of necessary building maintenance work could be reduced by improved methods of design, specification, construction and feedback of maintenance data to designers. In addition, effective maintenance management embraces many skills. These include the technical knowledge and experience necessary to identify maintenance needs and to specify the right remedies; an understanding of modern management techniques; a knowledge of property and contract law; and an appreciation of the relevant sociological and economic aspects.

The Building Conservation Trust[16] has described how its permanent exhibition at Hampton Court Palace shows quite dramatically how through neglect and deterioration every building faces certain death from progressive decay. The process can be accelerated by neglect or delayed by proper care. Proper maintenance is cheaper, quicker and easier than major repairs.

Chudley[17] has identified the principal criteria which could influence the decision to carry out maintenance work, such as cost, age and condition of property, availability of adequate resources, urgency, future use and possibly sociological considerations.

Assessing Maintenance Priorities

It is difficult to formulate a precise order of priorities of maintenance activities as they are so diverse and any assessment is likely to be a subjective evaluation. Some of the principal functions of maintenance are: to ensure the safety of occupants, visitors and the general public; to maintain services, such as heating, lighting, escalators and fire alarm systems; to maintain decorative surfaces and carry out adequate cleaning; and to prevent or diminish significantly deterioration of the fabric.

Some organisations have formulated maintenance priority guidelines which, in times of financial stringency, dictate how monies are to be spent. Typical is the following approach, adopted by one county council.

(1) Work required for health and safety, such as emergency exits and fire precautions.
(2) Work required to preserve the structure, such as essential roof repairs and external painting.
(3) Work required for occupational efficiency, such as increased lighting.
(4) Amenity work, mainly internal, such as interior decorations.[18]

The Local Government Operational Research Unit[19] identified three separate categories of building maintenance work—fabric maintenance; day-to-day repairs; and improvements and modernisation. The Unit endeavoured to establish techniques for determining whether a particular job should be done immediately or deferred.

For instance to draw up a long-term maintenance programme for the fabric of buildings, a maintenance manager must make many decisions. First, he must decide which of the various elements of the building, such as walls, floors, roof, windows and doors, merit detailed inspection. Upon inspection he needs some criteria for ranking them in order of priority coupled with a technique for assessing their condition. Finally, he must decide in each case whether work is necessary, and if so whether patching or replacement is more appropriate. To make these decisions he must not only know the cost implications of the various alternatives but must also know the minimum acceptable conditions of the elements, appearance being an important factor in deciding what type of repair should be undertaken. *Ad hoc* maintenance with an open-ended budget may seem attractive but it is unlikely to obtain full value for monies spent or an efficient maintenance system.

There is a need to improve the methods of managing and executing building maintenance. Maintenance budgets should be clear and well reasoned and supported by full information on the consequences of neglecting maintenance. Decision-making in building maintenance could be assisted by the application of operational research and computer-aided techniques. A prime aim should be to improve efficiency and productivity. There is a backlog of several years in the maintenance of many buildings and the more effective use of resources will help to reduce these arrears and assist the national economy.

Feedback from occupier to designer should be improved in order to assemble information on both the preference of the user, and the performance of materials, components and constructional methods. There is a general lack of essential basic data and appropriate recording systems. Design teams all too frequently neglect consideration of maintenance aspects and there is a great need to reduce the gulf between design and maintenance. Occupiers of new buildings should ideally be provided with maintenance manuals listing the materials and equipment used in buildings, together with precise details of the maintenance required for their most efficient and economic use, and this aspect is investigated in more detail in chapter 12.

Resources for Building Maintenance

It is generally recognised that the current rate of spending on building maintenance is at a depressed level, and that many necessary tasks are therefore being neglected. In consequence, considerably increased expenditure is likely to be needed in the future to rectify defects, which will by then be much more extensive and serious. Hence more should be spent on fabric maintenance now to save money in the future. However, maintenance policies of most organisations have to be tailored to their resources and needs.

To accelerate the rate of maintenance work requires an increased quantity of both labour and materials, both of which were in short supply in the early nineteen seventies. Statistics published by the Department of the Environment show that the construction labour force in Great Britain fell from 1 023 000 in 1976 to 776 000 in 1982. More serious still is the drop in the numbers of apprentices and skilled craft operatives. The failure of labour and material resources to match demand caused a sharp rise in building prices and erratic and unpredictable tenders in the early 1970s. This was followed by a substantial reduction in workload in 1974/75. There is a growing need to make employment in the construction industry more secure and attractive and to train persons to less depth in a number of trades. This latter proposal would certainly benefit maintenance work. Some have suggested a reduction in the apprenticeship period for building trades, by eliminating some of the tasks that are rarely performed, to help in recruitment.

There are difficulties in estimating the capital resources employed in maintenance as many contractors carry out both new and maintenance work and make fixed and working capital available to each as required. Official statistics show that in 1981, over 80 per cent of all contractors in Great Britain were small firms, employing less than 8 operatives, and they concentrated almost exclusively on repairs and maintenance. Larger firms were involved to a significant extent in refurbishment schemes. The trades engaged most intensively on maintenance work are painters and plumbers. No one measure of output seems both adequate and sufficiently widely used, be it turnover, materials used, facilities created or value of building stock maintained.

The distribution of construction resources as between new work and maintenance and their deployment within maintenance should ideally be determined in the way most advantageous both to the economy as a whole and to the needs of property owners and occupants. Unfortunately there are at present no reliable methods for determining the optimum maintenance cost in relation to the life or value of the building or of the value of the activities carried on within it. Moreover, because responsibility for building maintenance may be organised by one of a wide range of departments or disciplines—surveyors, engineers and architects—there is a general lack of concentrated experience of the special problems and techniques. The establishment of a separate building surveyors division in the Royal Institution of Chartered Surveyors has, however, helped to remedy this deficiency.

The Committee on Building Maintenance[2] believed that maintenance policies of the property owner should be consistent with his main objectives and financial position, taking into account any plans for new work and the effects of maintenance or its neglect on his operations and the morale of employees. Ideally, the principles of planned and preventive maintenance should be practised and reviewed periodically. These policies may well indicate how the actual work should ideally be performed—by directly employed labour, contractors or a combination of both. It should be borne in mind that caretakers and porters can with suitable training be usefully employed to carry out simple maintenance tasks such as internal decorating, re-washering taps and other valves, and clearing gutters and drains.

Nature of Maintenance

Maintenance comprises three separate main components—servicing, rectification and replacement.

Servicing is essentially a cleaning operation undertaken at regular intervals of varying frequency and is sometimes termed day-to-day maintenance. As more sophisticated equipment is introduced so more complicated service schedules become necessary. The frequency of cleaning varies—typical frequencies being: floors swept daily and polished weekly; windows washed monthly; flues swept every every 6 months; painting for decoration and protection every 4 years.

Rectification work usually occurs fairly early in the life of the building and arises from shortcomings in design, inherent faults in or unsuitability of components, damage of goods in transit or installation and incorrect assembly. Rectification represents a fruitful point at which to reduce the cost of maintenance, because it is avoidable. All that is necessary, at any rate in theory, is to ensure that components and materials are suitable for their purpose and are correctly installed. These seemingly simple requirements are not always easy to achieve. Frequently, the same component must fulfil many functions, such as weather-shield, load-bearer, thermal insulant, and still be of good appearance. A failure to perform any one of these functions satisfactorily can result in maintenance work. Typical examples are the failure of decorative floor coverings on solid concrete ground floor slabs due to damp penetration and the failure of joints between large slabs in wall cladding to exclude wind and rain. Rectification work could be reduced by the development and use of performance specifications and codes of installation.

Replacement is inevitable because service conditions cause materials to decay at different rates. Much replacement work stems not so much from physical breakdown of the materials or element as from deterioration of appearance. Hence the length of acceptable life often involves a subjective judgement of aesthetics of change. Furthermore, the measurement of the durability or length of life of a material is a very difficult technological problem, because of the complex nature of the environment and the difficulty of determining how much a material may change before it is discarded. Some indication can be obtained from observing materials in buildings, on exposed sites and in simulated exposure or use. A long history of accelerated tests shows the inadvisability of undertaking simulative tests without a detailed knowledge of the conditions being simulated. The frequency of replacement could often be reduced by the use of better quality materials and components, but the economics of this merit careful study.

Maintenance can also embrace *renovations* which consist of work done to restore a structure, service and equipment by a major overhaul to the original design and specification, or to improve on the original design. This may include limited additions and extensions to the original building. An element of improvement, or of new works, is frequently found under a heading of maintenance costs. This is to some extent unavoidable, since in replacing a fitting, such as a bath, the replacement will be of new design. An analysis of the cost of modernising a typical dwelling showed that 32 per cent was required to repair, maintain and replace to original standards, 39 per cent on upgrading to acceptable present-day standards and 29 per cent to improvements to above present-day standards.

One analysis of maintenance costs in buildings not more than 25 years old showed fair wear and tear accounting for 56 per cent, rectification of design or specification faults at 20 per cent, repairs due to faulty materials or workmanship at 12.5 per cent and the remaining 11.5 per cent was attributed to sundry causes.[20] The annual cost of maintenance is likely to increase in the future because of some new products by-passing the Agrément Board and the fact that many traditional products, whose properties and problems are largely known, are still misused. Finally, correct diagnosis of building defects is essential to ensure that the cost of remedial work is not excessive and that it is successful.

Research into Maintenance

The importance of research into various aspects of building maintenance was recognised by a former Minister of Public Building and Works when he established the Committee on Building Maintenance in 1965. Research and development problems in building maintenance spring essentially from the great diversity of the subjects. In the two decades after World War II, research in this field was mainly directed at properties of materials and few of the results were actually implemented. So much needs to be done concerning the relationship between design and maintenance, execution of maintenance, economic significance of maintenance and the actual performance of materials and components under varying conditions. There is also a great need for a continuing dialogue between research and development workers on the one hand and architects, surveyors, maintenance personnel and contractors on the other, to ensure the relevance of research to the realities of construction and the implementation of the results of research in practice. The Government has initiated a large amount of research in this field and has assisted in its dissemination by conferences and publications, primarily through the Building Research Establishment.

Concern has been growing at the very substantial sums of money, amounting to several hundreds of millions of pounds per annum in the public sector alone, which are being spent on correcting design and construction faults in housing. The Building Research Establishment mounted a major research project with the National Building Agency, to examine the quality of construction work in progress on a number of low-rise traditional housing schemes in England. The results of the three-year investigation were published in 1982[21]. They showed that just over one-half of the faults concerned the external envelope, with 20 per cent in walls, 20 per cent in roofs, and 13 per cent in windows and doors. One-quarter of all faults were infringements of the Building Regulations. Out of the total number of faults, 50 per cent were attributed to design, 41 per cent to site, and 8 per cent to materials. Figures 1.3 and 1.4 illustrate these aspects.

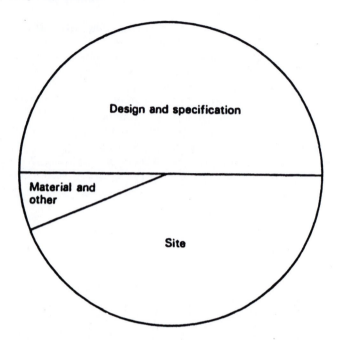

Figure 1.3 Origins of building faults (source: *Quality in Traditional Housing*)

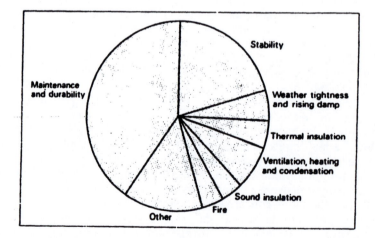

Figure 1.4 Attribution of types of faults by performance (source: *BRE News 58 (1982)*)

In 1986, the Building Research Establishment started providing advice for owners and their consultants on methods of inspection, maintenance and renovation of the different steel-framed house types. About 140 000 metal-framed houses have been built in Britain since the early 1920s, based on some thirty different proprietary systems. The BSIF system is the most common with over 30 000 houses built in England and Wales and over 4000 in Scotland. Some cases have been reported of serious deterioration of the main frame or failure of claddings and fixings, and there is a history of concern about thermal and fire performance in some of the house types; hence authoritative advice is welcome.

The Building Research Establishment included the creation of a Building Energy Efficiency Division in the 1986/87 programme, to give greater emphasis to this increasingly important area. The fire research programme was also recast to give priority to the scientific basis of fire testing, the relationship between smoke production and control, and fire problems in building services.

Advice on building maintenance is obtainable from a variety of different organisations and the following provides a selected list of the more important sources. They include the Building Conservation Trust, the Building Research Establishment, Building Maintenance Information, the Building Services Research and Information Association, the Construction Industry Research and Information Association, the National House-Building Council, the Ancient Monuments Society, the Historic Buildings Council, the Society for the Protection of Ancient Buildings, the Civic Trust and the Scottish Local Authorities Special Housing Group.

More needs to be known about the effect of weather on materials and components, criteria for visual tolerance and the cost of maintenance related to constructional elements, together with the nature of failures and their underlying causes. Costly items of maintenance such as condensation and mould growth, plumbing defects and painting cycles justify closer study. Finally, the variety of problems in maintenance management provides a potentially fruitful area for the application of operational research.[2]

Lives of Buildings

The lives of existing buildings are difficult to assess as all properties have, from the date of their erection, been the subject of varying amounts and standards of maintenance, besides being constructed to different standards. Most buildings are constructed with the intention that they should last at least 60 years and many exceed this period.

Stone[22] asserts that even cheaper buildings generally have a substantial life in the order of 50 to 60 years. Their possible physical life is often much greater but they may be demolished before the end of this period to permit a more profitable use of the site, or because it is found more economical to clear and rebuild rather than to adapt the building to meet changed requirements, because of physical or technical obsolescence as many of the building components become obsolete.

It is also possible to distinguish between 'structural life' and 'economic life'. Structural or physical life is the period which expires when it ceases to be an economic proposition to maintain the building, while economic life is concerned with earning power and is that period of effective life before replacement; replacement taking place when it will increase income absolutely. It is probable that optimum life is determined primarily by the earning power of the building, and only secondarily by the structural durability.

Changing social and economic conditions can have a considerable influence on the life of a building which can become ill-suited to present-day needs and its demise may also be accelerated by the significant ratio of land to building costs. Wherever possible the aim should be to extend the economic life of a building by making the structure adaptable and by careful management and control of the surroundings. Hence the actual physical life of a building is frequently much greater than its economic life, but the building is often demolished before its physical life has expired in order to permit a more profitable use of the site, or because it is found cheaper to clear and rebuild rather than to adapt the building to the changed requirements.[23]

Building for less than normal life saves little initial cost and, in any event, the services are unlikely to have a life of more than a generation. As a general rule the capital asset of a building is so valuable and is often appreciating, so that in practice maintenance is frequently directed to prolonging effective life. A limiting factor is the period ahead that one can plan in detail. This is seldom more than a decade.

Technology of Maintenance

The technology of maintenance is concerned with all the factors that influence and cause the need for maintenance work. The occurrence of defects in the fabric of a building can result from many unrelated design decisions—unsuitable materials, incorrect assessment of loads, inadequate appreciation of conditions of use and inadequate assessment of exposure. Exposure is influenced by rainfall, direction of prevailing winds, micro-climate, atmospheric pollution, and aspect and height of building. The durability of building materials is also influenced by frost action, crystallisation of salts, sunlight, biological agencies, abrasion and impact, chemical action and corrosion and incompatability of modern building materials. In addition to considering the physical and chemical properties of building materials, the designer should ensure that wherever practicable materials should be so used as to take full advantage of their potentialities. Where they are unlikely to last the life of the building, attention should be directed to convenience of replacement. Building operatives also need training in the use of new techniques, components and materials.

Defects may also occur from faults other than structural ones, resulting in inconvenience and discomfort to the occupant. Typical examples are overheating and glare due to excessive areas of glass, inadequate sound insulation, inconvenient layouts and very high operating costs.

Cracks in buildings normally result from failure or defective construction, and are almost invariably unsightly and unacceptable to occupants. If severe they may result in loss of stability. Furthermore, cracks frequently give rise to air infiltration, heat loss and reduced sound insulation, all of which result in reduced efficiency of the building. Cracking is generally caused by tensile stresses in excess of the tensile strength of the materials, produced by externally applied loads, or internal movements arising from temperature or moisture changes. It is essential that the cause of the failure is correctly diagnosed, otherwise there is the possibility of a component being reinstated at considerable expense and incorporating the previous deficiencies. Recognition of the location and extent of movements in building materials and components is essential for the satisfactory design of joints and fixings and the prevention of cracking.[24]

Other important concepts of the technology of maintenance can be illustrated by reference to roof construction. A good roof which is well maintained should last the life of the building and it is false economy to save money on the roof during construction, because, if it ever requires replacement, it will cause serious dislocation of production and other activities within the building. A leaking roof, apart from causing considerable inconvenience to users, can lead to accelerated deterioration of other parts of the building, such as ceilings, floors and walls, and can cause serious damage to decorations and electrical installations. Traffic over a roof should be kept to a minimum and, where it is essential, appropriate walkways and access ladders must be provided. To ensure that roofs are adequately maintained, they should ideally be inspected every three years, or alternatively one-third each year.

In the wake of new building techniques and innovations, building defects have multiplied. A partial collapse of a block of flats, a fall of a school roof, cladding defects on large housing estates and offices, and numerous failures to flat roofs and window joinery have been well documented. They have resulted in the expenditure of large sums of money to maintain the buildings in functional condition and, in some cases, for demolition and replacement only a few years after erection. Many of these failures resulted from the shortcomings of new materials, components and techniques not being fully understood.[25]

The term 'terotechnology' has been used to embrace the life cycle requirements of physical assets. It is a combination of management, financial engineering, and other practices applied to physical assets in pursuit of economic life cycle costs.[3] It is concerned with the specification and design for reliability and ease of maintenance of plant, machinery, equipment, buildings and structures with their installation, commissioning, maintenance, modification and replacement, and with feedback of information on design, performance and costs. It is a technology that takes into account the marketing and observance of design–maintenance–cost practice of all assets, the conservation of resources and the promotion of controlled and calculated life span of assets as against built-in or unpredictable obsolescence. Life cycle costs are defined in BS 3811[3] as "the total cost of ownership of an item of material, taking into account all the costs of acquisition, personnel training, operation, maintenance, modification and disposal, for the purpose of making decisions on new or changed requirements and as a control mechanism in service, for existing and future items."

Consideration of Maintenance at Design Stage

The importance of considering maintenance implications at the design stage of a building project is now generally recognised and the building maintenance conferences and seminars organised by the Department of the Environment have done much to draw attention to this. It is at the design stage that the maintenance burden can be positively influenced for better—or for worse. Skilful design can reduce the amount of maintenance work and also make it easier to perform—good maintenance begins on the drawing board. Ideally the design team should aim to produce a building which is attractive, functionally efficient and constructionally sound with a minimum of maintenance.[26]

The cost implications of building designs are often far wider than the effects on initial costs. For some types of buildings the equivalent of first costs is less than the running costs, and small changes in design have a much larger impact on running costs than on first costs.

Cost Yardsticks

The Government introduced cost yardsticks in 1967 with the aim of keeping building costs in the public sector within reasonable bounds and ensuring adequate cost planning of projects at the design stage. In the early 1970s opposition to these constraints increased rapidly, forcing the Government to rethink its policy. The Royal Institute of British Architects claimed that the yardstick system was "quite inadequate in balancing capital costs against subsequent costs in use— economies in finishes today will undoubtedly lead to inflated maintenance costs in future." The Association of Municipal Corporations believed that the yardsticks resulted in lower standards of design which led to serious maintenance problems, abortive work of the design teams and concentration of development on high-density residential developments with consequent lowering of environmental standards. The Royal Institution of Chartered Surveyors objected in 1973 to the use of *ad hoc* allowances to close the financial gap between the cost limit and the lowest tender on the grounds that this procedure was completely discretionary and destroyed the ability to cost plan construction projects, besides making it difficult to meet building standards, make best use of resources or reduce approval periods. There was unfortunately nothing built into cost limits to encourage the achievement of a satisfactory balance between initial and future costs. It was, however, not until 1981 that cost yardsticks were finally abolished.

Design Needs

It is important that the thread of 'building life' should flow through both design and construction processes, with effective lines of communication between client, designer, contractor and those charged with building maintenance. Outdated administrative procedures often result in the various parties to the building contract failing to appreciate the significance of other functions in the overall concept. This sometimes causes frustration and annoyance to maintenance personnel when taking over new buildings and finding themselves faced with bad details, poor choice of finishes, materials and components, and lack of basic information about the building and its services. Unfortunately, designers rarely have a long-term interest in the buildings they produce and hence they tend to become divorced from the maintenance problems that flow from bad design. There is a pressing need for maintenance surveyors and the like to be represented on design teams and for much increased feedback of maintenance and performance information from users and maintenance organisations to designers. When information on building defects and failures is fed back to the designer, it is sometimes inaccurate because the person diagnosing the fault has insufficient expert knowledge to assess the cause of failure. Furthermore, where manufacturers provide a good technical service, designers do not always make full use of it. A BRE report[21] found that architects sometimes had insufficient access to basic data such as British Standards and that there was a need for improved information on drawings and specifications.

Considerable loss of time and disruption of activities can stem from the failure of building components or alterations made necessary by poor design. Education for the designer in the appreciation of maintenance requirements and costs in use techniques could be most fruitful. It is emphasised that maintenance should not be the subject for thought after the erection of a building; it should be an integral part of the design process.[5] Designers could contribute significantly to a reduction in maintenance costs if they asked themselves four questions when designing each component or part of a building:

(1) how can it be reached?
(2) how can it be cleaned?
(3) how long will it last?
(4) how can it be replaced?

The designer requires a considerable amount of technological data to minimise the effect of a wide range of agencies, such as moisture, thermal and structural movement, corrosion, and insect and fungal attack, all of which combine to reduce the life of the building. Furthermore, the building owner has a right to obtain a building which will adequately perform the function for which it was designed. The primary functions are to withstand the effects of weather, to retain stability, give weather resistance, thermal and sound insulation, and to continue to function efficiently under the worst possible conditions, with a minimum of maintenance. The problem is aggravated by an increasing lack of concern among people as to the quality of the work they produce, the wider use of new and relatively untried materials and components and the acceptance of unrealistically low tenders.

High maintenance costs sometimes result from poor detailing at the design stage, including insufficient allowance for expansion or contraction, absence of weatherings or throatings, lack of or incorrectly placed damp-proof courses, unsound foundations, poor jointing between different materials or components, inadequate falls, incorrect choice or misuse of materials, and poor access or facilities for repairs.

On occasions, failures resulting from faults in design cannot be cured permanently and remain a continuing nuisance throughout the life of the building. The cause of the failure may result from insufficient consideration to performance in use aspects, such as the effect of sea air, industrial pollution or even the height of the building or its proximity to a number of high buildings. The latter situation may result in the creation of air currents which lead not only to an unpleasant environment around the building, but also to severe strains on the fabric at unexpected points.

It is also necessary to take account of the interdependence of the various elements of a building. For example, the heating, ventilation and insulation of buildings must be considered in relation to each other if excessive heat losses and condensation are to be avoided. Similarly the ventilation system may affect cleaning and the performance of internal finishes. The performance of a building can also be affected by external factors; for example, the provision or retention of trees can have an appreciable effect on the subsoil and the performance of foundations and services such as drains.

Peacock[27] has identified the principal failures of designers in minimising maintenance costs as (1) unsatisfactory detailing; (2) incorrect selection or specification of building materials, components and systems; (3) lack of standardisation; and (4) failure to appreciate how a structure will be used and maintained.

Effect of Metrication

Metrication of components and materials used in the construction industry has affected the maintenance of buildings to some extent. Nevertheless, to keep the changeover in perspective, it must be remembered that some materials and components have life cycles, according to their use and visual appeal. For example, kitchen and light fittings have changed radically in recent years, while ironmongery is constantly changing in style, so that fittings may become outmoded in a comparatively short time and are not replaceable. Hence the replacement of discontinued articles is not new.

Many materials or components were initially manufactured in both imperial and metric sizes and further changes came in a few cases with rationalised metric dimensions. In general, metric sizes are slightly smaller than the equivalent imperial sizes and so problems may arise in making future replacements of some products when existing stocks in imperial units have become exhausted.

The problems in future maintenance are likely to be comparatively small, since the ingenuity of maintenance staff will enable most of them to be overcome without too much difficulty. Sheet materials to determined sizes, such as laminated boards, plasterboard, hardboard and plywood present few problems as most used in maintenance are cut on fixing. Partition units can be cut or adjusted to size within cover strips.

Copper pipes have changed slightly in size but connectors are available for joining imperial and metric pipes and fittings. Lead pipes can be dressed and soldered to accommodate any variation in the size of brasswork.

Clay facing bricks are manufactured to sufficiently wide tolerances to avoid any problems. Tiles and paving slabs will cause some problems on account of the smaller metric sizes. Hence where part of a ceiling has to be renewed with discontinued size of removable tiles, it may be necessary to renew the whole ceiling and store the serviceable existing tiles for repairs to other ceilings. In some cases a thicker joint may suffice. In the case of wood floor blocks, specials may be needed for small areas of renewal. The introduction of a slightly smaller architrave or skirting as a complete length between breaks is unlikely to worry the maintenance staff or offend the eye. On the other hand, standard units with lapped or sliding joints such as roof tiles and gutters, will present no problems as they can be cut or lapped to the required dimensions.

Ready-made standard units will cause the greatest difficulty. New metric door set dimensions produce doors 12 mm narrower than the old 2 ft 6 inch door. Imperial doors were produced alongside metric doors for a considerable period while there was a significant demand, but eventually the time will come when it will be necessary to fit purpose-made doors or adjust or replace the frame, whichever is the more economical. Similar problems may arise with the replacement of imperial standard windows in years ahead. Standard joinery fittings will be wider and longer but slightly lower, requiring careful planning on replacement.

Sanitary fittings and drainage goods are unlikely to cause problems, but the size and threads of connections to heating equipment are changing and so cause difficulties. For example, an obsolete pattern of radiator can be renewed in a room with several matching radiators by taking one which is fixed in isolation and installing the new radiator in its place. The use of new threaded fasteners may involve re-tapping, reamering or re-drilling to accommodate them. With electrical threaded conduit and fittings, manufacturers produce special connectors as necessary.

A metric conversion table is provided in appendix 1.

Relationship of Capital, Maintenance and Running Costs

A building owner can reasonably expect a designer to provide a building which will satisfactorily meet his needs and will secure a reasonable balance between first and future costs. Hence it is important for design teams to develop more effective methods of predicting the functional and economic consequences of designs and so obtain best value for money. Unfortunately, the total occupancy costs of buildings are frequently difficult to assess and feedback of accurate information is very restricted. A variety of payments are made from different funds at different times and in different ways. Table 1.1 shows the breakdown of total costs and the wide scope of the activities involved.

Table 1.1   Breakdown of total costs

Total costs
(life span of building)

Initial costs		*User costs*	
Land	*Running costs*	*Occupational charges*	
Construction			
Professional fees	Maintenance	Rates	
	Operating services	Insurance	
	(operating and cleaning)	Modifications	
	Energy	and alterations	
		Estate control	
		(management)	

Source: *Building Economics*[23].

Difficulties in Assessing Total Costs

There are a number of problems in endeavouring to assess total costs or costs in use at the design stage, and the more important ones are now listed.

   (1) It is difficult to assess the probable maintenance costs of different materials, processes and systems. There is a great scarcity of reliable maintenance cost data tabulated in a meaningful way. It is not easy to predict the lives of materials and components in a variety of situations. Even the lives of commonly used materials like paint show surprising variations and are influenced by a whole range of factors, including type of paint, number of coats, condition of base, extent of preparation, method of application, degree of exposure and atmospheric conditions.

   (2) There are three types of payment involved—initial, annual and periodic. All three have to be related to a common basis for comparison purposes, and this requires a knowledge of discounted cash flow techniques which will be considered later in this chapter.

   (3) The selection of a suitable long-term interest rate is difficult; rates rose dramatically in the 1970s and fluctuated widely in the 1980s.

   (4) Inflationary trends may not affect all costs in a uniform manner, thus distorting significantly the results of costs in use calculations.

(5) Where the initial funds available to a building owner are severely restricted, it is of little consequence telling him that he can save large sums in the future by spending more on the initial construction.

Current and Future Payments

As mentioned earlier, a major difficulty in making costs in use calculations is that every building project involves streams of payments over a long period of time—usually the life of the building. The payments are of three main types

(1) present payments covering the cost of the site, erection of the building and architect's and surveyor's fees;

(2) annual payments relating to minor repairs, cleaning, and heating and lighting;

(3) periodic payments such as full internal redecorations possibly every 7 years, external redecorations every 5 years, and replacement of boilers and electrical wiring every 25 years.

All these varying types of payment have to be converted to a common method of expression to permit a meaningful comparison to be made between alternative designs. The process is often described as discounting future costs and is based on the premise that if the money were not spent on the project in question it could be invested elsewhere and would be earning interest. £100 invested today at 8 per cent compound interest will accumulate to £215.89 after 10 years (the multiplier of 2.1589 is found from a valuation table—the amount of £1 table).

In the reverse direction the 'present value of £1 table' shows that it will be necessary to invest £46.32 now at 8 per cent compound interest to accrue to £100 in 10 years' time. The 'present value of £1 per annum' or 'years' purchase' table shows that an expenditure of £1 per annum throughout a sixty year period—the period often accepted as the probable life of a building and also the central government loan period for buildings owned by local authorities—is equivalent to a single payment of £12.38 today, taking a compound interest rate of 8 per cent. Expressed in another way, if £12.38 were invested today at 8 per cent compound interest it would provide sufficient funds to be able to pay out £1 per annum for each of the 60 years. Hence it could be argued that it would be worth spending up to an extra £12.40 today on initial construction if this will reduce the expenditure on maintenance work by £1 per annum throughout the 60 year life of the building.

There are two possible approaches in making costs in use calculations.

(1) Discount all future costs at an appropriate rate of interest, probably between 5 and 8 per cent (long-term pure interest rate with no allowance for risk premium), and so to convert all payments to present value (PV) or present worth, using valuation tables. Variations in the discount or interest rate selected can have an appreciable effect on the calculations.

(2) Express all costs in the form of annual equivalents, taking into

account the interest rate and annual sinking fund. A building owner is entitled to interest on the capital he has invested in the project and requires a sinking fund to replace the capital when the life of the building has expired.

An example will serve to illustrate the discounting principle. A building is designed to last 60 years and can either be provided with a roof costing £3000 which will last 30 years and then need replacing, or be covered with a roof costing £4500 which will last the life of the building. It is necessary to determine which is the better proposition financially and to do this, payment in 30 years' time has to be converted to its present value. The present value of a payment of £3000 in 30 years' time is found from the present value of £1 table and is £3000 × 0.17411 = £522.33 taking an interest rate of 6 per cent. The calculations can be summarised as follows:

	Cost	Present Value
Roof A		
Initial construction	£3000	£3000
Replacement after 30 years	£3000	£ 522.33
Total cost	£6000	£3522.33
Roof B		
Initial cost	£4500	£4500

These calculations show roof A to be the better long-term proposition. It could however be argued that this is oversimplified as, for instance, it takes no account of the cost of demolishing the old roof or any temporary work that may be needed to protect the occupants and contents of the building during the reconstruction. The interest rate could also be criticised; also the hypothesis that the effects of inflation will be cancelled out through rising costs and values, so that the higher cost of building work in thirty years' time will be offset by the increased value, including rent, of the building itself, may not find ready acceptance. A range of costs in use worked examples covering building designs, services, external works and components are given in *Building Economics*.[23]

The following example compares the likely total costs of provision and maintenance of softwood painted windows and anodised aluminium windows. In common with government department practice, an effective life of 50 years has been assumed and a discounting rate of interest of 10 per cent. A total window area of 100 m² has been taken in both cases.

Softwood Painted Windows		*Anodised Aluminium Windows*	
Initial cost including glazing, fixing and ironmongery	£4500	Initial cost including glazing, fixing and ironmongery	£9500

Painting at 5-yearly intervals. £600 × 1.615 (Present Value of £1 at 5-yearly intervals up to 45 years at 10 per cent)	£970	Cleaning at 3-monthly intervals £150 × 9.915 (Present Value of £1 per annum for 50 years at 10 per cent) £1487
Total costs	£5470	£10987

As this calculation indicates, the total cost of the anodised aluminium windows is approximately double that of the softwood painted windows. Even costly stainless steel windows can rust and bronze turns green in time. In practice there are few if any maintenance-free materials.[18]

Initial and Future Cost Relationships

Most design decisions affect running costs as well as first costs, and what appears to be a cheaper building at the design stage may in the long term be far more expensive than one with much higher initial costs, although there can be exceptions to this general rule. Some idea of the relationship between initial and running costs can be obtained from an examination of table 1.2, from which it can be seen that running costs often amount to about one-half of the annual equivalent of first costs. The proportions do, however, vary considerably from one building to another of the same type and so the percentage figures listed can only form a rough guide.

Table 1.2   Breakdown of costs in use for various types of buildings

Type of annual cost	Houses	High flats	Industrial buildings	Schools	Offices
			(percentages)		
Maintenance	14	12	18	16	13
Fuel and attendance for heating and lighting	24	24	30	18	29
Initial costs (amortised)					
(a) Building	48	56	47	51	47
(b) Land and development	14	8	5	15	11
Total costs in use	100	100	100	100	100

Source: *Building Economics*[23].

Land costs are relatively low with industrial buildings which are erected on lower-priced land with a higher utilisation factor, whereas houses and schools have high land costs, being built to a low density on relatively highly priced land. Initial building costs represent the highest proportion of total costs with flats, and the least with industrial buildings. Heating costs of modern industrial buildings have been reduced by improved thermal insulation but are still high, and schools rank low with shorter periods of use. Lighting of offices is a relatively high cost item stemming from the high standards of illumination that are required. On maintenance and decoration, schools rank high with heavy wear and tear.

In studies of Crown office buildings[28] total costs in use comprised, at an interest rate of 10 per cent, approximately two-thirds capital or initial cost and one-third running and maintenance costs. The one-third sector consisted of three approximately equal parts: fuel, electricity and gas; cleaning; and repairs, decorations and minor new works. It should also be borne in mind that rates can amount to as much as two-thirds of the total running costs. The proportion of initial cost to the remainder is influenced considerably by the rate of interest used to amortise the initial cost for conversion to annual equivalent values. High maintenance costs can result from low first costs coupled with inadequate specifications, or through complex services and elaborate finishes.

It is worthy of note that in the study of Crown buildings, the costs were based on internal redecoration at eight-yearly intervals and washing down at the third and sixth years; external redecoration at six-yearly intervals; and replacement of bitumen felt roofing after twenty years, boilers after twenty-five, internal pipework, thirty, storage tanks, twenty-five, electrical installation wiring, fifteen, distribution switchgear, twenty-five, and passenger lifts, twenty years. While for houses, Knight[29] suggested renewal of 25 per cent of gutters after forty years, repointing of 50 per cent of brickwork after forty years, renewing external doors after thirty years, renewal of ironmongery after twenty to forty years, renewal of cupboards and sanitary appliances after twenty years, and hot and cold water services and electrical wiring every thirty years. The average annual charges for services maintenance in air-conditioned offices were approximately 60 per cent more expensive than heated offices.

Life cycle cost planning is concerned with examining the economics of the life cycle of a building to ensure that there is a balance between the use of capital resources in design and construction and the consumption of future resources by the building in use. A valuable local authority guide[30] contains useful examples of life cycle cost plans, lives of components and energy economics.

A useful guide to energy investment decisions in private housing[31] identifies the main factors to be considered in setting an economic evaluation framework as the dual impact of time horizon/discount rate and the price of the fuel saved. Appropriate time horizons and discount rates are largely determined by a householder's age, access to finance, tax situation, and whether or not he has a mortgage.

Economics of Maintenance

Economics examines the process whereby scarce resources or factors of production, such as land, labour and capital, are allocated among the various competing claims on their use. Because maintenance involves the use of resources, it follows that decisions have to be made as to the level and nature of maintenance expenditures.[32]

The interdependence and interrelationship of initial and user costs are of prime importance when planning maintenance expenditure. The relationship of one to the other is often in inverse proportions. A reduction in future maintenance costs may often be obtained by increasing initial costs; similarly, economies in initial costs may follow from the acceptance of an increased level of maintenance costs. Wright[32] has shown how decisions as to the ratio of initial costs to future (planned) maintenance costs are influenced by time preferences and commercial judgement. Speculative development with the objective of sale will generally show more regard to economies in initial costs than in user costs, although recognising that too high a level of user costs will jeopardise the opportunity for sale; purchasers, however, will show more concern for user costs.

Tax payments, reliefs, grants, allowances, subsidies, rates and the like should always be included in development and maintenance calculations. Because initial costs mainly constitute capital expenditure and because depreciation allowances are extremely limited, it may be worth while incurring additional (tax-deductible) maintenance and other running costs. Cash flow calculations are quite complicated and are more so when tax considerations are included with the distinct possibility of future changes.[32]

Provision of buildings with low maintenance costs will assist in reducing the demand for scarce building resources since such buildings often possess higher user and even environmental benefits, when viewed against the visual cost to society of deteriorating buildings. Surveyors, maintenance managers and other interested parties need to identify the items generating the highest maintenance costs, and to be constantly questioning the suitability not only of materials and components in meeting their functional requirements but also of the method by which they are assembled.

Property managers are concerned with the total occupancy costs of properties over their expected lives. When the cost of maintaining the facilities imposes an unduly heavy burden upon marginally profitable enterprises it is time to reappraise the use of the premises. They may be sold, demolished, let, or used for other activities for which a different standard and cost of maintenance renders the premises of economic value once again.[33]

A great dilema facing Britain in the mid 1980s was that the built environment was deteriorating at a rate far faster than it was being modernised, improved or replaced. Areas of economic decline, natural ageing, and obsolescence, ill-conceived post-war designed and constructed buildings, combined with the increased demands that modern society places upon its buildings, all contributed in varying ways to the total problem. The price for this neglect will be paid in the years to come—and there must be a day of reckoning—but it goes far beyond the cost in monetary terms. The other costs, in terms of human happiness, in quality of life, in productive capacity and international competitiveness will eclipse even this.

Were money to be made available, there is probably a limit of an extra £1 billion to £2 billion per annum that the construction industry could absorb without overheating. Unfortunately, this level of expenditure would not keep pace with the rate of deterioration. Investment in construction must be attractive in a time of high unemployment in that it is labour-intensive.

Under the right-to-buy legislation created by the Housing Act 1980, 880 000 houses had been sold in 1986 by local authorities to their tenants. Receipts exceeded £5.5 billion in initial payments and mortgage repayments, and yet local authorities could only spend 20 per cent of capital receipts on their housing programmes, which seems grossly inequitable. There are also other problems inherent in the sale of council houses, with a substantial number of mortgage default cases and neglect of maintenance because of former tenants becoming over-stretched. Atwell[34] has described how the 'cream' of local authority housing has been sold off and not replaced, and much of what remains are the unlettable and unsaleable flats and maisonettes.

Many families cannot be rehoused and waiting lists grow longer. There is an unquestionable need for a new housing-built-for-rent programme to replace some of the lost stock. There is also the moral question of selling off so much housing at large discounts that was originally built at public expense, and on which local authorities will be paying debt charges into the far distant future.

The removal of VAT from repairs would encourage property owners to maintain their buildings to a higher standard. As recommended in the RICS report on housing[6], the system of mortgage tax relief needs review to determine how best to ensure that, at the very least, only the less well off and first-time buyers derive substantial benefit.

References

1 Department of the Environment. Research and Development Bulletin. *Practice in Property Maintenance Management—A Review* . HMSO (1970)
2 Department of the Environment. Research and Development Bulletin. *Building Maintenance—The Report of the Committee.* HMSO (1972)
3 British Standards Institution. *BS 3811: 1984 Glossary of maintenance management terms in terotechnology*
4 E. D. Mills (Ed.). *Building Maintenance and Preservation.* Butterworths (1980)
5 Chartered Institute of Building. *Managing Building Maintenance* (1985)
6 Royal Institution of Chartered Surveyors. *Housing—The Next Decade* (1986)
7 Department of the Environment. *An Enquiry into the Condition of Local Authority Housing Stock in England.* HMSO (1985)
8 Royal Institute of British Architects. *Decaying Britain* (1985)
9 Leeds City Council. *A Review of the State of Repair of the Leeds Housing Stock* (1984)
10 Royal Institute of British Architects. *Local Authority Housing Repair and Maintenance Needs* (1985)

11 Audit Commission. *Capital Expenditure Controls in Local Government in England*. HMSO (1985)

12 National Economic Development Office. *Investment in the Public Sector Built Infrastructure*. HMSO (1985)

13 Institution of Civil Engineers. Education failure. *New Civil Engineer* (19 June 1986)

14 Joint Contracts Tribunal for the Standard Form of Building Contract. *Standard Forms of Building Contract* (1980)

15 A. Speaight and G. Stone. *The Law of Defective Premises*. Pitman (1982)

16 Building Conservation Trust. *Care of Buildings* (1985)

17 R. Chudley. *The Maintenance and Adaptation of Buildings*. Longman (1981)

18 I. H. Seeley. *Blight on Britain's Buildings*. Paintmakers Association (1984)

19 Local Government Operational Research Unit. Report C144. *Hospital Building Maintenance: Can Decision Making be Improved?* HMSO (1972)

20 E.A. Spencer. Building Occupancy Costs. *Third National Building Maintenance Conference, 1971*. Department of the Environment. HMSO (1972)

21 Building Research Establishment. *Quality in Traditional Housing. Vol. 1: An Investigation into Faults and their Avoidance* (1982)

22 P. A. Stone. *Building Economy*. Pergamon Press (1983)

23 I. H. Seeley. *Building Economics*. Macmillan (1983)

24 *BRE Digest 227*. Estimation of thermal and moisture movements and stresses: Part 1 (1979)

25 E. J. Gibson (Ed.). *Developments in Building Maintenance—1*. Applied Science Publishers (1979)

26 Chartered Institute of Building. *Maintenance Management—A Guide to Good Practice* (1982)

27 W. J. Peacock. The maintenance of buildings and structures—why should we care? *Municipal Engineer*. **3** (1986)

28 Department of the Environment. Research and Development Paper. *The Relationship of Capital, Maintenance and Running Costs—A Case Study of Two Crown Office Buildings*. HMSO (1970)

29 H. Knight. Capital cost and cost in use. *The Chartered Surveyor* (February 1971)

30 Society of Chief Quantity Surveyors in Local Government. *Life Cycle Cost Planning* (1984)

31 R. Wensley and J. Meikle. *Cost Effectiveness Criteria for Energy Investment in Private Housing*. Surveyors Publications (1985)

32 R. H. Wright. The economics of building maintenance. *The Architect and Surveyor* (January/February 1973)

33 J. P. Edwards. The economic significance of building maintenance to industry and commerce. *DOE Third National Building Conference 1971*. HMSO (1972)

34 D. Atwell. What future for decaying Britain? *Proceedings of Building 86 Seminar, London* (7 October 1986)

2 BUILDING MAINTENANCE PROBLEMS AND THEIR SOLUTION—I

Foundations, Shoring and External Works

This chapter is concerned with building defects that arise from site conditions and inadequate foundations, the various types of temporary works and the maintenance of external works, such as pavings and fences.

Site Conditions

Site Investigations

Site investigations should take place before carrying out new building work, including alterations and extensions. This aspect is becoming increasingly important since land that has not been used before is now being considered for building. Special measures may be needed to deal with difficult site conditions.

The local authority is usually the best source of information, but older editions of Ordnance Survey maps and old maps and records can give useful information on features which might cause problems, such as infilled ponds, ditches and streams, disused pipes, and sites of old buildings, services and workings. Slopes steeper than 1 in 10 may be subject to creep and this could result in heavy pressures on walls. In limestone or chalk areas, craters or gentle depressions usually indicate swallow holes formed by the collapse of sandy or loamy soils into the fissured rock below.

A polygonal pattern of cracks about 25 mm wide on the ground surface during a dry summer indicates a shrinkable soil. Shallow depressions around mature trees in open ground, repairs to paved surfaces close to trees in built up areas, and broken kerbs may indicate shrinkage due to drying. Larger cracks approximately parallel to each other normally result from deeper-seated movements such as caused by mining, brine pumping or landslips.[1]

Low-lying sites may be liable to flood, particularly where they are within the flood plain of a river, and the highest recorded flood levels should be obtained. It is highly desirable to keep all excavation work above ground-water level.

32

Some clays contain sulphates and these may cause corrosion of buried concrete, iron and steel. Where the presence of sulphates is suspected, the groundwater should be analysed.

Johnson[2] devised a comprehensive site visit checklist for use when inspecting a potential building site, to reduce the likelihood of missing any possible hazards. There is considerable merit in adopting a structured approach of this kind. Some of the more important matters to be investigated are now listed.

(1) Are there signs of damage to existing buildings?

(2) Are any existing buildings supported on special foundations?

(3) Are there signs of landslip or erosion of slopes, such as surface rippling or tension cracks on the surface (clay slopes of gradients greater than 1:10 can be subject to creep)?

(4) Is there evidence of imported soil, tipped material or rubbish?

(5) Stickiness when wet and cracking when dry may indicate a clay with shrinkage/swelling properties.

(6) Note the location, species, height, girth and condition of any trees.

(7) Bounciness underfoot or any evidence of past flooding can signify a high water table.

(8) How was the site used previously?

(9) If contamination is suspected, professional advice must be sought.

(10) Is the site situated in a known area of coal or mineral extraction?

(11) Will demolition prior to development affect the stability of adjoining buildings?

(12) Is there any evidence of existing services crossing the site?

Soils

Prior to designing foundations, it is necessary to identify the soils present on the site. Boreholes or inspection pits should be excavated on the site to obtain samples for soil testing, noting the depths in each case. Various soil characteristics are needed, including colour, smell and texture. Means of soil identification are detailed in BRE Digest 64.[1]

Soil conditions have an important influence on foundation design and the subsequent behaviour of buildings. Most soils consist of solid particles of varying shapes and sizes with the spaces between filled by water or air. Large particles, like sand, are held together mainly by their own weight and when loose have little strength whereas fine particles, like clay, hold more water in films which lie between the particles and bind them together. Clays shrink with drying coupled with an increase in strength, while on wetting they swell and lose strength.[3]

A foundation load increases external pressure on the soil, squeezing out water from between the soil particles. With larger particles, as in the case of sand, the water movements are rapid and the soil settles fairly quickly after the load is applied. By contrast, clays offer resistance to water expulsion, and settlement can continue for years after construction.

Foundation Problems

Movements resulting from Loading

The extent of foundation movement depends on the nature of the soil and the amount of imposed loads. Not even uniform ground uniformly loaded settles evenly and the complex properties of soil make it difficult to assess the degree of settlement of individual foundations. It has been suggested that with large structures compression of the foundations may continue for some 20 years after construction, although most significant movements take place within five years. Shallow foundations, such as strip, pad and raft foundations, subject to normal loadings increase pressure in the soil to a depth and breadth equal to one-and-a-half times the breadth of the foundation.[3] Differential patterns of loading are likely to lead to differential settlement only where there are changes in ground support. The principal exceptions are bays, rear additions and internal partitions, which may be built on smaller and shallower foundations. There is often differential movement between these features and the main enclosing walls.

Movements resulting from Other Causes

Movements of foundations can also result from seasonal weather changes, growth or removal of vegetation, earth flows and subsidence. The problems are accentuated with soils composed of fine particles and so different soil conditions are considered separately.

(i) *Clay soils* Clays, which shrink on drying and swell again when wetted, are commonly responsible for the movement of shallow foundations. Where clays are firm enough to support buildings of several storeys they are known as firm shrinkable clays.

The roots of trees and shrubs penetrate soil to considerable depths and extract moisture when rainfall is low in summer, causing drying out of the soil. Beneath large trees and shrubs in the United Kingdom permanent drying has extended to about 5 m and shrinkage of 50 to 100 mm has been measured at the ground surface (figure 2.1.1). To some extent the building protects the clay from seasonal drying and wetting, and movement is more likely under outer walls and corners. Shrinkage of clay occurs both horizontally and vertically, so there is a tendency for walls to be drawn outwards in addition to settling and for cracks to open between the clay and the sides of the foundations. These cracks permit water to enter during the following winter and to soften the clay against or beneath the foundations (figure 2.1.2). Buildings should not in general be erected closer to single trees than their height at maturity, or one-and-a-half times their height in the case of groups or rows of trees. New trees should not be planted nearer to existing buildings than these distances.[3] BRE 298[4] describes how the minimum distance between buildings and trees varies with the species of tree and that it can be half the height of the tree with lime, ash, elm, sycamore,

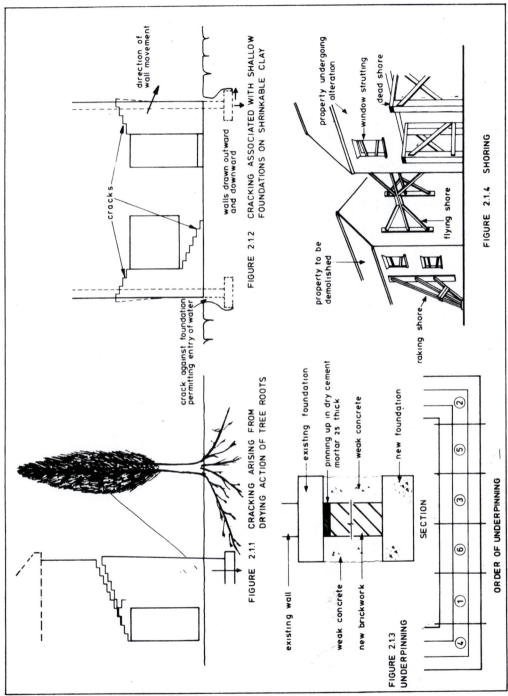

FIGURE 2.1.1   CRACKING ARISING FROM DRYING ACTION OF TREE ROOTS

FIGURE 2.1.2   CRACKING ASSOCIATED WITH SHALLOW FOUNDATIONS ON SHRINKABLE CLAY

FIGURE 2.1.3   UNDERPINNING

FIGURE 2.1.4   SHORING

Figure 2.1

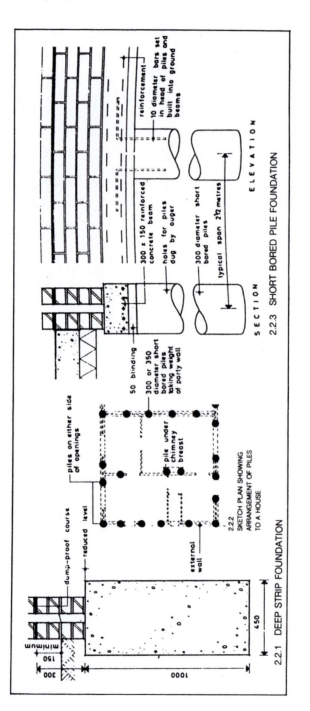

Figure 2.2 Foundation types

hawthorn, beech, birch and cypress. Constant pruning of growing trees is often necessary to restrict their height and avoid risk of damage to buildings.

When trees are felled to clear a site for building, a considerable time should be allowed for the clay to regain water previously extracted by the tree roots. Otherwise the clay as it swells may lift the building.[3]

Boilers and furnaces inadequately insulated from the clay beneath may dry and shrink the clay, resulting in the fracture of the concrete foundation slab through lack of support.[3] Short-bored pile foundations are often well suited for use in shrinkable clay, although narrow strip (trench fill) foundations may provide a satisfactory alternative, as illustrated in BRE Digest 241[5] and figures 2.2.1, 2.2.2 and 2.2.3.

(ii) *Sandy soils*   Dense beds of sand form excellent foundation soils, but underground water can wash out the finer particles, leaving coarser material in a less stable condition.

During severe winters in the United Kingdom frost may penetrate soil to a depth of 600 mm or more. Where the water table is close to the ground surface, the water can become frozen and cause lifting of the ground surface, known as 'frost heave'.

(iii) *Organic soils and made-up ground*   Peat and other soils containing a considerable proportion of organic matter in the form of decaying vegetation vary in volume as their water content changes, and are also readily compressible. Made-up ground often settles for many years unless it is good material, carefully placed and adequately compacted in thin layers. Indeed, poorly compacted fill is unsuitable for foundations and these need to penetrate the fill to firm strata beneath, often using a piled foundation. Steel drums, trunking, paint tins and even car bodies have been found in uncontrolled tips. The dangers to the structure of a building on unconsolidated fill or on soft ground are very serious. Further settlement and cracking may occur after repairs have been carried out; with settlements of this kind, movements are often extensive and their timing unpredictable.

Large-scale Movements

Some foundation movements occur in good foundation soil owing to natural or geological phenomena, artificial agencies or a combination of them. For instance, clay soils on slopes exceeding 1 in 10 are likely to move downhill, albeit slowly, while in chalk and limestone areas, cavities in the bedrock can be formed by underground streams or watercourses dissolving the rock. If a sandy overburden falls into the cavity, a vulnerable 'swallow hole' is formed at the surface.

Large settlements occur in mining areas as the ground subsides over workings. Normally the ground surface stretches as the front of a subsidence approaches and buildings start to tilt towards it. Subsequently the tilt decreases but settlement increases as the ground below is affected, which can result in fractures to the structure, sagging of arches, collapse of beams and fracture of pipe joints. Provided structural damage is not severe, the building will slowly return to an essentially vertical position but at a lower level. Small brick dwellings on comparatively thin reinforced concrete rafts can usually resist moderate movements without undue damage.[6]

Design of Foundations

With light traditional buildings, strip foundations, having a width equal to about twice the thickness of the loadbearing wall, will not impose a pressure in excess of the permissible bearing pressure on any soils, except very soft clays and silts, peats and made-up ground. For most soils, considerations of bearing pressure will only arise with heavier buildings where, for example, heavy load concentrations are imposed on pier or pad foundations. Permissible bearing pressures for different soils are listed in CP 101.[7]

The depth to which foundations have to be excavated is largely dependent on the following three factors.

(1) to secure adequate bearing capacity;
(2) in the case of clay soils to penetrate below the zone where shrinkage and swelling due to seasonal weather changes are likely to cause appreciable movement;
(3) in fine silts and sands, to penetrate below the zone in which trouble may be expected from frost.

The principal types of foundation and their main functions are well described and illustrated in *Building Technology*[8] and *House Foundations*.[9]

Concrete in Foundations

The strength of concrete is influenced by a number of factors

(1) proportion and type of cement;
(2) type, proportions, gradings and quality of aggregates;
(3) water content;
(4) method and adequacy of batching, mixing, transporting, placing, compacting and curing concrete.

The majority of concrete foundations contain ordinary Portland cement, although this may be varied in special circumstances. For instance, in sulphate-bearing soils and groundwaters[10] it is advisable to use a special cement, such as sulphate-resisting Portland cement to BS 4027[11] or supersulphated cement to BS 4248.[12] The customary minimum standard mix by volume of concrete is 1:3:6. BRE Digest 244[13] outlines the disadvantages of nominal mix proportions by volume. They fail to specify adequately the cement content of the mix, as the actual cement content of a m^3 of concrete made to a particular nominal mix varies with different aggregates and different water contents. There is considerable merit in specifying mixes in terms of cement content in kg/m^3 of fresh concrete or compressive strength at 28 days. Typical examples are grade C7.5 (7.5 N/mm^2) or 120 kg/m^3 of cement for plain concrete or grade C20 (20 N/mm^2) or 220 kg/m^3 of cement for reinforced concrete with dense aggregate. The water content should be kept as low as possible consistent with sufficient workability and water–cement ratios are usually in the range of 0.40 to 0.60. To secure satisfactory results, the concreting materials must be properly batched, adequately mixed, carefully transported and placed, adequately compacted and properly cured.

Premature failures of concrete foundations may arise through weak concrete or insufficiently mixed or unsuitable materials, or through the foundations being of inadequate cross-section or supported on a weak formation. Hence there are many ways in which the concrete may fail and ample precautions must be taken to ensure sound work. Concreting can proceed in frosty weather, provided rapid-hardening cement or a richer mix of cement is used, water heated and aggregates defrosted, and the concrete placed quickly and formwork left in position for a longer period.[8] Work can often be accelerated while at the same time ensuring a good-quality material by using ready-mixed concrete. The ready-mixed concrete suppliers also have good arrangements for defrosting aggregates and heating mixing water in cold weather. Ready-mixed concrete also has advantages on restricted sites or where large quantities of concrete are to be laid in a short time.

Modifications of Existing Foundations to take Increased Loads

It is often necessary, when carrying out alterations to existing buildings, to check on the adequacy of foundations. This check is carried out by digging down and exposing the foundations at suitable points and noting the size, depth and general condition of the existing concrete foundations. In a few cases it may even be found that the walls merely rest on brick footings with no underlying concrete foundations.

Where it is found that the existing foundations are insufficient to carry the additional loads, measures must be taken to increase the strength of the foundations. This is usually done by constructing new foundations under the existing foundations in lengths of 900 to 1200 mm as underpinning, and this type of work is described later in the chapter.

If existing foundations are of adequate thickness and quality, walls can be underpinned by installing raked mini-piles through the footings, generally alternating between the inside and outside of the building. The majority of small-diameter mini-piles are formed of driven steel casings filled with cement grout and provided with nominal reinforcement by a single central bar. In clay soils, casings are not normally necessary and augered, cast in grout or concrete piles can be used.

Settlement of Buildings

Causes of Settlement

Buildings may settle for a variety of reasons, including inadequate foundations, low-bearing or shrinkable clay soil, presence of large trees near the buildings, and the undertaking of extensive excavations or mining nearby. In many parts of the country, particularly in south-east England, settlement arises through foundations laid on shrinkable clay. This type of clay shows large surface cracks in dry weather and becomes very sticky in wet weather.

If it can be established that the cracks appeared or that they open and widen during dry weather in late summer and partially close in winter, and that windows and doors which jam in late summer become easier to open in winter, then the distortion can usually be attributed to shrinkage of clay below the foundations. Where there are fast-growing trees such as poplar, elm or willow within 30 m of the building, or vigorous shrubs or creepers within 1.50 m of it, then the drying action of the roots on the soil is likely to be substantial and windows and doors may remain jammed even in winter.

Assessment of Damage in Low-rise Buildings

It is necessary to examine all evidence of actual past and present damage to the building, having regard to its location, age and form of construction. The first step is normally to inspect all cracks, both internally and externally and any distorted doors or windows. It is generally necessary to establish whether the damage stems from foundation movement and whether it is likely to get progressively worse. BRE Digest 251[14] classifies damage to walls and makes recommendations as to remedial work as shown in table 2.1.

Remedial Measures for Settled Buildings

Remedial measures are generally difficult and expensive. Each case should be considered individually on its merits and where movements are detected in the early stages, further damage can sometimes be prevented at little cost. The repairs are best covered under four main headings.

(1) *Movements associated with direct climatic drying (no tree roots).*   Differential movements may be eliminated by underpinning all the external walls to a depth of at least 900 mm. It is unwise to underpin a fractured corner in isolation as this may accentuate relative movements in the remainder of the building. It is generally necessary to underpin the whole of the external walls, either by continuous underpinning placed in alternate sections or by separate blocks of underpinning so placed in relation to door and window openings as to take full advantage of the strength of the existing brick walls.

Table 2.1 Classification of damage to walls and appropriate remedial work

Category of damage	Degree of damage	Approximate crack width (mm)	Description of typical damage and ease of repair
1	Very slight	Up to 1	Fine cracks inside the building which can be remedied during normal decoration
2	Slight	Up to 5	Cracks may not be visible externally; doors and windows may stick slightly. Internal cracks easily filled and some external pointing may be required
3	Moderate	5 to 15 (or number of cracks up to 3 wide)	Cracks require opening up, external brickwork repointed and possibly small amount of brickwork replaced. Doors and windows sticking; service pipes may fracture; weathertightness often impaired
4	Severe	15 to 25 but also depends on number of cracks	Extensive repair work involving replacing sections of walls, particularly over doors and windows. Window and door frames distorted, sloping floors, leaning or bulging walls, some loss of bearing in beams, and service pipes disrupted
5	Very severe	Usually greater than 25 but depends on number of cracks	Requires a major repair job involving partial or complete rebuilding. Beams lose bearing, walls lean badly and require shoring. Windows broken with distortion. Danger of instability

Before any underpinning work is commenced the building should be carefully examined and any urgent repairs carried out. 'Tell-tales' should be fixed over existing cracks to show any further movement. In particular, the thickness and structural condition of the walls to be underpinned should receive special consideration, as well as the nature of the ground under the existing foundations. The wall to be underpinned should first be supported with flying or raking shores (figure 2.1.4), loads reduced as far as practicable, and holes then excavated alongside and under the existing foundations in a suitable sequence, as illustrated in figure 2.1.3, so that the sections being worked at any one time will be as far distant from one another as possible. The lengths of excavated sections are usually about 1.00 to 1.50 m, so that the existing wall can bridge the gap satisfactorily with a minimum of support. The sum total of unsupported lengths should not exceed one-quarter of the wall length or one-sixth in the case of heavily loaded walls.

When excavating each hole, sufficient space is allowed in front of the wall to provide adequate working space. No earth faces must be left unsupported overnight and it is customary to limit the lengths of poling boards to 900 mm. In some cases chemical consolidation of the ground or pressure grouting may be used to ensure stability of the ground. The foundations and brickwork are constructed in each section and the back timbering removed where possible and replaced with weak concrete. Steel dowels are often built into the ends of the concrete foundations to key the different sections together. The brickwork is normally constructed in English bond and toothings are left at the end of each section for subsequent bonding to adjacent sections. The top of the new brickwork has to be pinned up to the underside of the existing wall or concrete foundation. The most usual method is to ram dry cement mortar about 25 mm thick into the gap using a piece of board and a club hammer, while another alternative is to use manufactured keying blocks made of dense concrete. Alternatively, concrete 'legs' can be used in place of the lengths of brickwork.

Underpinning of foundations may prove too costly, particularly where the damage is not severe or the building is very old. In these circumstances it may be considered adequate to reduce further movements by surrounding the building with a relatively impervious apron of precast slabs or *in situ* concrete to a width of 1.50 m. After the laying of the apron, the ground should be left to absorb moisture for a winter before the cracks in the building are repaired. This does not provide a completely satisfactory solution and is very much a compromise.

(2) *Movements associated primarily with the drying action of tree roots.* The treatment in this case is usually more difficult. Where the trees have not reached maturity it is good practice to cut them down and kill the stump, probably using sodium chlorate. The ground under the fractured part of the building will slowly swell up during wet weather and tend to lift the building and partially close the cracks. The filling of the cracks should be delayed for at least one wet season to permit this movement to take place.

If the trees have reached maturity and the building is fairly old, it is unlikely that further movements will occur except in exceptionally dry spells. In this situation it would be best to leave the trees in position and merely fill up the cracks in the building. It is not often economical to underpin buildings badly affected by trees as it is frequently necessary to underpin to a depth approaching 3 m and the cost of this work may exceed the value of the property.

When repairing cracks to brickwork and similar walling materials, the horizontal gaps should first be wedged tight by driving in pieces of slate or tile at intervals to give support to the upper parts of the structure. The outside face of the cracks may then be filled with mortar and pointed. Plaster cracks should be cut back to a reasonable width and filled with gypsum plaster gauged with lime. The proportion of lime is varied to produce a plaster of similar hardness and suction to the existing material.

(3) *Mining subsidence.*   Ground subsidence often results from the extraction of minerals, particularly coal. A combination of horizontal movements at the surface coupled with vertical movements can cause serious damage to buildings. An extracted coal seam 1.30 m thick can cause subsidence at the surface of up to 1 m deep immediately above the seam and reducing on both sides.

A subsidence wave may first cause tension in buildings at the crest, followed by compression in the trough. The worst effects are with thick seams in shallow workings. Tensile stresses cause lengthening of structures with fractures in walling at butt joints and at the corners of window and door openings, followed by the fracture of pipe joints and displacement of beams. Compressive stresses may result in buckling of walls and the arching of pipes and paving materials. The use of flexible pipes or pipes with flexible joints, with provision for movement where they pass through the structure, is advisable in areas liable to subsidence. An ideal type of foundation in this situation is a reinforced concrete slab resting on a bed of friable material such as sand. Fibreboard, hardboard or wallboard should replace plaster as a finish to ceilings and partitions. Steel frameworks should have flexible joints such as pin joints, as in the Consortium of Local Authorities Special Programme (CLASP) buildings, which also contain coiled springs on diagonal members. All cladding units are hung free to slide against adjoining units. The extra cost of this precautionary work has been largely offset by the savings stemming from bulk purchases.

With dwellings, smaller units are more stable than larger ones; for example, semi-detached houses are less vulnerable than terraced houses. Outbuildings should ideally be independent of the main structures and projecting bays, porches and the like are best avoided. Breaks in long buildings should extend through the foundations. Paved surfaces should be of flexible materials, such as tarmacadam and asphalt. Door openings constitute points of weakness and are best located in short walls, avoiding front and back doors opposite one another and doors in adjoining dwellings from being placed side by side.

The traditional method of underpinning previously described is not well suited for dealing with buildings subject to mining subsidence, as it does not provide a solution to the problem of continued settlement after underpinning nor does it return the building to a truly level and upright position. The following methods are used in various situations, and these are well described and illustrated in *Repair and Renewal of Buildings*.[15]

(a) *Underpinning with concrete stools and ground beams.* This method is only really suitable where the settlement is small and unlikely to recur. The stools consist of short struts or columns, often of prestressed concrete about 225 × 225 mm in section and 450 mm high, spaced at about 900 to 1200 mm centres. A series of holes or pockets are cut into the wall to be underpinned and the stools with top and bottom steel distributing plates are inserted in the holes and packed around solidly with mortar. Once the mortar has set, the intervening brickwork is cut away to accommodate *in situ* reinforced concrete beams, with their tops normally two courses of brickwork below damp-proof course. Finally the tops of beams are pinned up to the brickwork above and their outside faces are normally rendered to give the appearance of a plinth. Bored piles or brick piers are often taken down to a firm base to provide support to the beams.

(b) *Underpinning by jacking.* Where a building needs relevelling and returning to a truly vertical plane, following extensive and possibly continuing settlement, jacks or rams replace the stools with the lift reproducing the settlement in reverse, and a part of the building which has for instance settled 200 mm must be pushed back into position at twice the speed of a part that has settled 100 mm, with an average jacking speed of about 25 mm per hour.

A contoured plan is prepared joining all points of equal settlement to produce 'jacking contours'. The apparatus controlling the jacks is based on a system of moving water levels with manually actuated jacks in small systems and power operated jacks in large ones. Finally the jacks are replaced by conventional stools pinned up to existing brickwork, followed by the insertion of ground beams and probably foundation piles or piers.

(c) *Pedatified foundations.* When erecting new buildings in areas where ground movements are continuing or are to be repeated, it is advisable to incorporate permanent jacking points often using pedatified foundations. One method, suitable for two-storey buildings, entails the construction of substantial foundation piers and bases which support a framework of ground beams. At each jacking point a steel prop or pedapyn passes through a vertical hole in the beam and rests upon a pier top. The building is levelled by means of inverted jacks with plungers resting on pedapyns and with casings bolted to beams. As the plunger is extended the casing of the jack rises, lifting the ground beam and the wall above.

(4) *Other forms of settlement.* Buildings may settle for a variety of other reasons and the cause of the settlement may sometimes be difficult to establish. A poorly constructed concrete foundation might disintegrate under load and would need replacement in short lengths, strutting the wall above while the replacement work is in progress, to prevent it slipping.

Settlement could result from soil being washed away from beneath foundations owing to leaking drains or water services. The first step must be to locate and rectify the defective service. Another possible cause is the lowering of natural groundwater level over a period of years. While the failure of an adjoining owner of land at a lower level to provide adequate retaining walls may result in landslips and consequent settlement of buildings at a higher level.

Buildings erected on unsuitable or inadequately compacted fill are likely to settle and where there are varying depths of fill, unequal settlement may occur unless the fill is well consolidated in layers not exceeding 300 mm thick. On most fills, pad or strip foundations are rarely suitable and raft or piled foundations are generally needed. Grouting of the fill would be one method of strengthening the base material with a view to preventing further settlement. A more expensive but sounder job could be obtained by underpinning with the supporting stools being taken down through the fill to a firm base below.

Shoring

Shoring may be needed to give temporary support to walls and floors during alteration work, demolition work or underpinning, or where a structure has become unsafe. In the absence of adequate shoring the buildings could collapse, possibly causing death or injury to persons in or near the building. Shoring may take a number of different forms—raking shores, flying shores, dead shores, window strutting and floor propping. Figure 2.1.4 illustrates the nature and uses of the main forms of shoring.

Shoring members are generally of timber, often pitch pine, with all needles, cleats and wedges preferably of hardwood. Each type of shoring is now considered in turn.

Dead Shores

The purpose of dead shores is to support dead and superimposed loads of a building, mainly while alteration and repair work is in progress. At the same time it is generally necessary to strut existing floors and roofs to relieve the walls of their weight, and a suitable framework could comprise 225 × 50 mm headboards and sole pieces, 225 × 75 mm dead shores and 150 × 25 mm braces. It may be possible to reduce costs, as timber prices have increased considerably, by using second-hand timber or combining lighter sections of timber to make up heavier ones, such as the use of three 225 × 75 mm members to build up one 225 × 225 mm. Another alternative is to use adjustable steel props which are very strong, easy to fix and may be hired at reasonable rates.

A dead shore, as illustrated in figure 2.3.1, supports a wall of a building while an opening is being formed in it. Needles may be of timber or steel with sizes depending on their spacing, distance apart of dead shores and the loadings. The dead shores must be of sound material fixed in an upright position, and in the case of timber members the minimum width should be at least one-twenty-fourth of the height and they must be securely fixed to base plates and needles, often by means of dogs. Clearance of 750 mm between the wall and dead shore is needed to give adequate working space.

Needles should not be located beneath window openings and the spacing can vary from 900 mm to 1.80 m, depending on the condition of the brickwork. Bracings are often of 225 × 25 mm timber on both faces of dead shores, fixed at an angle of about 45°. Sole plates must be placed on a firm base. The insertion of folding wedges between needles and dead shores (posts) permits tightening of the shores without exerting pressure on the structure.

Raking Shores

Raking shores may be used to provide temporary support to a wall which has become defective and unsafe, or as a precautionary measure while alteration work is being undertaken. The arrangement of the shores will depend on the

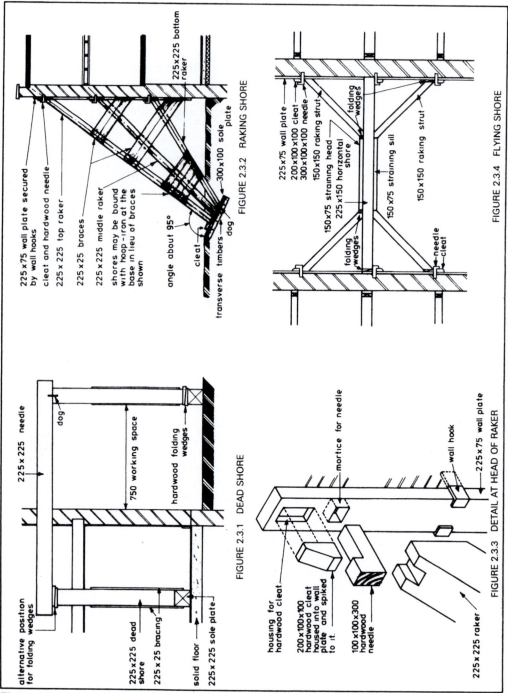

225 x 75 wall plate secured by wall hooks

cleat and hardwood needle

225 x 225 top raker

225 x 25 braces

225 x 225 middle raker

shores may be bound with hoop-iron at the base in lieu of braces shown

angle about 95°

cleat

transverse timbers

dog

225 x 225 bottom raker

300 x 100 sole plate

FIGURE 2.3.2   RAKING SHORE

225 x 75 wall plate

200 x 100 x 100 cleat

300 x 100 x 100 needle

150 x 150 raking strut

folding wedges

150 x 75 straining head

225 x 150 horizontal shore

150 x 75 straining sill

150 x 150 raking strut

folding wedges

needle

cleat

FIGURE 2.3.4   FLYING SHORE

225 x 225 needle

dog

750 working space

hardwood folding wedges

alternative position for folding wedges

225 x 225 dead shore

225 x 25 bracing

solid floor

225 x 225 sole plate

FIGURE 2.3.1   DEAD SHORE

mortice for needle

housing for hardwood cleat

200 x 100 x 100 hardwood cleat housed into wall plate and spiked to it.

100 x 100 x 300 hardwood needle

wall hook

225 x 75 wall plate

225 x 225 raker

FIGURE 2.3.3   DETAIL AT HEAD OF RAKER

Figure 2.3

height of the building, loads to be carried, extent of openings and space available adjacent to the building. They can consist of single rakers, or a number of them as illustrated in figure 2.3.2, which are inclined members, principally of timber, with the upper end terminating against the wall requiring support and the lower end supported by a sole plate bearing upon the ground. A wall which is out of plumb to the extent of 1 in 12 or more requires shoring immediately.

The angle of the shores with the ground should, as far as practicable, be between 45 and 75°. The spacing of flying shores varies between 2.50 and 5.00 m and it is advisable to position a shore at each end of the wall to be supported. The centre line of each raker or shore intersects the ends of suspended floor joists and centre lines of wall plates to provide maximum support at critical points of concentrated loads. Support from shores is spread over a larger area of wall by the use of wall plates or wall pieces, secured with metal wall hooks at about 2.50 m centres. With tall buildings the top shore may consist of two lengths—a 175 × 175 mm rider bearing on folding wedges at the top of a shorter and larger section back shore.

The connection between the head of a raker and the wall plate is usually formed with a needle and cleat (figure 2.3.3). A hole is cut in the wall and wall plate and a hardwood shouldered needle, probably about 300 × 100 × 100 mm, inserted. This is further supported by a hardwood cleat, often about 225 × 100 × 100 mm, which may be housed or nailed to the wall plate. The sole piece, which prevents the rakers from slipping, is normally about 100 mm thick and forms an internal angle of about 85° with the top raker. A grillage is sometimes provided below it. A crowbar can be used to lever a raker into position after which a dog is inserted to prevent any subsequent movement.

The usual order of erecting raking shores is as follows.

(1) Cut holes in the wall to receive needles, taking care not to disturb the surrounding brickwork or masonry.

(2) Wall plates are prepared, morticed for needles, notched for cleats, and then fixed to the wall with wall hooks.

(3) Where the wall is bulging badly it is necessary to place packing pieces behind the wall plate to provide an even bearing thoughout.

(4) The needles are prepared and inserted through the wall plate into the holes in the wall.

(5) A bevelled cleat is inserted in each housing in the wall plates and spiked.

(6) The heads of rakers are cut to the required angle and notched to receive the needle for a width of at least 75 mm.

(7) The sole plate is fixed in position and the bottom ends of the rakers tightened against it by levering them with a crowbar fitted into notches cut into the feet of rakers, after which dogs are inserted to prevent any subsequent movement.

(8) Bracing boards and/or hoop iron binding are then fixed for stiffening purposes.

Flying Shores

Flying shores are used to provide support between buildings, where an intervening building has been demolished, or across a narrow street or alley, when the consent of the highway authority will be needed in addition to that of the owner of the property from which support is required. The use of this type of shore is restricted to spans of 4.50 to 10.50 m at spacings of 2.50 to 4.00 m. A typical flying shore is illustrated in figure 2.3.4.

The horizontal member is termed a horizontal shore and varies from about 150 × 100 mm to 225 × 150 mm depending on the span. With tall buildings more than one horizontal member will be required. These members are supported at each end by a needle and cleat as described for raking shores. Raking struts, varying in size from 100 × 100 mm to 150 × 150 mm with the span, run from horizontal shores to the floors and ceilings in the two buildings. A length of straining sill and head nailed to the top and bottom of the centre part of the horizontal shore assist in stiffening the shore and providing support for the raking struts.

Sequence of Shoring Operations

Two examples of shoring work will serve to illustrate the sequence of operations.

(a) *Demolition of and rebuilding an intermediate three-storey terrace house.* It can be assumed that the adjoining houses are in poor condition so that maximum care must be taken. Following inspection, raking shores are erected to each upper floor and eaves opposite each party wall. Flying shores are then erected inside the building between the party walls, with the outside ones positioned close to the front and rear walls, with intermediate ones at intervals of 2.50 to 3.50 m, cutting holes in floors as necessary. Once the shoring is complete, demolition can proceed and the shores will remain in position until the new building work is virtually complete.

(b) *Inserting a shop front in an existing dwelling.* All window openings in the front wall are strutted to prevent deformation. Raking shores are then erected against the party walls and at intermediate points if the brickwork is in poor condition. Floor strutting consisting of 100 × 100 or 150 × 150 mm posts spaced about 1.20 m apart, wedged between 225 × 75 mm horizontal members (head and sill) at floor and ceiling level at each storey, will be provided, to relieve the front wall of floor and roof loads. This strutting is placed as near as possible to the wall, often 900 to 1200 mm from it, and may be sheeted to form a screen where the building remains in occupation during the alterations. Holes for needles are then cut and dead shores erected and wedged. The ground floor brickwork is cut away, piers built and steel beams inserted and wedged. After allowing about two weeks for the mortar to harden, the shoring is removed in the reverse order to which it was erected and needle holes in the brickwork made good.

Waterproofing Leaking Basements

One of the main requirements of basements is watertightness. Basements may be constructed in watertight concrete or in brickwork or other materials. These two broad categories will now be examined from the viewpoint of repairing leaks.

(1) *Watertight Concrete Basements*

There are difficulties in pinpointing leaks in concrete floor slabs and one good approach is to divide the slab up with small dams of bricks-on-edge in cement mortar (1:3) to locate leaks. Temporary steel tubes usually 20 mm diameter are then inserted in sockets set in the concrete, first for injecting a detergent solution to delineate the defective area and then for pressure grouting to seal it. Holes, 45 mm diameter, are drilled through the concrete with a portable percussion drill to take the steel sockets at a spacing of about 150 mm centres for fine cracks increasing to 600 or even 900 mm for wider cracks and areas of honeycombed concrete. The sockets are caulked into the holes with lead wool.

Various grouts are available including Portland cement and pulverised fuel ash, a chemical such as calcium chloride or sodium silicate, or resin. The grout is pumped in at a pressure of about 1.50 MN/m^2. After the grout has hardened satisfactorily, the tubes and sockets are removed and the holes made good.

(2) *Other Forms of Damp Basement*

Waterproofing a leaking basement with or without asphalt tanking can be both difficult and costly. The cause of the dampness must first be established and may arise from a number of causes, such as water pressure from waterlogged ground, damp penetration or condensation.

Porous drain pipes laid around the building to discharge into a natural watercourse or sewer can reduce the lateral water pressure arising in waterlogged ground. Latex-siliconate infusion by gravity or pump pressure is useful with porous basement walls free from cracks.

Another remedial measure is to apply dry linings to basement walls fixed on treated timber battens. Such a lining can prevent dampness and hygroscopic salts from damaging new decorations, but it will not provide a permanent cure. A polythene sheet vapour barrier should be fixed immediately behind the lining and it is desirable to ventilate the airspaces between battens to the outside air.

Bituminous dovetailed lathing (Newtonite) provides another useful alternative. It replaces the damp plaster and must be fixed with rust-resistant nails. A rotten suspended wood floor needs replacement with a concrete slab.

A waterproof rendering to the inner surface of a soundly constructed brick wall may stop damp penetration. Such renderings are usually based on Portland cement and sand (1:3) with or without admixtures. They must not be excessively wet or over-trowelled and should be applied in two coats not exceeding 50 mm total thickness. Angles and junctions in renderings should be coved.

Asphalt tanking requires a dry background which may be difficult to achieve in a damp or leaking basement. The cost of drying out will be very expensive and could cost as much as the tanking. The asphalt should be in three coats finishing not less than 30 mm thick for horizontal work and 20 mm for vertical work. Horizontal asphalt needs covering with at least 50 mm of concrete, while vertical work needs supporting by a half-brick or block wall. This reduces the size of the basement and obscures the source of any possible future leak. Some useful recommendations on asphalt tanking and the general protection of buildings against groundwater are given in CP 102.[16]

Maintenance of Paved and Grassed Surfaces

The cost and extent of maintenance work on paved surfaces varies with the type of construction.

Gravel or *hoggin* needs support at the edges from precast concrete kerbs or edgings, otherwise displacement of the material is likely to take place. This form of surfacing is not suitable for heavy loads. It is also expensive in maintenance as it will require periodic rolling and raking, coupled with occasional making up of depressions and application of weedkiller to retain the surface in good condition. This is one case where low initial costs may be more than offset by high maintenance costs.

Tarmacadam consists of graded stone coated with tar laid hot and compacted by rolling. It is normally laid in two coats with an average thickness of about 75 mm on a suitable base of hard material. Better-quality surfaces can be obtained by using bitumen macadam, hot rolled asphalt, mastic asphalt or fine cold asphalt. All these materials have the advantage of being easily laid to irregular shapes and varying falls and cambers. They provide a flexible paving which is desirable in areas liable to subsidence or where access may subsequently be required to underground services. They do however need support at edges, as described for gravel, and periodic surface dressing to seal the surface, with the frequency depending on the material and amount of wear. Some favour the use of 6 mm whinstone chippings for surface dressing on grounds of improved appearance and better riding qualities, but greater quality control of the binder viscosity is required to prevent the binder creeping above the smaller chippings. Attention to local weather forecasts determines whether an adhesive agent should be added to prevent stripping of chippings, should rain occur.

Defects in carriageway surfaces arise from various causes—exposed trench reinstatements, potholes and patches, crazing and cracking, edge failure, lack of roughness (skid resistance) and excessive undulation (ride quality). Some ranking procedure is necessary in determining maintenance priorities. Crazing and cracking warrant close investigation as they may indicate a need for strengthening or reconstruction, whereas in other cases a surface dressing may be economic and assist in arresting deterioration. Excessive undulation can also result in extensive maintenance work and an assessment of patches as a percentage of total paved surface will indicate the scale of the problem.

There are four main criteria for determining whether a road requires maintenance work:

(1) structural deterioration threatening to impair its load carrying ability;

(2) poor riding quality, when the riding comfort has deteriorated below an acceptable standard;

(3) slipperiness, when the surface no longer provides adequate resistance to skidding;

(4) other surface deterioration, such as surface ravelling, potholes and opening of concrete joints.

Concrete can be used for both roads and footpaths, but it must be of a suitable mix, often 1:2:4, or strength grade C7.5 (7.5 N/mm^2), laid on a waterproof membrane on a suitable base, and be adequately mixed, compacted and cured, with ample provision for expansion and contraction. It provides a hard-wearing surface, although irregularities sometimes occur at the joints and sun glare from the surface may be a disadvantage. It constitutes rigid construction and is not well suited for use in areas liable to subsidence or for subsequent service trench reinstatement. Joint fillers are often formed of softwood, medium density chipboard or fibreboard, with a groove at the top of the joint to take a pliable sealing material, an important function of which is to keep out grit. Periodic re-sealing of joints forms an important maintenance item with concrete roads.

The mean construction costs of concrete are likely to be more expensive than flexible construction, particularly for rural secondary and housing estate roads. When discounted maintenance costs over a 50 year period are taken into account, the differences between these classes of road are insignificant, although concrete roads normally have lower maintenance costs in the first 15 to 20 years. With major roads, concrete is likely to be the better long-term proposition.

Precast concrete paving flags are reasonably popular for footpaths because of their attractive appearance and relative ease of reinstatement. The main faults occurring with flag footpaths are trips (vertical misalignment of adjoining flags) and cracks, mainly owing to differential settlement, resulting in dangerous conditions for pedestrians. Foundation faults can also cause flags to rock. It is necessary for maintenance purposes to establish a maximum permissible height of trip, and 20 mm is a commonly adopted dimension.

Random events, such as vehicles mounting footpaths, contractors' damage and tree roots, result in broken and uneven paving. Hence frequent inspections and repairs are often necessary and shorter relaying cycles would not necessarily guarantee better paths.

Other pavings include cobbles, setts and bricks, and these are often laid in small areas for decorative purposes. If laid on a suitable bed the amount of maintenance should not be excessive.

Paved surfaces need periodic sweeping and road gullies require emptying from time to time to prevent their becoming silted up. In urban areas it is desirable to empty gullies six times a year and to sweep channels to town centre roads daily and channels to other roads weekly, often using a combination of mechanical and manual methods, with bonus schemes to secure maximum productivity.

Grass cutting is essential for reasons of amenity. The frequency is determined by the standard of maintenance adopted and weather conditions. The number of cuts per year could range from four to thirty, depending on the situation and standard of maintenance. Gang mowers provide the best equipment for mowing very large areas of grass, power mowers for medium areas and rotary mowers for semi-rough grassed areas. Trimming edges is labour-intensive and expensive and the ratio of edges to grassed area should ideally be kept to a minimum.

The provision of grass up to the face of buildings can result in damage to claddings by grass cutting machines. Without adequate drainage, rainwater running down the face of cladding may result in the grass adjoining the cladding becoming a strip of mud. Mud splashes up the side of the cladding are unsightly and increase maintenance costs. A concrete apron around the building will overcome these problems and also help window cleaners.

Maintenance of Fencing

Some types of fencing are particularly vulnerable from a maintenance aspect.

Interwoven fences with thin slats are liable to damage and fairly rapid deterioration, unless of oak or cedar, or of softwood which is kept regularly treated with preservative.

Oak close boarded fencing is much more durable but feather-edged softwood boarding is liable to warp and should be pressure-creosoted. The pales should be kept above ground level with horizontal gravel boards below them, which can be replaced if required without much difficulty. A capping should be provided at the top of the fence to protect the end grain at the tops of pales. Softwood posts are liable to rot at ground level and are better bolted to concrete spurs let into the ground. All nails should be galvanised to avoid rust stains on the woodwork.

Chestnut pale fences are not very attractive in appearance and maintenance costs can be high, mainly resulting from sagging of the galvanised wire supporting the pales.

Chain link fencing strained from concrete or steel posts is used extensively, as it provides a good boundary division. It is not very attractive and unless plastic-coated does rust over a period of time. It is important that the straining posts shall be well bedded in concrete and the chain link adequately strained.

Brickwork boundary walls sometimes become defective through loose or split copings. It is necessary to use engineering bricks set in cement mortar (1:3) for a brick-on-edge coping to provide adequate resistance to frost. Ends and angles may come loose unless the coping bricks are adequately anchored to the wall with galvanised steel brackets. When a boundary wall also acts as a retaining wall, frost action may cause spalling and hence the brickwork should be built in engineering bricks or be provided with a vertical damp-proof membrane.

References

1 *BRE Digest 64.* Soils and foundations: Part 2 (1972)
2 R. Johnson. *Foundation Problems associated with Low-rise Housing.* CIOB Technical Information Service Nr 61 (1986)
3 *BRE Digest 63.* Soils and foundations: Part 1 (1979)
4 *BRE Digest 298.* The influence of trees on house foundations in clay soils (1985)
5 *BRE Digest 241.* Low-rise buildings on shrinkable clay soils: Part 2 (1980)
6 *BRE Digest 67.* Soils and foundations: Part 3 (1970)
7 British Standards Institution. *CP 101: 1972 Foundations and substructures for non-industrial buildings of not more than four storeys*
8 I. H. Seeley. *Building Technology.* Macmillan (1986)
9 G. Barnbrook. *House Foundations: for the builder and designer.* Cement and Concrete Association (1981)
10 *BRE Digest 250.* Concrete in sulphate-bearing soils and groundwaters (1984)
11 British Standards Institution. *BS 4027: 1980 Specification for sulphate-resisting Portland cement*
12 British Standards Institution. *BS 4248: 1974 Supersulphated cement*
13 *BRE Digest 244.* Concrete mixes: specification, design and quality control (1980)
14 *BRE Digest 251.* Assessment of damage in low-rise buildings (1981)
15 Institution of Civil Engineers. *Repairs and Renewal of Buildings.* Telford (1983)
16 British Standards Institution. *CP 102: 1973 Protection of buildings against water from the ground*

3 BUILDING MAINTENANCE PROBLEMS AND THEIR SOLUTION—II

Wall Claddings, Dampness, Condensation and Smoky Chimneys

Defects in Wall Claddings

The Building Regulations[1] prescribe that materials used in building work shall normally comply with relevant British Standards or British Board of Agrément Certificates and Quality Assurance Schemes as recommended by the Approved Documents. Unfortunately, some modern buildings produce greater maintenance problems than many centuries-old buildings, even though the climatic conditions have undergone little change. There has been an evident failure to learn from experience and to make full use of known science and technology. When using large building units, sufficient allowance has not always been made for movement due to thermal and elastic changes and for creep in concrete under load. The effect of the rapid drying out of new buildings by central heating does not always seem to be appreciated. Insufficient attention to detailing and the failure to select suitable materials for use in exposed situations can have unfortunate effects.

Wall claddings suffer particularly from the following defects:

(1) inability to support imposed loads, resulting in differential movement, distortion or cracking;
(2) inability to keep out the weather;
(3) inability to insulate from cold with resultant condensation;
(4) deterioration of cladding materials.

The external face of a building has to resist the passage of rain and wind, since moisture penetrating to the inside can result in mildew or mould growth, or produce conditions conducive to corrosion of metals, fungal attack on timber, damage to decorations, and even detrimental effects on contents and the health of occupants. We are concerned not only with the wetting and drying of walling materials but also the chemical and physical changes in the structure caused by the complex action of moisture, frost, temperature changes, ultraviolet light, and formation and transference of salts.

Cracking of wall cladding usually results from climatic, physical or chemical changes or defective construction. It is almost invariably unsightly and unacceptable to the occupants. If severe, it may cause loss of stability,

Plate 1 Laminated brickwork

rain penetration, air infiltration, heat loss and reduced sound insulation, all of which cause a reduction in the efficiency of the building.[2] It often occurs when the tensile stress in a material exceeds its tensile strength, through externally applied loads or internal movements produced by temperature or moisture changes subject to external restraint.

Water penetration can be exceptionally critical in cold weather, particularly around vulnerable areas such as cornices and sills. Water expands by about one-tenth on freezing and can exert very great pressures within the fabric, causing it to lift and eventually break away. If water finds it way into the fabric through cracks or bad pointing, it will cause damage on freezing. Plate 1 shows laminated brickwork resulting mainly from frost action and the use of underburnt bricks, and there are signs of settlement in the brick pier.

Carbon accumulates to form soot deposits on the fabric which are both unsightly and harbour dangerous corrosive elements. Regular cleaning will move these deposits and prevent their build up, in addition to permitting a check of the condition of the fabric and enabling the necessary repairs to be undertaken.

Uncontrolled weathering and even normal use can lead to physical decay and deterioration, resulting in the need for an excessive amount of repair and renewal and often a change in the appearance of the building. This highlights the need for more detailed design at critical points of the structure and a better understanding of the nature and behaviour of materials and their use. Not only are traditional methods now being used in non-traditional ways, but designers are continually being faced with new materials, components and building techniques with insufficient back-up data. The traditional details of overhangs, cornices and drips protected wall surfaces and openings as well as enhancing the aesthetic qualities of the building.[3]

Brickwork

Clay Bricks

The majority of bricks in general use are made of clay. These are classified in BS 3921[4] according to variety—common, facing or engineering; quality —internal, ordinary or special; and type—solid, perforated, hollow or cellular. It is important to select the correct type and quality of brick for a particular situation. For example, internal quality bricks laid in weak mortar are suitable for internal walls where there is no early frost hazard, whereas ordinary bricks in medium-strength mortar are required for the outer leaf of cavity walls. Unrendered brickwork in parapet walls should contain special bricks laid in a strong mix of mortar, such as 1 part masonry cement: $2\frac{1}{2}$–$3\frac{1}{2}$ parts sand.[5,6]

BS 3921[4] specifies a minimum strength of 28 N/mm^2 for bricks, and this is sufficient for the loadings in low-rise housing and similar buildings. Higher-strength bricks should be specified only when they are required to meet structural needs, as strength is not necessarily an index of durability. Similarly, water absorption does not always indicate the behaviour of a brick in weathering. Low absorption—less than 7 per cent by weight—often indicates a high resistance to damage by freezing, although some types of bricks of much higher absorption may also be frost-resistant. Underfired bricks are likely to contain larger quantities of deleterious salts and present less resistance to surface attack by them.[7] Clay bricks expand slightly after leaving the kiln.

Other Bricks

Calcium silicate bricks of sandlime or flintlime to BS 187[8] in six acceptable classes and concrete bricks to BS 6073[9] can be used successfully in a wide range of situations subject to the selction of the appropriate class of brick and mortar.[10] To avoid shrinkage cracks, bricks should be kept dry prior to laying and a weak to medium mortar should be used as appropriate. Long lengths of external brickwork should be subdivided by vertical joints at intervals of not more than 7.5 m to 9 m to permit movement. A rigid joint filler, polythene or bituminous felt, should be inserted for the full thickness of the brickwork but kept back about 12 mm from the outside face. The joints must be sealed to ensure moisture exclusion and with facing work the joints are ideally filled with mastic.

Shrinkage cracking in brickwork may take two forms—running straight through the brick and joint in alternate courses or following a zigzag path along the joints. The latter type of crack is more easily made good and it is accordingly advisable to make the joint weaker in tension than the brick itself. Furthermore, with a weak mortar joint there is a greater chance of the bricks shrinking individually, without stressing the wall as a whole, with minute hair cracks forming around each brick rather than wider cracks at greater intervals.

Mortars

The principal requirements of mortars for brickwork and blockwork are good workability and plasticity but stiffening within a reasonable period, early attainment of strength, with a final strength adequate but not greater than bricks, and adequate durability. An excessively strong mortar concentrates the effects of differential movement by producing fewer and wider cracks and is liable to lead to increased efflorescence. Stronger mixes are preferable in cold weather to develop strength more quickly and so resist the effects of frost.[11] The selection of mortars is influenced by the type of construction and condition of exposure as shown in tables 3.1 and 3.2, extracted from BRE Digest 160.[11]

Cement mortar sets quickly and develops great strength, often more than is required, and is liable to craze, whereas *lime mortar* is extremely workable but is weak and slow hardening. Hence these mortars have been widely superseded by *cement:lime* (compo) *mortar* which is workable and sufficiently strong without the risk of drying shrinkage. A *plasticiser* may be added to cement mortar to produce an aerated or air-entrained mortar. The plasticiser entrains bubbles of air in the mix, increasing workability and permitting the use of weaker mortars in place of lime. *Masonry cement mortar* usually consists of a mixture of Portland cement with a very fine mineral filler and an air-entraining agent. It has good working properties. *Special mortars* are used in certain cases, as where soil has a high sulphate content, or high early strength or resistance to heat or chemicals is required.[11]

Table 3.1   Mortar mixes (proportions by volume)

	Mortar group	Cement : lime : sand	Masonry-cement : sand	Cement : sand, with plasticiser
Increasing strength	i	1 : 0–$\frac{1}{4}$: 3	—	—
but decreasing ability	ii	1 : $\frac{1}{2}$: 4–4$\frac{1}{2}$	1 : 2$\frac{1}{2}$–3$\frac{1}{2}$	1 : 3–4
to accommodate movements	iii	1 : 1 : 5–6	1 : 4–5	1 : 5–6
caused by settlement,	iv	1 : 2 : 8–9	1 : 5$\frac{1}{2}$–6$\frac{1}{2}$	1 : 7–8
shrinkage, etc.	v	1 : 3 : 10–12	1 : 6$\frac{1}{2}$–7	1 : 8

Direction of changes in properties

equivalent strengths within each group

increasing frost resistance

improving bond and resistance to rain penetration

Where a range of sand contents is given, the larger quantity should be used for sand that is well graded and the smaller for coarse or uniformly fine sand.

Because damp sands bulk, the volume of damp sand used may need to be increased. For cement : lime : sand mixes, the error due to bulking is reduced if the mortar is prepared from lime : sand coarse stuff and cement in appropriate proportions; in these mixes 'lime' refers to non-hydraulic or semi-hydraulic lime and the proportions given are for lime putty. If hydrated lime is batched dry, the volume may be increased by up to 50 per cent to get adequate workability.

It is important to use an appropriate mortar, properly batched and adequately mixed, using the same mix throughout and taking adequate precautions against frost.

Defects in Brickwork

Brickwork defects arise in a variety of ways, of which the most common are efflorescence, stains, sulphate attack, frost action, settlement, lack of stability, use of unsound materials or poor workmanship, corrosion of iron and steel, drying shrinkage, growth of lichens and moulds, fumes from cavity insulation, and need for repointing.

Efflorescence

This consists of deposits of soluble salts formed on the surface of new brickwork, and it usually appears as loose white powder or as feathery crystals, or more occasionally as a hard glossy deposit penetrating the brick faces. It can occur on internal as well as external surfaces, causing damage to decorations where applied before the walls have dried out. Efflorescence is generally a temporary spring-time occurrence appearing as new brickwork dries out for the first time. It sometimes reappears in the second spring of a

Table 3.2   Selection of mortar groups

Type of Brick:	Clay		Concrete and calcium silicate	
Early frost hazard[a]	no	yes	no	yes
Internal walls	(v)	(iii) or (iv)[b]	(v)[c]	(iii) or plast(iv)[b]
Inner leaf of cavity walls	(v)	(iii) or (iv)[b]	(v)[c]	(iii) or plast(iv)[b]
Backing to external solid walls	(iv)	(iii) or (iv)[b]	(iv)	(iii) or plast(iv)[b]
External walls; outer leaf of cavity walls:				
—above damp-proof course	(iv)[b]	(iii)[d]	(iv)	(iii)
—below damp-proof course	(iii)[e]	(iii)[b, e]	(iii)[e]	(iii)[e]
Parapet walls; domestic chimneys;				
—rendered	(iii)[f, g]	(iii)[f, g]	(iv)	(iii)
—not rendered	(ii)[h] or (iii)	(i)	(iii)	(iii)
External free-standing walls	(iii)	(iii)[b]	(iii)	(iii)
Sills; copings	(i)	(i)	(ii)	(ii)
Earth-retaining walls (back-filled with free-draining material)	(i)	(i)	(ii)[e]	(ii)[e]

[a]During construction, before mortar has hardened (say 7 days after laying) or before the wall is completed and protected against the entry of rain at the top.
[b]If the bricks are to be laid wet, see text.
[c]If not plastered, use group (iv).
[d]If to be rendered, use group (iii) mortar made with sulphate-resisting cement.
[e]If sulphates are present in the groundwater, use sulphate-resisting cement.
[f]Parapet walls of clay units should not be rendered on both sides; if this is unavoidable, select mortar as though *not* rendered.
[g]Use sulphate-resisting cement.
[h]With 'special' quality bricks, or with bricks that contain appreciable quantities of soluble sulphates.

building's life but on a reduced scale. It is unsightly but usually harmless and shortlived unless water is able to percolate into the brickwork, or soluble salts such as magnesium sulphate crystallise just inside the surface pores.

The salts may come from the brickwork, as most clay bricks contain water-soluble salts, from soil in contact with the brickwork particularly in the absence of an effective damp-proof course, or by contamination with seawater or spray as with unwashed sea sand. Bricks can be tested for efflorescence in the manner described in BS 3921.[4]

Efflorescence can be minimised by effective damp-proofing, avoiding the use of facing bricks with a high soluble salts content in very exposed positions, using suitable mortar, keeping bricks dry and covering new brickwork at the end of each day's work.

Fortunately, surface efflorescence is normally washed away by rain and no special treatment is needed. To accelerate removal, the brickwork can be dry-brushed periodically until the soluble salts cease to crystallise. In sheltered situations, it may be necessary to remove the efflorescence by periodic washings with a hose with deposits of salt brushed off between washes.

Where efflorescence persists there is likely to be abnormal water penetration of the brickwork and constructional faults, such as leaking rainwater pipes and defective damp-proof courses, must be repaired. It is usually destructive only in exceptional cases where the soluble salts crystallise just below the brick surface which, if weak, may crumble,[6] sometimes referred to as cryptoflorescence. Individual disintegrated bricks should be cut out and replaced with bricks having a low sulphate content. Where disintegration occurs over a wider area, the best remedy is probably to render the face of the brickwork, after raking out joints to a depth of 10 mm and dry brushing off the efflorescence. The rendering should be weaker than the brickwork; for example, 1:1:6 mix of cement:lime:sand for strong bricks and 1:2:9 for weak bricks.

Efflorescence on internal plaster should be lightly brushed off before decorations are applied. In rare cases where spalling of the plaster occurs, the source of dampness must be remedied, the plaster removed and it is advisable to provide a capillary break between the brickwork and the plaster. One suitable method is to nail corrugated pitch or bitumen lathing to the inside face of the wall; an alternative is to use impregnated timber battens and a lining.

Stains

The worst stains appear in the absence of projecting features resulting from rainwater carrying deposits on to the wall face, as shown in plate 2. White stains under concrete and limestone components, such as string courses and copings, generally result from lime being deposited on the brickwork by rainwater. The normal remedy is as follows:

(1) thoroughly wet brickwork with clean water;
(2) carefully brush on diluted hydrochloric acid, starting with a small area;
(3) when stains have dissolved, thoroughly wash wall with clean water and a bristle brush;
(4) after removal of stains, flashings should be provided to prevent further percolation and staining.

Green stains caused by the corrosion of copper or bronze are very difficult to remove, and it is advisable to prevent rainwater draining from these metals discharging over the brickwork.[12]

Plate 2 Stained brickwork

Sulphate Attack

Sulphate attack on brickwork is the result of the reaction of tricalcium aluminate present in all ordinary Portland cements, with sulphates in solution. Its effect is an overall expansion of the brickwork, which can be followed in more extreme cases by progressive disintegration of the mortar joints. Except for earth retaining walls, where the attacking sulphates could emanate from groundwater, the source of sulphates is usually the clay bricks, with the sulphates transferred from bricks to mortar joints by percolating water, usually rainwater.[13]

Sulphate attack first becomes evident through horizontal cracking on the inner face of the wall, which with cavity walls may be concentrated near the roof. In long stretches of brickwork some oversailing of the damp-proof course is likely. Subsequently, mortar joints become white and a narrow crack may occur in the middle of the joints. Later still, the surface of the mortar joint spalls off and the mortar reduces in strength, while advanced stages of attack are accompanied by spalling of facing bricks. The expansion due to sulphate attack can be distinguished from drying shrinkage as it normally takes at least two years to develop.

In the past sulphate attack on unlined chimney stacks serving slow-burning appliances, resulting from condensation from flue gases, were quite common, but the provision of flue liners in new chimney stacks will prevent this. An ash-blinded sub-base and failure to link damp-proof courses in walls and floor, led to severe sulphate attack on brick walls in a Scottish bungalow. Sulphate attack on external renderings to brick walls usually results in a predominance of horizontal cracking associated with the expansion of mortar joints, as distinct from the fine map cracking emanating from the drying shrinkage of rendering.[13]

Ideally, bricks of low sulphate content should be used but this is rarely practicable. Alternatively, steps should be taken either to increase the resistance to sulphate attack or to limit the extent to which the brickwork becomes and remains wet. The sulphate resistance of mortars can be increased by using either richer mixes (1:5–6, cement:sand with plasticiser is excellent) or sulphate-resisting or supersulphate cements. Excessive wetting of the brickwork can be avoided by improving design details and ensuring a generous overhang at eaves and verges, adequate flashings, and damp-proof courses and special precautions at parapets and free-standing walls (low sulphate bricks, good copings with adequate overhangs and drips, damp-proof courses under copings and at base of walls, expansion joints not more than 12 m apart and suitable mortar mixes). Brick earth retaining walls should only be built of special quality bricks laid in a sulphate resisting mortar.[13]

Affected brickwork should be dried out and moisture excluded as far as practicable, as well as remedying poor design features, such as correcting the detailing on parapets and forming one or two expansion joints. Where attack is rather more severe but without visible damage to mortar, surface waterproofer may be applied when the brickwork is reasonably dry and, if successful, repeated at intervals of a few years. In cases of advanced attack, with brickwork expansion and severe damage to mortar, some form of cladding should be applied. Weatherboarding or tile hanging both form suitable treatments, which are equally applicable to failed renderings. When rebuilding parts of the structure it is essential to use materials suited to the prevailing conditions.[13]

Frost Action

In Great Britain, frost failures are usually confined to partly built unprotected brickwork or to brickwork subject to conditions of severe exposure, such as free-standing walls, parapets and retaining walls and, occasionally, brickwork below damp-proof course. Bricks in these positions should have good frost resistance and the work should be adequately protected from frost during construction by taking all necessary precautions and particularly laying loose bricks on top of the wall, overhanging 50 mm on each side, and covering the wall with polythene sheeting or other covering.

Frost can cause spalling of the face of bricks and disintegration of mortar. On occasions bricks may become detached from the mortar. Brick-on-edge copings often split if they are not frost-resistant and require replacing by engineering bricks laid in cement mortar (1:3). Stronger mortar mixes are needed when there is a danger of frost. While weak mortars are susceptible to frost attack, the stronger less-flexible mortars, containing a high proportion of cement, are vulnerable to shrinkage and movement-induced cracking, which allows water to penetrate into the cracks which, in turn, can freeze and cause disruption if it is unable to drain away freely.[14]

Settlement

The normal slight overall settlement of a building should not disturb the brickwork but differential settlement, often resulting in cracked walls, may occur where there are abrupt changes in ground conditions over the site, or where there is inadequately consolidated fill under foundations. Another common occurrence is the differential settlement of bay windows caused by the foundations being taken to a shallower depth than the house.[15] The effect of shrinkable clay, fast-growing trees and mining subsidence on foundations has been described in chapter 2. Repairs to cracked brickwork are examined later in the chapter.

Stability of Brick Walls

BS 5628[16] makes recommendations for the design of loadbearing walls consisting of bricks or blocks. To utilise the full capacity of high-strength bricks (70 N/mm^2 or more), cement mortar (1:3) is needed. For lower-strength bricks or blocks, mortars with an increased proportion of lime can be used without any great loss in masonry strength. Walls of a given thickness and material strength tend to fail at lower loads as their height increases. In design, the slenderness ratio of effective height to effective thickness is important. BS 5628[16] recommends that for walls set in Portland cement mortars the slenderness ratio should not exceed 20, for walls less than 90 mm thick in buildings of more than two storeys, and 27 in all other cases. The strength of a 260 mm cavity wall with both leaves loaded is approximately 16 per cent less than that of a 215 mm solid wall.[17] Approval Document A1/2 of the Building Regulations 1985[1] prescribes minimum thicknesses of external, compartment and separating walls.

Where floors and roofs span parallel to a wall, it is essential to provide straps or ties if the floor is to act as a lateral support. If the straps are not positioned or fixed correctly or are of inadequate design, this can result in instability of the walls. The Building Research Establishment investigated the collapse of gable end walls of large buildings and found the causes to be under-design, high wind pressures and lack of intermediate support.[18]

Cracked brickwork may not always be unstable. Tests at the Building Research Establishment showed that the capacity of a 215 mm brick wall to carry vertical loads was reduced by no more than 30 per cent by a stepped or slanting crack up to 25 mm wide, provided it was not accompanied by considerable transverse movement. On the other hand the resistance to side loading of a half-brick wall with sound joints but with a visible bulge could be impaired considerably. With cavity walls the effects of leaning or bulging, and of eccentricity of loading, are more serious than with solid walls, and wall ties play an important part in securing stability.

Zinc-coated steel wall ties for cavity walls complying with the original BS 1243 were found to offer inadequate resistance to corrosion, and it became necessary to revise the standard in 1981 to require a considerable increase in the thickness of zinc or the use of plastic coating in addition to zinc.[19] In some instances the use of black ash and permeable mortars has accelerated the corrosion process.

Techniques were developed in the early 1980s for the reinstatement of cavity walls by the insertion of new wall ties without recourse to demolition and reconstruction, and these can be used to advantage where steel wall ties have corroded badly. Alternative methods are described and illustrated in BRE Digest 257[20], and they include the insertion of resin-grouted offset or replacement ties, the installation of all-metal or metal/plastics expansion grip fixings in staggered positions, special screw-in ties grouted to the outer leaf, and the use of a partial tying method and filling the cavity with heavy-duty polyurethane foam, which will increase thermal insulation qualities, but the long-term structural performance of the foam is unknown.

Symptoms of wall tie failure and remedies vary with the type of tie. The substantial thickness and bulk of the vertical twist tie means that it will increase in volume significantly as it corrodes. The resultant expansion will cause horizontal cracking along brickwork courses in tie positions, accompanied by expansion of the walls. Where ties corrode within the inner leaf, floors and roofs may lift and inner walls can be disrupted. Shear cracks are likely to appear on walls bonded to the inner leaf.

Where there is doubt as to tie failure, a few bricks should be removed and a tie examined. When the less bulky butterfly ties corrode there are likely to be few, if any, external signs of disruption. The ties may be traced with a metal detector and replaced one by one by removing a few bricks, pulling out the ties and inserting new ones. A simpler and quicker approach is to insert new ties close to the old ones, using one of the methods described earlier.

The significance of a defect must be judged in relation to the whole building—loadings, transverse support, openings and piers—all are important. It is also necessary to keep a sense of proportion—a wall which is out of plumb not more than 25 mm or bulges not more than 12 mm in a normal storey height would not usually require repairing on structural grounds. The following remedies could be applied to unstable walls.

(1) Insert tie rods through the building in the thickness of a floor, or at roof level, anchoring the suspect wall to another wall or structural member that is sound. This is generally the cheapest and most effective method.

(2) Build a buttress or buttresses keyed into and thrusting against the unstable wall and carried to a firm base, possibly involving underpinning.

(3) Demolish walling that is bulged or out of plumb and replace by new brickwork, preferably built in cement: lime: sand mortar, 1:1:6.[21]

Cracks

Cracks which do not impair structural stability may appear in brick walls. A distinction may be made between cracks that run more or less diagonally, following horizontal and vertical mortar joints alternately, and those that pass straight down through vertical joints and the intervening bricks and mortar beds. The latter form of cracking may involve cutting out bricks. Fine cracks (up to 1.5 mm wide) in joints between absorbent bricks are usually best left unfilled as they are unlikely to be harmful. With non-absorbent bricks, it may be advisable to rake out the defective joints and repoint with 1:1:6 cement:lime:sand mortar.

Wider cracks (1.5 to 3.5 mm wide) will need to be repaired with the method varying according to the type of mortar in the existing wall. With weak mortar joints, the joints are raked out deeply on both sides of the wall, and filled and pointed with cement:lime:sand mortar not richer than 1:3:12. With strong mortar joints it is customary to cut out the bricks adjoining the crack and rebond using a 1:1:6, cement:lime:sand mortar. The same procedure would be adopted where there are cracked bricks. It is important not to use an excessively strong mortar which is likely to shrink. Where cracks may continue to widen with further movement, the cracks are best sealed with an oil-based mastic.

When examining cracks, care should be taken to record precisely the direction of the cracks, whether or not they extend through the wall, whether they taper off in any direction and whether they are progressive. Horizontal cracks require very careful consideration particularly to determine whether the part of the building above the crack has risen or whether the part below has fallen. Cracks of similar appearance can be due to different causes, which need identifying, and this occurs particularly in the case of parapet walls, where cracking may be the result of expansion due to frost, thermal movement, sulphate attack or movement of the adjoining roof slab. Distinction should be made where possible between tensile cracks, compressive cracks with small pieces of brick squeezed from the surface and localised crushing, and shear cracks identifiable by relative movement along a crack or points on opposite sides of it. Staveley and Glover[22] advocate accurate monitoring of crack damage, which is usually undertaken using calibrated tell-tales, being rigid indicators fixed over cracks and showing measured horizontal and vertical movements over a period of time.

Unsound Materials

Occasionally defective brickwork results from unsound bricks or mortar. Bricks with a high absorption rate used in parapet or freestanding walls or below damp-proof course are liable to spalling through periodic saturation and frost action, entailing replacement with more durable bricks. Mortars may be much stronger than the bricks with the likelihood of bricks cracking rather than mortar joints in the event of movement. Imperfectly slaked lime in a mortar can produce effects ranging from minor pitting of the mortar to general expansion with deformation and cracking of the brickwork. Repointing of brickwork may be required after a period of 25 to 40 years depending on exposure and type of mortar. The old mortar should be raked out to a depth of at least 20 mm, the joints brushed and moistened, and the mortar used for repointing should not be appreciably stronger than the original bedding mortar. Gauged mortar (1:1:6) is commonly used for this purpose.

Corrosion of Iron and Steel

Iron and steel embedded in brickwork may corrode and cause opening of brick joints or cracking of brickwork and also rust staining. Ferrous metals embedded in brickwork should be protected from rusting by coating with bitumen, anti-corrosion paint or metallic zinc paint, or another suitable method. Remedial action consists of removing brickwork to expose metal, clean metal, prime with rust-inhibitive primer and paint with bitumen paint, and when rebuilding take steps to reduce moisture penetration.[21]

Drying Shrinkage and Expansion on Wetting

Concrete or calcium silicate brickwork may crack, especially at window and door openings, owing to excessive drying shrinkage. The risk is minimised by using well-matured, dry bricks and laying in a weak mortar. Conversely, some clay bricks undergo slight expansion when first wetted and this can cause movement and cracking of brickwork. Both of these defects occur early in the life of the building and are unlikely to be progressive. Where bricks are cracked, they should be cut out and replaced. Solar expansion is unlikely to be a problem except in multi-storey structures where allowance will need to be made for possible movement.

Lichens, Moulds and Other Growths

These organisms are rarely destructive but they do produce disfiguring stains on brickwork and other wall surfaces. Such growths can be prevented or destroyed by applying one of the toxic washes recommended in BRE Digest 139[23], during a dry spell after partially removing any thick surface growths. The effective life of the treatment depends on the porosity of the surface and the extent to which it is washed by rain; periods of one to three years are common.

Climbing and other plants growing on walls can cause damage to walls, but much depends on the condition of the wall and the extent to which the growth of the plants is controlled. Each case has to be decided on its merits, balancing the effect on the appearance of the building against possible damage. Ivy, with its aerial roots, can penetrate cracks or open joints and cause damage. Virginia creeper will not harm a sound wall but plants like climbing roses, jasmine and honeysuckle are usually supported on wires or trellis and the fixings may cause damage. All plants should be trimmed to below eaves level and be kept clear of window or door frames.

Fumes from Urea–Formaldehyde (UF) Foam in Cavities

The cavities in the hollow walls of many dwellings in the United Kingdom have been filled with urea–formaldehyde foam to improve their thermal insulation qualities. Where the inner leaf is vapour-permeable, particularly when coupled with an impermeable outer leaf and a wide cavity (100 mm or more), fumes of formaldehyde gas may penetrate the building and cause irritation to the eyes and noses of occupants.

Stonework

Defects in Stonework

Limestone is generally one of the least durable of stones and offers least resistance to weather. *Sandstones* are harder and more durable than limestones and are more difficult to work and clean. In polluted atmospheres they tend to blacken more readily. *Granite* weathers extremely well and is extremely durable. Carved work is often of relatively soft stone.

Strength. Building stones are normally of adequate strength to carry imposed loads and rarely need special consideration except for lintels and civil engineering work. On the other hand fixings of stone claddings have not always proved satisfactory, particularly where reliance has been placed on mortar pats.

Moisture Resistance. Slates and granites absorb very little water but limestones and sandstones may absorb up to 20 per cent. In addition some sandstones are subject to appreciable moisture movement. Penetration of damp through stone is unlikely, except in situations such as window mullions or jambs, or from the catchment surfaces of sills, copings, cornices or string courses. Penetration through joints is more likely to be a problem, particularly with impervious stones such as granite. There is a need for cavities and adequate damp-proof courses.

Compatibility. Damage can result from the use of different types of stone in direct contact with one another. An acid atmosphere can attack limestones, forming soluble salts which if washed on to the surface of sandstones can cause decay.

Durability. The durability of a stone is influenced by its chemical composition and structure and the performance of stones in a particular locality gives a good guide. For example, a calcareous stone is liable to attack in a polluted atmosphere containing sulphur dioxide. The principal form of decay is the formation of a skin of soluble salts often accompanied by blisters. The salts can result from decomposition of stone by atmospheric pollutants, materials in contact with stone, or from groundwater. Most mortars contain alkali salts and need care in selection. Brickwork containing soluble salts is a possible hazard where natural stone is backed by brickwork and it should be separated by a suitable waterproof membrane. Stone may occasionally be damaged by frost action in exposed situations, such as cornices, string courses and copings.

Plates 3 and 4 show how stonework can spall and laminate in aggressive conditions, while plate 5 illustrates a stone chimney stack with badly weathered stones and open joints, which permit the entry of rainwater and can result in further decay.

Plate 3 Spalling stonework

Plate 4 Defective stonework

Plate 5 Dilapidated stone chimney stack

Repairs to Stonework

In some cases a soft stone in decayed condition can be cut back to expose a new, sound face. This method cannot easily be adopted where there are elaborate mouldings and difficulty is experienced in dealing with door and window openings and slender columns. With plain wall surfaces in soft stone it does however provide a relatively simple method of restoring a stone facade at reasonable cost.

The choice between replacement of damaged blocks with new stone and the execution of 'plastic' repairs depends on the extent of the damage and the character of the building. Plastic repairs skilfully undertaken may permit the original appearance of the stonework to be secured more quickly and completely, and this method avoids the disturbance of surrounding stone-work. Stone used for replacement purposes should be similar in colour, type and texture to the original. Where only a limited amount of the original stone is sound it will probably be advisable to replace all the old stone with new. On occasions it is possible to use similar sound stone from old buildings that are being demolished.

Plastic repairs are usually less costly than replacement with new stone, but workmanship and supervision needs to be of the highest standard. The principal materials used for this purpose are

(1) mortars based on Portland cement, lime and sand possibly 1:2:9 and often containing pigments or 1:8 cement:sand with a vinsol resin plasticiser;[24]

(2) mortars based on zinc or magnesium oxychloride cement with sand or crushed stone aggregate;

(3) crushed stone or sand with an organic binder, often based on cellulose.

Various precautions need to be taken, such as cutting the stone back to a sound surface, using an adequate thickness of plastic material, obtaining sufficient key and building up large areas gradually.

Colourless Treatments for Stonework

These treatments are of two main categories—water repellents and preservative treatments.

Colourless water-repellent liquids are intended to improve the resistance to rain penetration of stonework without markedly changing its appearance. They line the pores and inhibit capillary absorption, but may in the process increase the degree of penetration through cracks or defective joints. Hence, pointing should be examined and cracks made good prior to treatment. Soluble salts in a stone wall could be trapped and cause spalling of the treated surface. Most water repellents are silicone-based in accordance with BS 3826[25] and there are various classes for different types of stone. They should be applied to dry surfaces by brush or spray during dry spells, and they require renewal from time to time.[26]

The indiscriminate application of *colourless stone-preservatives* often results in disappointment. Some powdering often takes place on the surface of the stone and the treatment requires periodic renewal. The most common application is silicone-based masonry water repellents. It is advisable to investigate the performance of preservatives on similar stones exposed under similar conditions before arranging treatment. Stone decay usually takes place very slowly, so even after some years an untreated building may look no worse than a treated building.[24]

Cleaning of Stonework

Deposits of dirt spoil the appearance of stonework, retain harmful chemicals and hide decay. The choice of cleaning method is important as an unsuitable one can result in damage. Consideration should be given to the type and condition of surface to be cleaned, and the cost, speed and convenience of the cleaning method. Before letting a contract, it is advisable to test the proposed cleaning method on typical parts of the building.[27]

Water washing with a fine mist spray softens the deposits of dirt, beginning at the top of the building so that surplus water runs down and pre-softens dirt below. It is often necessary to assist the removal of dirt with brushes of bristle and non-ferrous or stainless steel wire. Abrasive stones may be needed to clean projecting features. It is one of the cheaper, least harmful but slower methods, well suited for cleaning limestones and marbles. Poultices of wet powdered clay are sometimes applied.

Dry-grit blasting uses abrasive non-siliceous grit blown under pressure to scour away dirt. Various sizes of nozzle are used according to the nature of the stonework and the delicacy of the work. Protection must be provided against dust and re-bounding grit by close sheeting. It is particularly suitable for sandstones, granites, slates and harder stones generally, is a fast method but high in cost, and there is a risk of damage to surfaces being cleaned. Noise may be a serious problem.

Wet-grit blasting is similar to the previous method except that water is introduced into the air/grit stream, thereby reducing the visible dust. The uses, merits and demerits are similar to dry-grit blasting.

Mechanical cleaning makes use of conical-shaped carborundum heads of various sizes and textures, grinding and buffing discs, and rotary brushes, all used with power tools. They spin off the dirt and weathered face in one operation. Operatives need to take special precautions and exercise great skill to avoid causing damage. It is a fast method, useful with hard stones, but the cost is also high. Hand tools may be used to supplement power tools for cleaning intricate carved work.

Chemical cleaning generally makes use of hydrofluoric acid as it leaves no soluble salts in the stonework, but it is dangerous in inexperienced hands and every precaution must be taken to prevent contamination. Any proprietary chemical cleaner should desirably be supported by an appropriate Agrément certificate. It is a moderately fast method at relatively low cost for use with harder stones.

Steam cleaning uses mains water pumped to a flash boiler and the steam generated is fed to a lance and played on to the stone surface assisted by brushes and abrasive stones. It has however little to commend it compared with the other methods, apart from moderate cost, and is seldom used.[27]

Other Claddings

Light claddings

Light claddings have been used extensively in the last three decades to form continuous envelopes or 'curtain walls' suspended from the loadbearing structure or as panel infillings in the spaces between members of a structural frame. These components save weight, space and building time, extend the range of architectural expression and exploit new materials. A danger with new materials is the absence of adequate experience on which to assess performance accurately.

A light cladding does not require high compressive strength, but must be able to resist wind loads. Provision must be made for diverting heavy rainwater runoff from joints, windows and doors. Thermal movement can be extensive and there may be differential movement between panels and frames, and compressible horizontal joints should be used in the cladding.[28] It is desirable to separate the external waterproof skin from the inner insulating layer, and an intervening ventilated air space helps to lower the temperature of the external skin in summer and to exclude water and reduce condensation. Joints between panels need to be flexible as well as watertight; they may be formed of plastic compounds or mastics, mechanical joints or those combining a mechanical outer barrier with an internal airtight seal. Unfortunately, all too often excessive reliance was placed on new jointing techniques which could not accommodate the diverse movements of the components which occurred in practice.

Limited aesthetic scope and performance failures led to the reduced popularity of curtain walling in the early 1970s. Also the horizontal emphasis of buildings made other constructional methods more popular, such as ribbon walling on dwarf brick walls built within slabs. The scope for using alternative construction methods was also on the increase.

In the late 1970s designers had a better understanding of the problems of water ingress caused by pressure differentials and produced pressure-equalised and self-draining walls—systems no longer dependent on the then fallible mastics for sealing the structure. Glass and spandrel panels became more attractive; the solar control bronzes were just not more efficient, they looked better and encouraged improved specifications. Panel systems were introduced using much larger panels to accept part of the loading. These panels were usually flat with minimal mullion and transom intrusions.[29]

By the mid 1980s, slopes, angles, and curves on plan and elevation became available to the specifier. Finishes in polyester powder coat aluminium neoprene or stainless steel provided a variety of colour and vision. Well engineered and well insulated panels were available to complement the ever-increasing choice of tinted and mirror glass. Many systems accepted triple glazing in addition to sealed-unit double glazing.

The material used for curtain walling had traditionally been aluminium, with its very favourable strength/weight ratio. A uPVC system was introduced in the mid 1980s specifically for the refurbishment of two-storey modular schools, hospitals and offices built in the 1960s. This had the advantage of excellent grid thermal insulation and reduced structure-borne sound with little maintenance. Hence curtain walling can provide cost-effective and visually attractive ways of upgrading buildings to the comfort and low-maintenance cost requirements of the 1980s.[29]

Concrete Panels

Precast concrete cladding may be subject to cracking and crazing which besides being unsightly may permit water to penetrate to the reinforcement and cause corrosion. Fixings must be of adequate strength and durability and be adjustable to accommodate the dimensional deviations arising from the building and manufacturing processes.[30] Problems arise from variations in dimensions of panels and the need to obtain jointing which will successfully withstand varying conditions of temperature and moisture. The range of sealants available enables the designer to specify the most appropriate product for a given set of conditions, with adequate regard to cost-effectiveness.[31] There is also a need for adequate water shedding drips and projections to prevent unsightly surface staining.

Concrete has been subject to two major forms of chemical failure, namely carbonation and alkali–silica reaction. Carbonation is a form of deterioration which attacks exposed concrete. All Portland cement contains a proportion of calcium hydroxide (free lime). The hydration, or hardening of the cement in concrete, does not change the state of the free lime. Subsequently, concrete is in contact with carbon dioxide in the atmosphere. This reacts with the free lime to form calcium carbonate, thereby reducing the alkalinity of the concrete. The alkalinity is necessary to protect mild steel reinforcement in concrete from oxidation or rusting. If the alkalinity of the concrete is destroyed by carbonation, the steel is liable to rust in the presence of moisture and air. The same concrete which is vulnerable to carbonation is the type of concrete which is inclined to be porous and therefore most likely to expose the reinforcement to these two elements. The resulting rust can cause progressive cracking and spalling of the concrete.

Carbonation starts at the surface of the concrete and proceeds inwards at a decreasing rate determined by the type of concrete, its quality and its density. It was originally thought that there was little risk of carbonation becoming a problem if the reinforcement had sufficient cover of concrete. In practice, the cover has often proved to be inadequate and the concrete too permeable, producing a fatal combination. This adverse condition was sometimes aggravated by the addition of chloride additives to the concrete mixes.

Where the carbonation has not reached the steel reinforcement, overcladding the structure to prevent extensive rain penetration or treatment with a coating which is resistant to water ingress and carbon dioxide diffusion, can reinstate its life expectancy. If carbonation has reached the steel, the area of carbonated concrete requires cutting out, then the rusting steel should be cleaned and treated and the wall reinstated with a water-repellent mix before the whole wall surface is overclad or resurfaced.[29]

The other concrete problem is alkali-silica reaction, which has been termed 'concrete cancer'. This defect is caused by a chemical reaction between the alkalis normally present in concrete and certain forms of aggregate quarried in various parts of the United Kingdom. For this defect to occur, moisture has to be present. Alkali-silicate reaction is most likely to occur in exposed concrete and the best means of protection is to shield it from excessive wetting by rain or condensation.[29]

Table 3.3 outlines the more common defects occurring in precast concrete claddings and the likely causes and recommended remedial action.

Generally accepted methods of repairing concrete are now described. The use of polymer-modified cementitious materials is generally considered to be the most suitable method of patch repair after removing rust from the steel. However, recasting or sprayed concrete, possibly with polymer additives in the mix, is likely to be used for large volume repairs. Epoxy and polyester resins are also used in some specialised areas, such as crack injection. Methods for dealing with efflorescence, and the removal of stains and growths from concrete are detailed in Cement and Concrete Association publications.[32,33]

Aluminium Sheeting

The appearance of aluminium as manufactured is satisfactory for many situations, although dulling of the surface and subsequent pitting is likely to occur. The original condition can be preserved by regular washing or abrasive cleaning. The frequency of this treatment varies from once every few months to once a year, depending on the composition of the alloy and local atmospheric conditions. Surfaces sheltered from rain need more frequent cleaning than rain-washed areas to maintain the same appearance. Various treatments can be adopted for decorative purposes or to give protection against aggressive conditions, including conversion coatings, painting and lacquering, stove enamelling, vitreous enamelling and anodising.

Table 3.3   Defects in precast concrete cladding

Defects	Causes and remedial action
1 Cracks, uniform in width or closed; at a later stage, spalling especially at corners or edges of cladding units, possibly also misalignment of faces	Cladding units under compression owing to inadequate allowance for differential movement between cladding and supporting structure. Check provision and effectiveness of horizontal 'soft' compression joints and any corrosion of reinforcement
2 Cracking or spalling of cladding units in a regular pattern which seems to indicate position of fixings; no evidence of compression	Possible corrosion of ferrous fixings; check drawings or specification. If serious, remove unit, otherwise keep under observation
3 Misalignment between faces of adjoining units	If excessive, suspect either potential compression failure or fixings absent or defective
4 Iron stains on surface of precast concrete units, random occurrence	If units unreinforced likely to be iron-bearing aggregate. If reinforced, see 5
5 Iron stains on surface of precast concrete units, occurrence suggests a pattern	If units reinforced, likely to be corrosion of reinforcement. If adequate cover, suspect also possible use of excessive calcium chloride in manufacture, especially if cracking or spalling
6 Sealant extruding considerably from joint and edges of cladding units in contact	Insufficient allowance for movements, risk of future displacement, cracking or spalling of cladding. If apparent within first 5 years, keep under annual observation
7 Sealant not deformed, suggesting no compression occurring at joint	Compression joint may be inoperative owing, for example, to concealed presence in it of mortar or other incompressible material
8 Sealant wrinkled at an angle to the line of the joint	Sign of differential movement between cladding units in the direction of the line of joint; seek cause and assess consequences
9 Sealant split or adhesion lost.	Greater movement at joint than can be accommodated, especially by aged sealants; likely to lead to rain penetration

Source: BRE Digest 217[28].

Plastics

A variety of plastics are used for wall cladding and their main disadvantage is that of combustibility. Phenolic resin laminates are used extensively for curtain walling and can be expected to remain structurally sound under normal weathering conditions for upwards of twenty years. The natural surface gloss soon disappears but they can be painted. Others have a decorative melamine formaldehyde face with good weathering qualities.

With glass-fibre reinforced plastics (GRP) cladding panels, even slight distortion of a nominally flat surface is noticeable. A textured surface will help to mask it but at the cost of increased dirt retention. Bright colours are less stable; darker colours such as greys, browns and near-blacks fade less but have higher surface temperatures. Damaged portions can be patched on site but they stand out and it is better to replace a complete panel. Badly exposed glass fibres must be scrubbed off before any new surface treatment is applied. Acrylic and polyurethane paints can be applied to surfaces that have deteriorated.[34]

Glass-fibre reinforced cement (GRC) is a composite material consisting of a cement matrix reinforced by a small proportion of glass fibres. One of the first applications was for cladding, providing a lightweight construction combined with freedom of design of panel shape and the choice of a wide range of durable finishes. The panels may be either single skin or of sandwich construction with an insulating core. The fibre reinforcement enhances the ultimate tensile and flexural strength of the matrix and greatly increases its toughness, although these properties change with time, depending on the environment. GRC undergoes drying shrinkage on exposure to low humidity/high temperature conditions.[35]

Timber

The most satisfactory form of weatherboarding is rebated shiplap boarding, preferably treated with preservative and backed with bitumen felt. Cedar boarding, even where heartwood, tends to weather very badly if left untreated, resulting in a streaky appearance. It is advisable to apply a suitable preservative regularly; medium to high build exterior-quality wood stain applied every 2 to 3 years provides an effective treatment. Timber weatherboarding fixed vertically is most vulnerable to rot in the end grain of its lower edges, and water should be allowed to drain freely from them. BRE Digest 286[36] gives guidance on natural finishes for exterior timber.

Structural Frames

All large buildings are subject to movement due to compression of foundations, shrinkage of concrete, thermal movement, variable loadings and wind pressure. Cracking of reinforced concrete columns or beams can take the form of surface cracks, which are influenced by the effective concrete cover to the steel reinforcement, and internal cracking where the member is subject to bending. The latter case is more serious and may result in a breakdown of the adhesion bond around reinforcing bars.

The roof to the assembly hall of the Camden Girls' School collapsed in June 1973. It consisted of prestressed concrete roof beams with insufficient bearing, insufficient structural cross-tying of the building, inadequacies in prestressing wires, conversion of high alumina cement and corrosion of continuity reinforcement. This incident has highlighted the need for greater

Plate 6 Corroded steelwork

care in design and execution of structural work, periodic inspection of existing structures and the strengthening of weak points.

All buildings incorporating high alumina cement in their structures must now be regarded as suspect following building failures, reports of the Building Research Establishment, and circulars issued by the Department of the Environment which required local authorities to check every roof or floor member over 5 m in length containing this cement. Conversion or degradation of the cement hydrate can occur as a result of the penetration of heat and water, high water/cement ratio, unsuitable aggregate or high setting temperature. As described earlier, failures can result from carbonation and alkali–silica reaction. The method of repair of defective concrete will be influenced by the nature of the distress, environment, cost and desired appearance.[37]

Steel frames can have defects resulting from a variety of factors such as faulty material, faulty design, overloading of the structure, bad workmanship, bad erection and corrosion. Faults resulting from poor workmanship or erection often cause excessive stresses and could in extreme cases result in failure of the structure. Typical examples are (1) stanchions that are erected slightly out of position on their foundations and then subsequently pulled over by floor beams at first floor level, and (2) grouting under stanchion bases where the full bed of cement mortar only occurs around the edge of the stanchion base. Periodic inspections and the carrying out of remedial works are important. Badly corroded steel members, as shown in plate 6, can result in serious structural conditions.

Weathering steels such as 'Cor-ten' as used on the Wills factory at Hartcliffe, Bristol, are low alloy steels which, although not rustless, are designed to form protective corrosion when exposed to the wet and dry cycles of normal weather. They are available in thin metal sheets, plate, rolled and hollow sections. The uncertain finish, varying from earthy brown to purple-grey depending on conditions and length of exposure, can be very attractive. The run-off rainwater from the steel during the initial weathering period will cause rust-coloured stains on other materials, such as concrete, and must be taken into account. Provided the design problems are successfully overcome, weathering steels should outlast exposed painted galvanised steel.[38]

Timber-framed Housing

Gordon[39] has described how timber frame has changed from the early 100×50 mm uninsulated studding without a vapour barrier, to insulated studding without vapour barriers, to insulated studding with vapour barriers, brick-clad with unventilated cavities, through to engineered sheathed structures with non-permeable sheathing to brick-clad ventilated cavities with permeable sheathing. In 1986 it was being suggested that the insulation should be on the outside of the studs to create a warm frame, and the inside of the studs covered with internal vapour barriers, now called vapour checks. The frame itself progressed from non-preserved, partially preserved, mainly preserved and finally totally preserved.

A BRE report in 1983[40] advocated that a more radical approach to timber-frame construction should be adopted, which would reduce the inherent potential for interstitial condensation. The permeability of the sheathing materials was considered and the need for an upper limit for vapour resistance established. In its summary of potential workmanship errors, it cited sixteen major areas that required particular attention by site operatives and supervisors, ranging from insufficient anchorage at the base, resulting in structural weaknesses, inadequate glueing which could cause failure in stressed skin floors, insufficient fixing between components causing structural weakness and the incorrect placing of fire stopping and vapour barriers. The latter could result in severe interstitial condensation.

The BRE subsequently surveyed ten housing sites and issued a further report in 1985[41]. The faults identified by BRE related to the four main areas of fire, strength and stability, durability, and differential shrinkage. Overall, almost two-thirds of the 433 faults were found to be either universal or common. Because of their importance, each of these areas is now considered in some detail.

1. *Fire.*   One in six of all the faults found on timber-framed building sites related to fire. There was an alarming "universal inability to provide effective barriers against fire." Surveyors and other professionals should look out for cavity barriers made of polythene-wrapped mineral wool,

timber battens, or both, which fail to close cavities and are not fitted around openings such as those for balanced boiler flues.

2. *Strength and stability.* Design problems became apparent, as well as site incompetence. The anchoring of the framework to foundations was "often poor and sometimes inadequate." Plasterboard, used to provide racking resistance, was often inadequately nailed; wall ties were found nailed to no more than the sheathing, and poorly fixed into the outer brick skin. Surveyors should consider how such defective workmanship would affect a house's performance over a period of years, and look for tell-tale defects.

3. *Durability* accounted for over a quarter of all faults, and needs particular attention when second-hand timber-framed houses are being surveyed. Untreated timber was found in vulnerable locations. Too often, breather papers were torn or missing altogether, and in general were considered too vulnerable and susceptible to damage. Repairs to damaged papers were mostly inadequate, with no lapping to cloak cavity trays and damp-proof courses. A major problem with timber frame in the United Kingdom results from the mild but wet conditions which prevail. It is, therefore, very disturbing for BRE to have found "widespread inadequate provision for the exclusion of rainwater." Surveyors should be on the alert for signs of rot.

4. *Differential shrinkage.* Site practice indicates that builders often appear oblivious of the critical need to allow for relative movement in timber-frame construction. On one 150-house site, the feet of the trussed rafters were built into the top of the brick cladding. This means that both brickwork and rafters "would be stressed in a way for which they were certainly not designed." More widespread was the absence of clearance, or sufficient clearance, at critical points, such as near sills. Lack of provision for vertical relative movement between frame and cladding was "very common."

Surveyors will, however, experience problems in establishing the type and position of moisture and vapour barriers, and the adequacy of fire stopping in existing timber-framed dwellings.[42] Faults will normally be investigated by optic probe or by opening up. As to the scale of the problem, it was estimated that more than half-a-million timber-framed houses had been built by 1985.[43]

The BRE subsequently issued a statement concerning the 1985 report,[41] of which the following is an extract. "Faults were defined in the report as departures from good practice, of a kind which are commonly found in all forms of house construction and it is clearly desirable that they are avoided; but they only rarely lead to significant failures in service. Whilst design and construction practice needs to be improved, the evidence available to BRE indicates that the performance in service of timber-framed housing is no less satisfactory than that of traditional construction." This can probably be substantiated in the light of the numerous constructional faults that occur in cavity wall construction daily.

It appears that the National House-Building Council accepted timber-framed houses in large quantities prematurely, before the necessary vital research was carried out to adapt the system to the United Kingdom climate and changing social attitudes. In the public sector, the design and build contractors convinced local authorities that the speed of erection and high insulation values were to their advantage. The tragedy is that public confidence in timber-framed housing suffered badly just at the time when most of the answers for sound construction were becoming known. Gordon[39] highlights the lesson learned—that all major innovation should be thoroughly researched and approved before being implemented on a large scale.

Dampness Penetration

Causes of Dampness

Damp penetration is one of the most serious defects in buildings. Apart from causing deterioration of the structure, it can also result in damage to furnishings and contents and can in severe cases adversely affect the health of occupants. The main sources of dampness in buildings have been identified by Oxley and Gobert[44] as direct penetration through the structure, faulty rainwater disposal, faulty plumbing, rising damp and dampness in solid floors. These aspects are now examined in some detail.

Water Introduced during Construction

In building a traditional three-bedroomed house, about 7000 litres of water are introduced into the walls during bricklaying and plastering. The walls often remain damp until a summer season has passed. As the moisture dries out from inner and outer surfaces it is liable to leave a deposit of soluble salts or translucent crystals. A porous wall with an impervious coating on one surface will cause drying out on the other surface. Typical moisture contents of some of the more common building materials are plaster: 0.2 to 1.0 per cent, lightweight concrete: up to 5 per cent; and timber: 10 to 20 per cent.

Penetration through Roofs, Parapets and Chimneys

Tiled roofs may admit fine blown snow and fine rain, particularly in exposed situations. Both tiles and slates must be laid to an adequate pitch and be securely fixed. It is wise to provide a generous overhang at eaves. Parapets and chimneys can collect and deliver water to parts of the building below roof level, unless they have adequate damp-proof courses and flashings. Leakage through flat roofs is more difficult to trace and needs to be distinguished from condensation.

Faulty Plumbing or Rainwater Goods

Dampness may result from leaks in a plumbing system, although this must not be confused with condensation on cold pipes. Rainwater goods which are cracked or have defective joints over long periods can cause damp penetration and deterioration of the structure.

Penetration through Walls

Penetration occurs most commonly through walls exposed to the prevailing wet wind or where evaporation is retarded as in light wells. On occasions the fault stems from excessive wetting from a leaking gutter or downpipe. There must be a limit to the amount of rain that a solid wall can exclude. In the wetter parts of the country (for example, south-west, western and north-west England, rising to a maximum in north-west Scotland) and on exposed sites, one-brick thick solid walls may permit penetration of water. In this connection Lacy has devised indices of exposure to driving rain.[45] The greatest penetration is likely to occur through the capillaries between the mortar joints and the walling units. The more impervious the mortar and the denser the bricks or blocks, the more serious is the penetration likely to be. Dense renderings can prevent moisture drying out more effectively than preventing its entry, and this tendency is accentuated in cracked renderings with moisture penetrating the cracks by capillary action, becoming trapped behind the rendering and subsequently drying out on the inner face of the wall.

Cavity walls when properly detailed and soundly constructed will not permit penetration of rain. Penetration when it occurs is usually the direct result of faulty detailing at openings or mortar droppings on wall ties. Finally, disintegration of brickwork may be caused by the action of sulphates or frost when the bricks are saturated.

Plate 7 shows the rapid and extensive deterioration of a reinforced precast concrete sill, where rainwater has penetrated a joint causing rusting of the steel reinforcement and spalling of the precast concrete.

Rising Damp

In older buildings damp may rise up walls to heights in excess of 1 m because of the lack of damp-proof courses. The height of damp penetration depends on several factors, such as the pore structure of the wall, degree of saturation of soil, rate of evaporation from wall surfaces and presence of salts in the wall. In newer buildings rising damp may occur through a defective damp-proof course, the bridging of the damp-proof course by a floor screed internally or by an external rendering or pointing, path or earth outside the building, or mortar droppings in the cavity. Damp may also penetrate a solid floor in the absence of a damp-proof membrane. These sources are well illustrated in BRE Digest 245.[46]

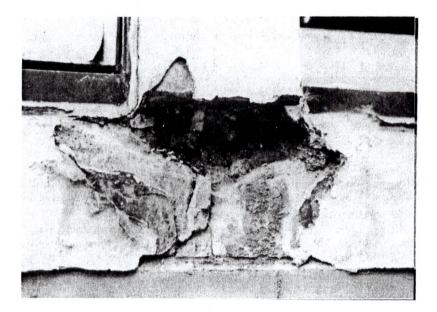

Plate 7 Defective precast concrete sill

BRE Digest 245[46] describes how measurements of surface moisture are in themselves no positive indication that a genuine rising damp problem exists. The electric meters commonly used by surveyors are responsive to both the amount of moisture present and to the salt concentration, and cannot distinguish between the two. Such meters often given high readings on the walls of old properties where an accumulation of salts inevitably appears on internal surfaces. This does not mean that the property necessarily has a dampness problem, as high readings can be obtained from a wall where the concentration of salts is high, even if the wall is virtually dry. Nevertheless, as preliminary surveying instruments, these meters have a valuable role to play and will identify areas where further investigation is necessary.

To obtain more conclusive proof of the presence of rising damp the most satisfactory approach currently available is to obtain samples of the bricks, blocks or mortar at some depth in the wall. From these samples an accurate measurement of the moisture content of the wall can be obtained, together with an indication of the influence of any hygroscopic salts that may be present. The sampling technique creates a certain amount of dust and slight damage to decorations. The testing of the samples can be undertaken using the direct weighing and drying method or a carbide meter. A wrong diagnosis could, however, lead to considerable unnecessary expenditure.

It is unusual for walls with moisture contents less than 5 per cent at their base to be severely affected by rising damp, though hygroscopicity caused by salts which may have accumulated over many years can cause damage to the plaster and decorations. Comparing the hygroscopic moisture content (HMC) with the registered moisture content (MC) indicates which factor is determining the dampness at any position. An HMC higher than the MC indicates that the dampness results from moisture from the air rather than from the ground or some other source, while an MC higher than the HMC indicates that water is coming from some source other than the air, such as rising damp or rain penetration.[46]

Some building materials possess an HMC of up to 5 per cent even without the presence of salts from external sources. Although only a rough indicator, the BRE believes that the 5 per cent threshold does represent a reasonable general guide as to whether or not some kind of remedial treatment is needed. Rising damp is a seasonal phenomenon, increasing in winter with rising water tables and falling in summer. This seasonal effect must be taken into account in any diagnosis since the problem could disappear in the summer months and return in the winter.

Curing Rising Damp Penetration

Some causes of rising damp can be dealt with quite cheaply and easily. For instance, it may be possible to lower earth, rockeries or pavings which extend above the damp-proof course, or to remove rendering or pointing which bridges the damp-proof course. The erection of porches, sun lounges and other additions, often by do-it-yourself enthusiasts with little constructional knowledge, can also result in the damp-proof course being bridged and damp penetrating the main building. Where there is no damp-proof course or where it is defective, more expensive measures are required and these are now described.

Where the wall is of stable brickwork or coursed stonework and not unusually thick, the best method is to saw a slot in a mortar bed joint with a tungsten-carbide tipped chainsaw, normally just below the level of an existing suspended wooden floor or as close as possible to the top screed of an existing solid floor. A damp-proof membrane is then inserted in the slot, wedged with temporary wedges and varying the type of mortar filling to the remaining gap according to the strength of the wall. The membrane is normally inserted in 1 m lengths and usually consists of flexible black polythene. The membrane can be extended internally to form a vertical damp-proof course between the solid floor and the horizontal damp course. It is, however, difficult to apply this method to party walls and it is costly, time-consuming and causes considerable disturbance.

BRE Digest 245[46] recommends that any new plastering which follows the damp-proofing work should act as a barrier to residual salts and moisture in a wall. Ideally it should have as high a vapour permeability as possible to help the evaporation of residual moisture and should be weaker than the background to which it is applied. Some tests by BRE on a 1:3 cement/sand undercoat finished with class C gypsum plaster gave satisfactory results.

BRE[46] have recommended that non-traditional methods of damp-proofing should be considered only if they have been awarded an Agrément certificate. The only method satisfying this requirement in 1986 was *chemical injection*, which was also covered by a British Standard code of practice.[47] Chemical injection systems involve the use of silicone or aluminium stearate water repellents, either injected at high pressure or transfused into the wall under gravity or at low pressure. The high-pressure systems are the more commonly used.

BRE[46] point out that water repellents are pore liners rather than pore blockers and hence allow the passage of some water vapour while preventing the rise of liquid moisture. The repellents are not intended to provide a damp-proof barrier against a substantial positive pressure of water and should not be used in basements subject to high water tables and penetrating damp. Another limitation is that repellents in common use for damp-proofing work are not durable over long periods in highly alkaline conditions, although this is not usually a problem in older buildings with well-weathered lime/sand mortar joints, unless the walls are exceptionally thick.

The repellents are injected or transfused into closely spaced holes in brick or mortar courses at dpc level and the treatment must be carried out at different depths in construction other than a single leaf wall, to ensure penetration through the entire thickness of the structure. As rising damp is a seasonal occurrence, injections are best carried out in late summer when water tables are at their lowest and the walls are relatively dry.

Another damp-proofing technique is *electro-osmosis*, which is of two types; passive and active. A continuous copper strip is inserted into drilled holes at an appropriate level and, in the case of the passive system connected to deep sunken earthing rods, and is designed to reverse the electrical potential of the masonry and thus stop the upward movement of water from the ground. The active system is connected to a power source, usually from within the property, and suitably stepped down to the necessary low voltage. There have been some reported failures with this technique and it does have inherent weaknesses such as the possible corrosion of the metal strip and the accidental breaking of the circuit. It was not recommended by the BRE in 1986.

Another approach would be to reduce the amount of moisture in the wall by diverting groundwater through land drains or French drains, and possibly to increase evaporation from the wall such as by removing dense external renderings and defective plaster. These measures would not however remedy a severe case of rising damp.

Where treatment of the source is virtually impossible or a building subject to rising damp has only a limited life, a relatively cheap remedial method is to conceal the rising damp in the walls by battening out or lining the inside face of the wall to prevent dampness and salts affecting decorations. The lining may consist of wallboard on pressure-impregnated battens or a proprietary system. The main disadvantage is that these remedies will reduce the evaporation rate from a wall and cause the dampness to rise further.

Vegetation on the external faces of walls can cause problems. For example, creeping plants can damage masonry, lichens are useful for dating old stone but emit acid which corrodes metal flashings and gutters, while moss although picturesque, encourages insects, reduces evaporation from porous masonry and may block gutters. All plants should be trimmed to below eaves level and kept clear of window and door frames.[42]

Damp-proofing Solid Floors and Dry Rot in Suspended Timber Floors

Dampness may penetrate ground floors and the various remedial measures are described in chapter 4.

Treatment of Damp Walls due to Rainwater Penetration

The treatment of cavity walls may entail clearing mortar droppings from wall ties or above the damp-proof course by cutting out bricks in the outer leaf. Alternatively it may be necessary to rectify defective damp-proof courses around window or door openings.

A one-brick thick solid wall is unlikely to withstand severe weather conditions satisfactorily and it may therefore be necessary to apply a suitable external finish. The main disadvantage is that the attractive colour and texture of good brickwork will be lost.

The most common external finish is rendering or roughcast. *Rendering* may consist of two coats of Portland cement and sand (1:3 or 1:4), possibly incorporating a waterproofing compound, finished with a float to a smooth finish and often painted with two coats of emulsion or stone paint. This produces a rather dense coat which is liable to develop hair cracks. Moisture enters the cracks, becomes trapped behind the rendering and evaporates from the inner surface. Hence it is advisable to use a rendering of cement:lime:sand in the proportions of 1:1:6 or 1:2:9. These produce porous finishes which absorb water in wet weather and permit free evaporation when the weather improves.[48] It will be necessary to rake out the brick joints to form a key for the rendering, which should comply with BS 5262.[49]

With *roughcast* or *pebbledashing* there is less risk of the external coat cracking or breaking away from the wall. In roughcast the coarse aggregate, usually gravel, is mixed into the second coat which is applied to the wall, whereas in pebbledashing, the chippings are thrown on to the second coat while it is still 'green'. In both cases the appearance is rather unattractive but the need for periodic decoration is eliminated.

Common brick surfaces can be decorated in a variety of different ways to make them watertight. *Masonry paints* have good water shedding properties and are available in a wide range of colours. In more recent times, *chlorinated rubber paints* have been applied to external brick and block walls with satisfactory results. *Bituminous paints* give an almost impervious surface coating but their use restricts future treatment owing to the bitumen bleeding through other applications. *Silicones and other colourless water repellents* are useful on exterior wall surfaces in good condition, when the appearance of the brickwork or stonework is to be retained. The permanence of this protection is variable, depending on the type of treatment selected and the condition of the masonry and the degree of exposure. Thick, textured sprayed coatings, often based on alkyd resins with mica, perlite and sometimes fibres, are widely used and can have a life of up to ten years.[50]

Vertical tile hanging on the external elevations of buildings can be very attractive if skilfully fixed, but it is expensive. It is extremely durable provided the fixing battens are pressure-impregnated with preservative. With plain tiling all tiles must be nailed but a 40 mm lap is sufficient. Angles may be formed with purpose-made tiles or be close cut and mitred with soakers, while vertical stepped flashings are usually introduced at abutments to form a watertight joint.[51]

Weatherboarding provides a most attractive finish and may be either of painted softwood, or of cedar boarding treated periodically with a suitable preservative such as high solids exterior wood stain.[36] Feather-edge boarding is used extensively but probably the best results are obtained with ship-lap boarding. The boarding is usually nailed through a felt backing to vertical impregnated batterns. A variety of *sheet claddings* is also available in steel, aluminium, asbestos cement and plastics.

Condensation

Nature of Condensation

In years past the major causes of dampness in buildings were rain penetration and rising ground moisture, but condensation has become an even greater cause in post-war dwellings. Warm air can hold more water vapour than cold air and when warm moist air meets a cold surface it is cooled and gives up some of its moisture as condensation. Air containing a large amount of water vapour has a higher vapour pressure than drier air and hence moisture from the wetter air disperses towards drier air. This has special significance since (1) a concentration of moist air as in a kitchen or bathroom readily disperses throughout a dwelling, and (2) moist air at higher pressures inside buildings tries to escape by all available routes to the outside, not only by normal ventilation exits but also through the structure when it may condense within it.[52]

Condensation takes two main forms—(1) *surface condensation* arising when the inner surface of the structure is cooler than room air, and (2) *interstitial condensation* where vapour pressure drives water vapour through slightly porous materials, which then condenses when it reaches colder conditions.

The term *relative humidity* (rh) expresses as a percentage the ratio between the actual vapour pressure of an air sample and the total vapour pressure it could sustain at the same temperature (per cent rh at °C). Air is described as saturated when it contains as much water vapour as it can hold.

It is then at 100 per cent rh. If moist air is cooled, a temperature will be reached at which it will become saturated and below which it can no longer hold all of its moisture. This temperature is the *dew point*.[52]

The occurrence, persistence, extent and level of condensation are influenced by a number of factors, of which the most important are probably:

   (1) number of occupants of the property;
   (2) type of dwelling and construction;
   (3) heat levels maintained in the property;
   (4) type of heating;
   (5) length of time the property remains unheated;
   (6) degree of insulation;
   (7) amount of ventilation; and
   (8) prevailing weather conditions.[53]

A BRE report[54] estimated that 1.5 million dwellings in the United Kingdom are seriously affected by dampness caused by condensation, and a further 2 million have slight condensation problems.

Causes of Condensation

There are two main reasons for the increase in the frequency and severity of condensation—(1) changes in living habits, and (2) changes in building techniques. More housewives now go out to work, often resulting in dwellings often being left unoccupied, unventilated and unheated for much of the day. Moisture-producing activities such as cooking and clothes washing tend to be concentrated into shorter periods of time. Furthermore, washing and drying of clothes are often carried out within the main dwelling area instead of in a separate washhouse or fairly isolated scullery. Unflued paraffin and bottled gas heaters are still used quite extensively for background heating and they emit considerable quantities of water vapour. Furthermore, occupants have become more sensitive to slight dampness in their dwellings and endeavour to maintain a high standard of decoration, so that local deterioration assumes greater importance.[55]

Structurally, probably the most significant change is the disappearance of open fires and air vents which provided valuable ventilation routes. Modern windows reduce ventilation rates and this may be further accentuated by draught-proofing by occupants. Solid floors without an insulating floor finish or screed are slow to warm up, and modern wall plasters and paints are less absorptive. Flat roofs and newer forms of wall construction also need careful design if they are not to lead to increased condensation.

Surface condensation can lead to unsightly and unpleasant blue, green and black mould growth on walls, ceilings, fabrics and furnishings, which produces many complaints from occupants. On paint it may show as pink or purple staining.[56] Condensation within the fabric is slower to show but may be much more serious in the long term.

Diagnosis

Rising damp can be distinguished from condensation by the pattern and positioning of staining, while moisture penetration through cavity brickwork across wall ties also shows pattern staining. Gutters and downpipes must be checked for cracks, defective joints, blockages and the resultant water penetration. Roofs can also be checked for defects and here again the type and position of staining is often a useful guide. Less obvious causes of dampness are slight weeping at pipe joints and wastes, and pinhole leaks in pipes, where the pipes and fittings are concealed.

Drying out of construction moisture can lead to defects similar to those resulting from condensation and it is desirable to allow drying out to finish before carrying out remedial measures. As this can take up to three years, occupants are only likely to accept this advice with reluctance.

Condensation frequently occurs as occasional damp patches in cold weather, although a sudden change from cold to warm humid weather may also cause condensation. Apart from investigating the damp conditions, attention should also be directed to the heating arrangements, possible use of portable oil or gas-fired appliances, ventilation, arrangements for drying clothes, means of dispersal of moisture from the kitchen, form of construction of floors, walls and roof, and whether there is any uninsulated pipework. Measurement of temperatures and humidities will show whether conditions favourable to condensation exist at the time of measurement. Suitable charts and useful calculations are contained in the DOE publication *Condensation in Dwellings—Part 1*.[52] A sling or whirling hygrometer is useful for this purpose and consists of both wet and dry bulb thermometers.[55] Protimeters can also help in indicating the amount of moisture held beneath the surface of any material, subject to the limitations described earlier in respect of electric meters. An investigator also needs the capacity to assess the reliability of information supplied by occupants.

Condensation problems will lead to damp patches that are more diffuse and without the definite edges that occur with other causes. Impermeable surfaces, such as gloss paint or vinyl wallpaper, can be covered with a film or droplets of water. Trouble starts in areas that are usually cold, such as inside exposed corners, wall to floor junctions or solid lintels, or poorly ventilated, such as kitchen cupboards, wardrobes or behind furniture. Spores from moulds and other fungi can germinate over a whole range of temperatures (0 to 20°C) given suitable conditions (supply of food, oxygen and liquid water), resulting in deterioration of decorations, a musty smell and possible health hazards.[57]

Remedial Measures

The principal remedial measures consist of improved ventilation, insulation or heating, or a combination of them. If the relative humidity is excessive, the amount of moisture must be reduced or temperatures raised. Alternatively the moisture vapour should be removed at source, preferably by mechanical means.

Ventilation is normally the cheapest solution and can be very effective provided it does not result in unpleasant draughts, otherwise it is likely to be rendered ineffective by occupants. This is particularly important in kitchens. A limited amount of ventilation is essential to keep relative humidities below 70 per cent. This can be achieved by the installation of trickle ventilators in bedrooms and extractor fans in kitchens and bathrooms. Automatic controls, such as humidistats, can improve the effectiveness of fans at little extra cost. Care should be taken to make the fans as unobtrusive as possible, both in visual and acoustic terms.[56] Ventilated hoods above cookers are valuable in quickly trapping and removing the steam generated by cooking.

Insulation is generally more expensive than ventilation but more acceptable to the occupier. The main aims are to keep surface temperatures above dew point, improve U-values and secure better value for money from heating and improved comfort conditions. (U-value is the rate of transfer of heat through an element of a building.) This is normally done by filling cavities with suitable insulant or fixing a plasterboard–insulation composite board with an integral vapour check internally.[56]

Heating is the most effective measure of all but also the most expensive, and may be opposed by occupants faced with increased running costs. The aim is to raise air and surface temperatures and so reduce the relative humidity. Living rooms, even if heated only during evenings, seldom suffer from surface condensation, while bedrooms, which are often very poorly heated, often present problems.[52]

Condensation should not be a problem in *bedrooms* if the structure has a U-value of not greater than 0.60 W/m^2 °C, with reasonable ventilation—one to two air changes per hour—and where the room temperature is at least 5°C above outside air for most of the time. In the case of poor ventilation, heating should be improved to give 8 to 10°C above outside air in very cold weather. The prevalence of condensation in bedrooms is due mainly to lack of adequate heating and ventilation, and the best remedy is to provide reasonable heating with some night ventilation.[57] Casual heat gain from other rooms can be important, such as, for example, a bedroom over a well-heated living room. Conversely, a bedroom at one end of a flat may obtain little benefit from other heating and may also have a high heat loss if it is an external corner room with two exposed walls. Similarly, top floor rooms may suffer high heat loss through poorly insulated roofs.[55]

Clothes drying cupboards should be heated and well ventilated. *Cupboards* on external walls, especially clothes cupboards in poorly heated bedrooms, often suffer from condensation. They will benefit from high and low level internal vents and in extreme cases low wattage tubular heaters should be installed.

With *bathrooms*, rapid ventilation provided by opening windows after bathing is usually adequate, and particularly so if there is a heated towel rail. Bathroom doors should be well-fitting and kept closed. Internal bathrooms with fan ventilation rarely give trouble if the fan is functioning satisfactorily. Separate WCs are rarely heated and condensation often occurs on cold fittings. In extreme cases the provision of a low wattage tubular heater could be considered. In a similar manner, condensation may drip from cold storage cisterns located in cupboards. A drip tray or insulation at least 12.5 mm thick to the underside of the cistern should prove effective. Condensation on cold water pipes may also cause drips and the pipes should ideally be insulated.

There is normally sufficient air movement in halls, passages and stairways to prevent condensation but additional heat may be needed in extreme cases. *Living rooms* which are heated to above 18°C for several hours a day rarely suffer from serious condensation. If it does occur, it may be caused by an adjacent kitchen without a fan, very poor ventilation of the living room or poor structural insulation value. In the latter case some background heating for longer periods should be provided or, alternatively, a low thermal capacity lining should be fixed to the room face of the structure.[55]

The measures necessary to prevent *interstitial condensation* can be determined by calculation. Unless there is a vapour barrier on the room side of an external structure, water vapour will enter and condense when it reaches colder conditions towards the outside. With flat roofs, a vapour check at ceiling level may be formed of gloss paint or vinyl-faced paper. With some composite forms of walling, a vapour barrier is needed on the inside face, often in the form of polythene sheeting on impregnated battens with a dry lining, or an insulated lining with an integral vapour barrier. Anti-condensation paints can be used in certain situations but their use on a wide scale is rarely justified. Finally, occupants of dwellings should be informed of what is and what is not reasonable so far as living patterns are concerned, as quite trivial changes in living habits may bring a major improvement.[56] Plate 8 shows the results of condensation on the interior face of a glazed door to a school classroom, which has necessitated the replacement of the decaying door. Plate 9 shows mould growth in a dwelling caused by condensation on walls affected by penetrating damp.

Electric dehumidifiers that operate on a closed refrigeration cycle both dry and heat the air. They are most effective in warmer dwellings where condensation problems are caused by high vapour pressures but they tend to be relatively obtrusive and noisy in operation.[56]

Chimney Problems

Smoky Chimneys

Causes. One of the most numerous complaints in the past, and often one of the most difficult to cure satisfactorily, is that of smoky chimneys. A normal open fire requires 110 to 170 m³ of air per hour for it to burn satisfactorily; air is drawn into the flue from the room and this must be replaced by further air drawn in from outside the room. There are many

Plate 8 Rotting door resulting from condensation

Plate 9 Mould growth on walls of dwelling

factors or combinations of them which can prevent chimneys functioning satisfactorily, and the more important ones are now described.

(1) Blockage of the flue by soot or debris.

(2) Air starvation—insufficient air to carry the smoke into the flue.

(3) Adverse flow conditions resulting from poor design of the fireplace through which the smoke passes. For example, the throat over the fire may be too large or badly formed, thus reducing the velocity of the smoke and gases; the fireplace opening may be too large in relation to the flue size (more than six times area of flue); flue pipe may project too far into flue and so restrict the free flow of smoke and gases; or the fireplace opening may be too high (possibly 610 mm), thus assisting smoke to enter the room.

(4) Unsuitable size of flue; flues larger than 230 mm square may cause smokiness because they never really get warm.

(5) Air leaks into flue through defective brick joints and cracked parging, cooling gases and reducing draught.

(6) Poorly constructed flue—offset too low, bend too abrupt or traversing length too long—resulting in poor chimney draught and a smoky fire.

(7) Unsuitable chimney pot—possibly round base pot on a square flue causing obstruction to flow.

(8) Steady downdraught due to chimney top being in a high-pressure zone, such as a chimney pot lower than a nearby ridge and on the windward side.

(9) Downdraught due to doors, windows or ventilators being in a low-pressure zone, occurring mainly with short chimneys serving inset open fires in bungalows and the top two storeys of blocks of flats.

(10) Intermittent downdraught caused by downward striking wind currents near chimney tops where there are higher buildings, trees or hillsides nearby.

Remedies

The various remedial works are now described.

(1) Ensure that the chimney has been swept—excessive quantities of soot may result from the use of unsuitable fuel—and that the flue is free from debris by lowering a metal coring ball down the flue. Check on the size of throat, shape and position of lintel, and similar matters.

(2) Open the room door or window. If smoke ceases to enter the room the trouble is probably due to air starvation and the removal of draught-proofing and/or provision of ventilators or underfloor ducts are likely to cure the problem.

(3) Where the throat is too large, a thin sheet of metal can be wedged across the front of the throat, reducing the entry aperture to about 100 × 250 mm. If this produces an improvement, a variable throat restrictor can be fitted. Where the fireplace opening is too high, place a thin piece of metal, about 75 to 100 mm high, across the top of the opening to reduce its height to 510 to 560 mm. If a greater depth than 100 mm is involved, this solution

would be undesirable as it could adversely affect the heating of the room. If smoking ceases, a permanent metal canopy can be installed.

(4) Excessively large brick flues can be lined with fireclay or refractory concrete pipes with spigot and socket or rebated joints, or other suitable lining.

(5) A flame of a lighted candle will be drawn in by an air leak in a flue and defects in the chimney can be located by a 'smoke bomb test'. Defective joints are made good with asbestos rope dressed with fire cement.

(6) The run of a flue can be checked with a chimney sweep's rods. In bad cases, the chimney breast is opened up and the faulty section rebuilt.

(7) An unsatisfactory round base pot can be replaced with a square base pot.

(8) With steady downdraught due to the chimney pot being in a high pressure zone, a door or window opened on the windward side will probably balance the pressure and restore updraught. The chimney can be extended temporarily with sheet metal pipes and, if satisfactory, the chimney stack can then be raised.

(9) By producing sufficient smoke in the room, the movement of air can be traced and in this instance it will show air being drawn out of the room. Remedies include a draught-inducing cowl, a throat restrictor, or finally an openable room heater or a free-standing convector fire.

(10) With intermittent downdraught, the usual remedy is to fit a draught-inducing cowl or a 'dovecote', consisting of a concrete capping at the top of the chimney supported by small piers at each corner and gaps in between.

Domestic Boiler Chimneys

Many pre-war dwellings were built with parged brick flues serving solid fuel boilers, and sulphates, acids and water vapour forming the products of combustion have often attacked the parging and brick joints. This has resulted in staining of brickwork and decorations, distortion and cracking of brickwork in the chimney, and general expansion of the brickwork, sometimes allowing the pot to sink into the chimney stack. Furthermore, these chimneys were often built alongside external walls with long lengths of exposed chimney which aggravated condensation problems.

All defective brickwork must be cut out and replaced with new materials, and the flue needs lining, usually either with a flexible metal lining or an *in situ* lightweight concrete lining. The flexible metal linings are made of stainless steel, but their use is restricted to oil and gas-fired flues, and they are pulled down existing flues by means of a cord attached to a temporary tapered plug. The top of the lining is secured by a sealing plate bedded upon the stack upon which the terminal construction is built. The air space between the lining and the chimney fabric can be filled with loose insulation.

Lightweight concrete linings can be cast *in situ* using a temporary inflatable rubber core centred by metal spacers to form the flue. The bottom is sealed with a metal plate and the annular space around the rubber core filled by pumping lightweight concrete through a hose. When the concrete has set, the tube is deflated and removed.

Clay, refractory concrete and asbestos cement rigid pipe linings are only really feasible where the existing chimneys are in straight runs.

New brickwork must be allowed to dry out before plastering. Brick joints should be raked out to a depth of at least 15 mm, and the plaster often consists of cement and sand (1:3) undercoat finished with a suitable gypsum plaster. Where staining is severe, adhesive-backed metal foil or foil-backed plasterboard coated with bitumen paint may be used to enclose the brickwork prior to plastering.

Closing Fireplace Openings

Occupiers of dwellings sometimes request permission to block up fireplace openings. A number of precautions need to be taken to avoid later trouble.

(1) The chimney must first be swept, as if soot is left in the flue it will become damp and the salts and tarry matter it contains may pass through the brickwork and cause stains and damp patches on the chimney breast.

(2) An unused flue provides an easy means of access for rainwater, and it is advisable to provide some form of capping, such as a half-round ridge tile, over the head of the chimney to prevent direct entry of rain.

(3) Ventilation is needed and so the flue should not be sealed at either top or bottom, and a ventilator should be provided in the blocked up opening.

(4) Where the opening is blocked up with bricks or blocks and plastered over, this entails the use of a large quantity of water, and the surface should not be decorated until it has had ample time to dry out. The use of incombustible sheet materials will overcome this difficulty and they should preferably be easily removable to allow for the extraction of any debris that may accumulate behind them.

References

1   *The Building Regulations 1985*: SI 1985 Nr 1065. HMSO (1985)
2   *BRE Digest 75*. Cracking in buildings (1977)
3   E. D. Mills (Ed.). *Building Maintenance and Preservation*. Butterworths (1980)
4   British Standards Institution. *BS 3921: 1985 Specification for clay bricks*
5   *BRE Digest 164*. Clay brickwork: Part 1 (1980)
6   *BRE Digest 165*. Clay brickwork: Part 2 (1974)
7   W. H. Ransom. *Building Failures: Diagnosis and Avoidance*. Spon (1981)
8   British Standards Institution. *BS 187: 1978 Specification for calcium silicate (sandlime and flintlime) bricks*
9   British Standards Institution. *BS 6073: Part 1: 1981 Specification for precast concrete masonry units*

10 *BRE Digest 157.* Calcium silicate brickwork (1981)
11 *BRE Digest 160.* Mortars for bricklaying (1973)
12 Department of the Environment/Department of Transport. Staining of brickwork. *Construction 33* (April 1980)
13 *BRE Digest 89.* Sulphate attack on brickwork (1971)
14 Institution of Civil Engineers. *Design Life of Buildings.* Telford (1985)
15 E. J. Gibson (Ed.). *Developments in Building Maintenance—1.* Applied Science Publishers (1979)
16 British Standards Institution. *BS 5628: Code of practice for structural use of masonry. Part 1: 1978 Unreinforced masonry*
17 *BRE Digest 246.* Strength of brickwork and blockwork walls: design for vertical load (1981)
18 *BRE Digest 281.* Safety of large masonry walls (1984)
19 British Standards Institution. *BS 1243: 1978 Specification for metal ties for cavity wall construction*
20 *BRE Digest 257.* Installation of wall ties in existing construction (1982)
21 *BRE Digest 200.* Repairing brickwork (1977)
22 H. S. Staveley and P. V. Glover. *Surveying Buildings.* Butterworths (1983)
23 *BRE Digest 139.* Control of lichens, moulds and similar growths (1982)
24 *BRE Digest 177.* Decay and conservation of stone masonry (1984)
25 British Standards Institution. *BS 3826: 1969 Silicone based water repellents for masonry*
26 British Standards Institution. *BS 6477: 1984 Specification for water repellents for masonry surfaces*
27 *BRE Digest 280.* Cleaning external surfaces of buildings (1983)
28 *BRE Digest 217.* Wall cladding defects and their diagnosis (1978)
29 Curtain walling/cladding choices. *Building Products* (Autumn 1985)
30 *BRE Digest 235.* Fixings for non-loadbearing precast concrete cladding panels (1980)
31 M. Wheat. What's new in sealants. *CIOB Technical Information Service Paper Nr 27* (1983)
32 D. D. Higgins. *Efflorescence on Concrete.* Cement and Concrete Association (1982)
33 D. D. Higgins. *Removal of Stains and Growths from Concrete.* Cement and Concrete Association (1982)
34 *BRE Digest 161.* Reinforced plastics cladding panels (1974)
35 *BRE Digest 216.* GRC (1978)
36 *BRE Digest 286.* Natural finishes for exterior timber (1984)
37 Chartered Institute of Building. *Building Maintenance Management* (1985)
38 DOE. Weathering steels. *DOE Construction.* HMSO (1972)
39 M. Gordon. Stopping the rot. *Building Design* (11 April 1986)
40 BRE. *Timber Framed Housing: A Technical Appraisal* (1983)
41 BRE. *Quality in Timber Framed Housing* (1985)
42 I. H. Seeley. *Building Surveys, Reports and Dilapidations.* Macmillan (1985)

43 Timber and Brick Homes Information Council. *Timber and Brick Homes Handbook* (1985)
44 T. A. Oxley and E. G. Gobert. *Dampness in Buildings*. Butterworths (1983)
45 *BRE Digest 127*. An index of exposure to driving rain (1971)
46 *BRE Digest 245*. Rising damp in walls (1981)
47 British Standards Institution. *BS 6576: 1985 Code of practice for installation of chemical damp-proof courses*
48 *BRE Digest 196*. External rendered finishes (1976)
49 British Standards Institution. *BS 5262: 1976 Code of practice for external rendered finishes*
50 *BRE Digest 197*. Painting walls. Part 1: Choice of paint (1982)
51 I. H. Seeley. *Building Technology*. Macmillan (1986)
52 DOE. *Condensation in Dwellings, Part 1: A Design Guide*. HMSO (1970)
53 A. Benster. Effective treatment of dampness in building. *Public Service and Local Government* (December 1979)
54 C. H. Sanders and J. P. Cornish. *BRE Report. Dampness: One Week's Complaints in Five Local Authorities in England and Wales* (1982)
55 DOE. *Condensation in Dwellings, Part 2: Remedial Measures*. HMSO (1971)
56 *BRE Digest 297*. Surface condensation and mould growth in traditionally-built dwellings (1985)
57 *BRE Digest 176*. Failure patterns and implications (1975)

4 BUILDING MAINTENANCE PROBLEMS AND THEIR SOLUTION—III

Timbers, Floors, Roofs, Sound and Thermal Insulation, and Vibration

Timber Defects

Nature of Timber

Botanically trees are grouped into two classifications

(1) *Broad leaved trees* (hardwoods) are generally hard, tough, dense and dark-coloured with acid, aromatic or even poisonous secretions, although not all hardwoods are hard. The medullary rays in hardwoods are usually more clearly visible than in softwoods. Typical examples of hardwoods are oak, teak, mahogany, walnut, elm, iroko and sapele.

(2) *Needle leaved trees* (softwoods) are coniferous with cone-shaped seed vessels and narrow, needle-shaped leaves. They are usually elastic and easy to work, and have resinous or sweet secretions. Some softwoods such as pitch pine are quite hard. Typical examples of softwoods are European redwood, yellow pine, Douglas fir, spruce, Western hemlock and Western red cedar.

Trees are generally felled between mid-October and mid-January and are converted into suitably sized timbers by using a variety of sawing techniques. The 'quartering' method is the most expensive but shows the grain to best advantage, while the 'slash' method is extremely economical. The 'through and through' method, with all cuts parallel, reduces waste to a minimum but some boards will twist on shrinking.

Seasoning of Timber

In 'green' timber large quantities of free water are present in the cell cavities and the cell walls are also saturated. Seasoning consists of drying out the free water and some of the water from the cell walls, which on withdrawal causes the timber to shrink, with the object of reducing the moisture content to a level consistent with the humidity of the air in which the timber will be placed. The importance of seasoning can scarcely be exaggerated. It is vital that timber is dried to an appropriate moisture content, and care in drying will be wasted unless the timber is adequately protected in transit and

98

storage and after fixing in a wet building. There are two principal methods of seasoning timber.

(1) *Air seasoning* whereby the 'green' timber is stacked with laths or 'stickers' between the timbers to allow the passage of air and to assist in the evaporation of moisture from the timber. A suitable roof is needed to protect the timber from sun and rain. Air seasoning is unlikely to reduce the moisture content below 17 per cent even under ideal conditions and may take up to 2 years. Timber used internally in centrally heated buildings should not have a moisture content exceeding 10 per cent to prevent shrinkage and warping.

(2) *Kiln seasoning* is normally carried out in a forced draught compartment kiln, in which the air is heated by steam pipes and humidified by water sprays or steam jets. The temperature, degree of humidity and rate of air flow are all controlled from outside the kiln.

A maximum moisture content of 22 per cent should be specified for green timber. For structural timbers the moisture content in service is likely to vary between 12 per cent in continuously heated buildings and 20 per cent in unheated buildings.[1]

Preservation of Timber

Few timbers are resistant to decay or insect attack for long periods of time, and in many cases the length of life can be much increased by preservative treatment. The need for preservative treatment is largely dependent on the severity of the service environment.[1] The principal protective liquids are toxic oils, such as coal tar creosote; water-borne inorganic salts such as copper/chrome and copper/chrome/arsenic,[2] and organic solvent solutions, such as copper and zinc naphthenate,[3] all of which are suitable for both exterior and interior use.[4]

Preservatives can be applied by non-pressure methods such as brush application, spraying, immersion and steeping.[5] The double vacuum method uses a closed cylinder. A vacuum is created, the organic solvent preservative flows in, the vacuum is released and the preservative is taken into the wood. Sometimes the cylinder is pressurised to increase penetration, then a second vacuum is created to drain surplus preservative. For lasting preservation, a pressure method is preferable. In the 'full-cell' process, the timber is placed in a closed cylinder and a partial vacuum applied to draw out air from the cells, hot preservative admitted, air pressure applied for one to six hours and a partial vacuum reapplied to remove excess liquid. The 'empty-cell' process is cleaner and more economical and a deeper penetration can be obtained with only limited excess preservative. The timber is subjected to air pressure, the preservative admitted and a higher pressure applied causing the liquid to penetrate the timber and compress air in the cells. When the timber is extracted, the air trapped in the cells forces out excess liquid leaving the cells empty but impregnating the cell walls.[4]

Present-day experience contradicts the extreme view that preservation treatment of building timber is never justified. The annual expenditure of some £300 million on remedial treatment and consequent repair following insect and fungal attack indicates the size of the problem, and newer and non-traditional forms of construction generate additional maintenance problems. Recent trends of lower timber quality, higher timber prices and increased costs of remedial and repair work in relation to initial costs, strengthen the economic case for timber preservation. Preservation forms a second defence if the protection afforded by design and maintenance proves inadequate.

Some timber components justify treatment more than others. Where timber is exposed to a continually hazardous environment unprotected by design, it should be treated, and where experience shows a high risk of failure in loadbearing structures, preservation is essential, especially in such cases as joists, firring and decking of sealed flat roofs. Furthermore, preservation is desirable where there is a high risk of failure through faults in design or workmanship and where remedy is difficult and expensive, where there is a high risk of condensation, and in loadbearing concealed frames of external walls. Preservation is clearly optional where there is a low risk of attack or where remedial action or replacement is simple, as in the case of timbers in normal pitched roofs and unsealed flat roofs of unoccupied buildings, and ground floor joists. Preservation of floor boards, skirting boards and internal joinery is not usually necessary.

Strength of Timber

There are thousands of different species of timber but relatively few produce timber for structural use. For structural use, where appearance is generally unimportant, strength properties and durability are usually the prime considerations when making the choice of species.[1] Tables of reliable strength properties for each timber species and stress grade combination are contained in BS 5268.[6]

Timber Defects and their Rectification

The strength and usefulness of timber can be affected by a wide variety of defects, some of which can occur during natural growth, others during seasoning or manufacture, while others result from attack by fungi or insects. The principal defects are listed and described.

A.   Defects arising from natural causes

(1) *Knots.*   These are portions of branches enclosed in the wood by the natural growth of the tree; they affect the strength of the timber as they cause a deviation of the grain and may leave a hole. This is particularly important where they occur in critical parts of structural timbers subject to high tension, and it is sometimes helpful to turn the timber so that the defect is on the compression edge. Large knots on edges or arrises which penetrate deeply into structural timbers are particularly harmful. There are several types of knot. A *sound* knot is one free from decay, solid across its face and

at least as hard as the surrounding wood. A *dead* knot has its fibres interwoven with those of the surrounding wood to an extent of less than one-quarter of the cross-sectional perimeter; a *loose* knot is a dead knot not held firmly in place. *Rind gall* is a surface wound that has been enclosed by the growth of the tree.

(2) *Shakes.* These consist of a separation of fibres along the grain due to stresses developing in the standing tree, or during felling or seasoning. A *cross* shake occurs in cross-grained timber following the grain; a *heart* shake is a radial shake originating at the heart; a *ring* or *cup* shake follows a growth ring; and a *star* shake consists of a number of heart shakes resembling a star.

(3) *Bark pocket.* Bark in a pocket associated with a knot which has been partially or wholly enclosed by the growth of the tree (inbark or ingrown bark).

(4) *Deadwood.* Timber produced from dead standing trees.

(5) *Resin pocket.* Cavities in timber containing liquid resin (pitch pocket or gum pocket).

B. *Defects due mainly to seasoning*

(1) *Check.* A separation of fibres along the grain forming a crack or fissure in the timber, not extending through the piece from one surface to another.

(2) *Ribbing.* A more or less regular corrugation of the surface of the timber caused by differential shrinkage of spring wood and summer wood (crimping).

(3) *Split.* A separation of fibres along the grain forming a crack or fissure that extends through the piece from one surface to another.

(4) *Warp.* A distortion in converted timber causing departure from its original plane. *Cupping* is a curvature occurring in the cross-section of a piece and a *bow* is a curvature of a piece of timber in the direction of its length.

C. *Defects due to manufacture (including conversion)*

(1) *Chipped grain.* The breaking away of the wood below the finished surface by the action of a cutter or other tool.

(2) *Imperfect manufacture.* Any defect, blemish or imperfection incidental to the conversion or machining of timber, such as a variation in sawing, torn grain, chipped grain or cutter marks.

(3) *Torn grain.* Tearing of the wood below the finished surface by the action of a cutter or other tool.

(4) *Waney edge.* This is the original rounded surface of a tree remaining on a piece of converted timber.

D. *Fungal attack*

(1) *Fungal decay.* This is decomposition of timber caused by fungi and other micro-organisms, resulting in softening, progressive loss of strength and weight and often a change of texture or colour. Fungi are living plants and require food supply, moisture, oxygen and a suitable

temperature. A fungus is made up of cells called *hyphae* and a mass of hyphae is termed *mycelium*. It also contains fruit bodies within which very fine spores are formed. When timber is infected by spores being blown on to it, hyphae are then formed which penetrate the timber and break down the wood as food by means of enzymes. The Building Research Establishment[7] distinguishes two main forms of decay in wood according to the colour of the decayed timber (brown rots and white rots).

(2) *Dote*.   The early stages of decay are characterised by bleached or discoloured streaks or patches in wood, the general texture remaining more or less unchanged. This defect is also known as doaty, dosy, dozy and foxy.

(3) *Dry rot*.   This is a serious form of timber decay caused by a fungus, *Serpula lacrymans*. The fungus develops from rust-coloured spores, which can be carried by wind, animals or insects, and throw out minute hollow white silky threads (hyphae). The fungus also produces grey or white strands 2 to 8 mm thick which can travel considerable distances and penetrate brick walls through mortar joints. The strands throw off hyphae whenever they meet timber and carry the water supply for digestion of the timber, which becomes friable, powdery and dull brown in colour, accompanied by a distinctive mushroom-like smell. Often the timber shrinks and splits into brick-shaped pieces formed by deep longitudinal and cross cracks.[8] Table 4.1 shows the main characteristics of *Serpula lacrymans*.

The fungus needs a source of damp timber with a moisture content above 20 per cent. Damp, still air will encourage the establishment and spread of the fungus, particularly if these conditions are maintained for long periods. To eradicate an outbreak of dry rot, all affected timber and timber for 300 to 450 mm beyond must be cut away and burnt, preferably on the site. Surrounding masonry should be sterilised with a suitable fungicide such as sodium pentachlorophenoxide or sodium 2-phenylphenoxide. All apparently sound timber which is at risk should receive three brush coats of a suitable preservative, such as type F in BS 5707.[9] Where it is likely that damp conditions will persist as in a damp cellar, replacement timber should be pressure-impregnated with a copper/chromium/arsenic preservative conforming to BS 4072.[10] Similar treatment with creosote to BS 913[11] is effective but it emits a strong odour and stains. Where dampness is not expected to persist, an organic solvent type of preservative can be applied by immersion or the double vacuum process in accordance with BS 5707.[9] Levels of treatment are detailed in BS 5268[6] and BS 5589.[12]

The cause of the outbreak must be established and rectified, usually by preventing damp penetration and improving ventilation. The most vulnerable locations are cellars, inadequately ventilated floors, ends of timbers built into walls, backs of joinery fixed to walls and beneath sanitary appliances. The design of timber floors to prevent dry rot is covered in BRE Digest 18.[13]

Dry rot frequently starts near a damp wall, behind woodwork or under a floor. Look particularly for irregularities or waviness in the surface of panelling, skirtings, window linings or other woodwork, as they may be caused by dry rot. A knife blade will easily penetrate and be withdrawn from infected timber. Attacked wood becomes light in weight, crumbles under the fingers, has a dull brown colour and often breaks up into pieces by splitting along and across the grain. Sometimes the silky grey or white fruiting bodies can be seen on skirting boards, panelling and even outside walls near ventilators. Spore dust in the form of reddish-brown powder may penetrate between floor boards or cracks in woodwork, and the distinctive smell may also be detected. Plates 10, 11 and 12 illustrate the nature and serious consequences of outbreaks of dry rot in diverse situations.

The causes of dampness must be positively established and rectified. Typical sources of dampness are buried or obstructed airbricks, flower beds or rockeries formed against walls above damp-proof course, blocked or defective gutters, downpipes and gullies, broken pavement lights and windows, badly fitting cellar flaps, missing weatherboarding, damaged flashings, soakers and valleys, dislodged pointing, cracked masonry and

Plate 10 Dry rot to underside of floor

Platge 11 Dry rot mycelium

plumbing leaks. Regular inspections of existing buildings should prevent these defects occurring for more than short periods. Table 4.2 lists the measures needed to control an outbreak of dry rot.

(4) *Coniophora puteana (cellar fungus).* This is a fungus which attacks wet timber and is mainly found in badly ventilated basements and bathrooms. The fruit bodies are in sheets, yellow to brown in colour.

(5) *Wet rot.* Originally this was defined as chemical decomposition generally arising from exceptionally wet or alternate wet and dry conditions, as occurs with timber fencing posts at ground level. The decay may be

Plate 12 Dry rot attack behind panelling

accelerated by fungus attack. More recently the term has been applied to fungus attacks other than dry rot, for instance *Coniophora puteana*, *Poria vaillantii* (white strands) and *Paxillus panuoides* (pale yellow strands and reddish-brown wood).[7] Other fungi associated with wet rot are *Phellinus megaloporous*, *Letinus lepidous*, *Poria xanthus* and *Trametes serialis*. Dealing with wet rot is a relatively straightforward process, embracing removal of the source of dampness and subsequently drying out. Timbers which have become structurally unsound require replacement or strengthening.

Table 4.1 Recognition of *Serpula lacrymans* (dry rot)

Usual effect on the wood	Strands	Mycelium	Fruit-bodies and spores
Wood becomes light in weight, crumbles under the fingers and has a dull brown colour. It shrinks and splits into cubic pieces	Strands grey or white, 2–8 mm thick, become brittle when dried	In damp dark places, soft white cushions or silky growths; in drier places, thick silver-grey sheets or skins usually showing patches of lemon yellow and tinges of lilac	Fruit bodies fleshy, soft, but rather tough; shaped like pancakes or brackets. Spore-bearing surface rusty red with shallow pores or ridges and furrows; margin white. Spores often settle on horizontal surfaces as a layer of rust-coloured dust

Source: BRE Digest 299[8].

Table 4.2 Measures for controlling an outbreak of dry rot

Primary measures	Locate and eliminate sources of moisture
	Promote rapid drying of the structure
Secondary measures	Determine the full extent of the outbreak
	Remove rotted wood
	Contain the fungus within the wall
	Treat remaining sound timbers with preservatives
	Use preservative-treated replacement timbers
	Introduce support measures

Source: BRE Digest 299[8].

E. *Insect attack*

BRE Digest 307[14] describes how most of the insects which cause damage to timber are beetles. The adults lay eggs on the surface of the wood, in splits or in bark, and these hatch into active, grub-like larvae which eat their way into the wood and form tunnels. The damage to timber is caused largely by the feeding and tunnelling of the larvae. The tunnels usually become filled with excreted wood pellets known as 'bore dust'. The size, shape and cross-section of the tunnels, and to a lesser extent the characteristics of the bore dust, help to identify the type of insect.

When the larvae are fully grown, which may take from one to five years, they pass through a pupal stage to emerge from the infested wood as adult beetles, leaving emergence holes. The beetles do not themselves cause further damage although they can spread the infestation by egg-laying.

BRE Digest 307[14] lists the damage features used in the identification of wood-borers as:

1. type and condition of the wood (hardwood or softwood, sound or rotted);
2. size and shape of holes on the wood surface;
3. colour, shape and texture of bore dust (this may be ejected through holes, forming small piles on or beneath the wood); and
4. size and shape of tunnels within the wood, exposed by probing or by wear on floorboards.

BRE Digest 307[14] categorises wood-boring insects into three groups, according to whether insecticidal treatment is usually needed; where treatment is necessary only to control associated wood rot; and where no treatment is needed, as shown in table 4.3. The characteristics of the more common types of wood-boring insects are now described and a more comprehensive list is contained in table 4.3.

Common furniture beetle. Commonly referred to as 'woodworm', and can attack hardwoods and softwoods in buildings throughout Britain. Attack is most common in damp locations, such as timber ground floors or damp roof voids. Problems rarely arise in very dry situations, and existing infestations tend to die out where effective central heating is introduced. Imported tropical hardwoods, modern plywoods, chipboards and most other wood-based products are immune from attack.

The larvae live for two to five years and create a network of tunnels, circular in cross-section, each up to 2 mm diameter. The tunnels are loosely packed with cream-coloured bore dust, which is gritty in texture and consists mainly of lemon-shaped pellets when viewed with a ×10 hand lens. The beetles emerge during late spring and summer leaving circular emergence holes, 1 to 2 mm in diameter[14] (see plate 13).

Ptilinus (Ptilinus pectinicornis). Closely related to the common furniture beetle, but is encountered less frequently. It occurs in such hardwood timbers as beech, elm and willow, and is mainly found in air-drying timber selected for furniture manufacture. The tunnels are very densely packed with a compacted bore dust which is not easily removed.[14]

House longhorn beetle. Numerous infestations have been found in pre-1950s buildings in north Surrey. Building Regulations and the former Building Byelaws have required pre-treatment of roof timbers in new buildings in this area since 1962. Isolated cases of house longhorn beetle damage have been reported elsewhere. Attack occurs only in softwoods and predominantly in roof timbers. Where damage is severe, a survey of structural timbers may be necessary.

Larvae live for three to seven years and create oval tunnels (6 to 10 mm wider dimensions); these usually coalesce to form a powder-like mass beneath a thin, intact external veneer of wood which may develop a corrugated surface. The cream-coloured bore dust, with easily visible coarse, sausage-shaped pellets, may be emitted from cracks in the surface. In an active attack a scratching sound may be heard as the larvae feed in warm weather. The adult beetles emerge in the summer leaving ragged, oval emergence holes roughly the same size as the tunnels.[14]

Powder-post beetle. Found world-wide, this insect derives its name from the severity of the damage it causes, often reducing the sapwood to powder. Most tropical hardwoods are attacked as well as some coarse-pored European hardwoods, such as oak, elm, ash and chestnut. Timbers become less susceptible with age and are normally immune after 10 to 15 years. Infestation commonly occurs in plywoods manufactured from susceptible timbers.

The life cycle is one to two years and damage occurs rapidly. The larvae create a meandering network of tunnels of circular cross-section (about 1.5 mm diameter) coalescing to form a loose, powdered mass beneath a thin, intact, wood veneer. The bore dust is very fine with a talc-like feel. Adult beetles emerge throughout the year in heated buildings, leaving circular emergence holes about 1.5 mm diameter.[14]

Death-watch beetle. Commonly found throughout southern England, less frequently in the north and is non-existent in Scotland. Infestations are most common in large dimension hardwood such as oak and elm, in which there is some wood rot. It generally requires damp conditions. However, the beetle can attack slowly in relatively dry timbers where rot has ceased. Softwoods may be attacked if they are adjacent to infested hardwood or very occasionally, if rotted, in damp ground floors. If damage is severe a structural survey may be necessary. Infestations are most common in older buildings such as churches and manor houses. Timbers in contact with damp masonry are particularly vulnerable. Infestations may die out if effective drying methods are introduced.[14]

The larvae may live for up to ten years and excavate a series of bore dust-filled circular tunnels (about 3 mm diameter), eventually coalescing to give a honeycomb appearance. Where large timbers are attacked, internal cavities may develop with few visible signs externally. The bore dust, which is gritty, contains numerous flattened disc-like pellets and these are visible to the naked eye. Adult beetles emerge during the early spring forming circular emergence holes 3 mm in diameter[14] (see plate 14).

Prevention and detection of insect attack. Timber can be protected against all wood-boring insects, as well as rot, by treating with a preservative before use. In certain districts where the house longhorn beetle is prevalent, this is a requirement of the Building Regulations.

Early detection of insect attack can reduce both expense and inconvenience, for once established in a building it can spread rapidly. Ideally, timber and furniture should be inspected annually, preferably during warm summer weather when the wood-boring insects are most likely to be active, looking particularly for emergence holes and small piles of powdery sawdust-like material (bore dust) emitted from workings. Particular attention should be directed to damp or inconspicuous places such as under stairs, in roof spaces, under sanitary fitments, backs and undersides of furniture and the like. Dampness, incipient decay and excessive sapwood all render timber more susceptible to attack.

An extensive attack of furniture beetle in roof timbers is illustrated in plate 13, and a severe attack of death watch beetle in floor joists in plate 14.

Table 4.3 Common wood borers: recognition and significance

Type of borer	Timber	Emergence hole shape and size (mm)	Recognition of damage			Treatment
			Tunnels	Bore dust	Persistence	
Damage Category A: Insecticidal treatment usually needed						
Common furniture beetle	Sapwood of softwoods and European hardwoods	Circular 1–2	Numerous, close	Cream, granular, lemon-shaped pellets (× 10 lens)	Long-term except in very dry situations	Insecticidal
Ptilinus pectinicornis	Limited number of European hardwoods	Circular 1–2	Numerous, close	Pink or cream, talc-like, not easily dislodged from tunnels	Long-term except in very dry situations	Insecticidal, but replacement, may be more cost effective
House longhorn beetle	Sapwood of softwoods	Oval 6–10 often ragged	Numerous, often coalesce to powdery mass beneath	Cream powder, chips and cylindrical pellets	Can continue until sapwood consumed	Insecticidal
Powder-post beetle	Sapwood of coarse-pored hardwoods	Circular 1.5	Numerous, close	Cream, talc-like	Can continue until sapwood consumed	Replacement may be more cost effective
Death-watch beetle	Sapwood and heart-wood of decayed hardwoods. Occasionally softwoods	Circular 3	Numerous, close, eventually form a honeycomb appearance	Brown, disc-shaped pellets	Long-term, except in very dry situations	Insecticidal
Damage Category B: Treatment necessary only to control associated wood rot						
Weevil	Any, if damp and decayed	Ragged 1	Numerous, close, breaking through to surface in places	Brown, fine lemon-shaped pellets (× 10 lens)	Dies out on drying	Dry out and replace damaged timber
Wharf borer	Any, if damp and decayed	Oval 6	Numerous, close, often coalescing to form cavities	Dark brown, mud-like substance. Bundles of coarse wood fibres	Dies out on drying	Replace damaged timber
Leaf cutter bee/solitary wasp	Any, if badly decayed	Circular 6	Sparse network	Brown chips, metallic fragments, fly wings, barrel-shaped cocoons of leaves	May continue for several years	Replace damaged timber
Damage Category C: No treatment needed						
Pinhole borers	Any in log form	Circular 1–2	Across grain, darkly stained	None	Dies out on drying	None
Common bark-borer	Bark of softwoods	Circular 1.5–2 some in bark, few in sapwood	Network between bark and underlying wood	Cream and brown round pellets	Dies out when bark consumed	Remove bark edges
Wood wasps	Sapwood and heartwood of softwoods	Circular 4–7	Few, widely spaced	Coarse, powdery	Dies out after drying	None
Forest longhorns	Any	Oval 6–10 on bark edges only. May be larger in some hardwoods	Few, widely spread. Sections on sawn surfaces oval 6–10 mm	None or rarely, small piles of coarse fibres	May continue on bark edges, otherwise a few insects may survive for a few months after sawing of logs	None
Shipworm	Any	None but tunnels sectioned by sawing	Circular up to 15	None but tunnel may have white chalky lining	Dies out immediately after removal from sea	None

Source: BRE Digest 307.[14]

Plate 13 Furniture beetle attack on roof timber

Treatment of infested timber. There are many proprietary preparations available for treating infested timber, most of which contain chemicals such as chlorinated naphthalenes, metallic naphthenates and pentachlorophenol. Insecticides brushed over the surface and worked into all cracks and crevices during spring and early summer will destroy eggs and newly-hatched larvae, but may not reach larvae located deeper in the timber. A single thorough application normally eradicates common furniture beetle in softwood, but death-watch beetle in oak may need several applications.

In furniture, interior joinery and small timber sections, an injector should be inserted into emergence holes at about 150 mm centres and all timber in the vicinity not already painted, varnished or polished should be brushed

Plate 14 Death watch beetle attack on floor joists

with insecticide. Care should be taken to penetrate all open joints, splits and shakes and to treat all inconspicuous areas. This treatment should be repeated at least once each year during the months when insects are active until there is no sign of continued activity.

Structural timbers under attack but not seriously weakened should be thoroughly brushed or sprayed. Holes, 3 to 5 mm diameter, should be drilled at vulnerable points, such as connections of members or where they are built into walls, for the injection of insecticide, particularly in the case of death-watch beetle attack. Structural timbers which have been seriously weakened by infestation should be removed and immediately burnt. Replacement timber must be suitably impregnated. Gloves should be worn when applying insecticide and care taken not to inhale the fumes; ventilation should be good, smoking and naked lights avoided.

Floors

Suspended Timber Floors

The most serious defect that is likely to occur in a suspended timber ground floor is an outbreak of dry rot. As described earlier in the chapter this would entail the removal and destruction of all infected timber and also adjoining timber, together with the treatment of infected masonry. It may be necessary to improve ventilation by inserting new airbricks in external walls below floor level, preferably 215 × 140 mm terracotta at 3 m centres. Cast iron ventilators will corrode unless painted regularly. Airbricks are needed on all walls to avoid stagnant corners and where suspended floors adjoin solid ones, it is advisable to lay air ducts under the solid floor. Sleeper walls under suspended floors must be honeycombed to permit a free flow of air under the timber floor. Internal partition walls will also need openings in them below floor level.

Any sources of damp penetration must also be dealt with, such as the replacement of defective damp-proof courses and lowering of outside ground or paved levels where they adjoin or extend above damp-proof course. All new timbers must be impregnated and it is advisable to treat existing timbers that are being retained. Where a defective suspended timber floor is constructed below ground level, the best remedy is probably to replace it with a solid floor incorporating a suitable waterproof membrane. In old buildings there may be no concrete oversite or damp-proof courses and full remedial work could be very expensive.

A suspended timber floor should not move perceptibly when walked upon nor should furniture or ornaments placed on them vibrate. These deficiencies can result from inadequately sized joists or insufficient support to or fixing of the joists.

Where floorboards have curled across the grain resulting in wide open joints (1.5 mm or more gap), the fault may be due to various causes, such as:

(1) use of unsuitable, wet or insufficiently seasoned timber;
(2) boards insufficiently cramped on laying;
(3) insufficient nailing.

A large gap between the bottom edge of a skirting and floor boards generally stems from excessive shrinkage, both in the floor and the skirting. The squeaking of floorboards causes annoyance to occupiers and is usually due to loose fixing or the accidental contact of nails or screws with other metal components. Loose boards should be fixed more securely by nailing or screwing. In industrial buildings, badly worn floorboards may be lined with good-quality plywood.

Joists to upper floors are rarely attacked by dry rot but outbreaks can occur in damp cupboards or other enclosed spaces or through leaks in rainwater, waste or service pipes. More common faults in upper timber floors are sagging and springiness resulting from overloading, inadequate size of joists, lack of strutting or joists badly notched for services.[15] Uneven and springy floors supporting ceilings which are uneven or cracked at edges could result from the underlying courses of masonry not being level, use of packing under hangers, wrong grade of hangers, hangers not fixed tight to wall or gaps too large between joist ends and back plate.[16] Herringbone strutting should be provided to all floors with spans exceeding 3 m to stiffen the joists. A slightly springy upper floor can be strengthened by taking up floorboards and inserting solid strutting between joists, with each row of strutting wedged against the wall at each end. Another approach is to bolt new joists alongside existing ones, keeping the new joists shallower and packing at their ends to avoid disturbing the ceiling. An old floor can be strengthened by screwing chipboard slabs over the existing floorboards.

Chipboard is attractive as a flooring material because of its relatively favourable cost and flat surface compared with tongued and grooved boarding. It is, however, important that flooring grade bd to BS 5669[17] is used. BRE has found joints between boards not nogged, tongues and grooves in them not glued and lack of support at floor perimeters resulting in sagging and breakage. Boards can spring if plain nails are used instead of ring-shank nails.[18] BRE Digest 239[19] recommends the application of a polyurethane type finish in kitchens and bathrooms, where there is high relatively humidity.

Solid Floors

There is always some risk of rising damp with concrete floor slabs supported on the ground. It is therefore customary with new floors to insert an effective damp-proof membrane to prevent possible damage to the floor finish. Different floor finishes offer varying degrees of resistance to dampness. For example, pitchmastic and mastic asphalt flooring both provide effective damp-proof membranes in themselves, while concrete, terrazzo and clay tiles transmit rising damp without dimensional, material or adhesion failure. Thermoplastic and PVC (vinyl) asbestos tiles may suffer dimensional and adhesion failure under severe conditions, while magnesium oxychloride, PVA emulsion/cement, rubber, flexible PVC flooring, linoleum, cork carpet and tiles, wood blocks in cold adhesives, wood strip and board flooring, and chipboard are all particularly susceptible to damage in damp conditions.[20]

Concrete beds may also be adversely affected by soluble salts in the ground or hardcore below. In severe cases it may be necessary to replace the concrete slab and the fill beneath, with a sheet of polythene between them.

A settling floor slab may be underpinned with mini-piles and any voids beneath grouted through holes at about 1 m spacing in the slab, but the grout must not penetrate the inner skin of a cavity wall. BRE Digest 313[21] recommends that mini-piles should be placed at spacings not exceeding 1.5 m, with maximum distances from slab edges of 300 mm. Where a floor slab settles so badly that it has cracked, mini-piles should be installed each side of the crack at a distance from it not greater than 300 mm.

Solid Floor Coverings

A *granolithic* finish is best laid monolithically with the concrete base, that is, within three hours of laying the base, to avoid subsequent cracking and lifting. In some situations, as with suspended floors, monolithic construction is impracticable and the finish with a minimum thickness of 40 mm should be applied to a hardened base, in bay sizes not exceeding 15 m² The base should be mechanically roughened, cleaned, wetted and covered with a slurry of neat cement or of cement and sand to ensure a good bond with the finish. Even the best granolithic concrete will wear and produce some dust, and this can be substantially reduced by applying two or three coats of sodium silicate solution or other surface hardener. When repairing small patches of granolithic concrete, the defective material should be cut out in a rectangle with clean, straight edges, the exposed concrete covered with cement slurry and the recess filled with new granolithic. In extreme cases it may be necessary to take up all the old granolithic and replace with new, taking all the precautions previously described.

Terrazzo is best laid immediately after the screed has been placed and it should not be richer than 1 part of cement to $2\frac{1}{2}$ parts of aggregate by volume. To reduce the risk of shrinkage cracking, the flooring should be divided into panels not exceeding 1 m^2, with the lengths of the sides in a ratio not greater than 3:1. The panels are normally separated by strips of metal, ebonite or plastics set into the screed before placing the terrazzo.

Alternatively terrazzo may be provided in the form of tiles, similar to concrete floor tiles, usually bedded in mortar on a wetted concrete base or screed. Occasionally the tiles lift owing to shrinkage of the base and to avoid this the bedding mortar is best separated from the base with polythene or building paper. Properly designed and laid terrazzo flooring has good wear resistance and attractive appearance, and is easily cleaned. The surface should be kept dry and free from soap or wax. Disinfectants containing phenols and cresols may react with iron in the cement to produce indelible pink stains.

Stores, wash-houses, garages and external paved areas are frequently finished in concrete. These may crack or wear unevenly. Cracks may be undercut on each side with a cold chisel to form a key for the cement mortar which is worked into the crack, after cleaning and wetting. Rough patches should be cut out and the exposed surface brushed and wetted, and covered with fine concrete to a minimum depth of 20 mm floated to a smooth finish.

Clay tiles may fail owing to arching or ridging as the tiles separate cleanly from the bedding, when the newly laid screed shrinks but the tiles remain constant. Where the tiles are firmly bonded by their bedding to the screed surface, considerable stresses may develop which are eventually relieved by areas or rows of tiles lifting. This normally occurs during the first year after laying and thin tiles rise more readily than thick ones. This defect can be avoided by introducing a separating layer of polythene or building paper over the base as described for concrete tiles. If the base is subject to high temperatures, as around boilers, bituminous bedding is recommended. Expansion joints are needed around the perimeter of a clay tiled floor.[22] If an old tiled floor becomes uneven or tiles are loose and cracked, the defective tiles should be removed, the surface of the base hacked to form a key, brushed and wetted, and covered with bedding mortar to receive the new tiles. Treatment with linseed oil or polishes is not recommended as they tend to make the tiles slippery.

Thermoplastic and *PVC tiles* need polishing, preferably with emulsion polishes, to retain their initial appearance. Excessive use of polish should be avoided as it leads to slipperiness and high dirt retention. Worn or dirty coats of polish can be removed by washing with a solution of neutral detergent and subsequently rinsing with clean water. These tiles can be marked by black rubber in footwear, castor tyres and protective thimbles on the legs of metal furniture. These markings are best removed by rubbing with scouring powder and fine steel wool.[23] Thermoplastic tiles are less resistant to oils and grease than PVC flooring.

Linoleum may be washed with warm water using mild soap or neutral detergent. Harsh, alkaline or abrasive cleaners and scrubbing brushes should be avoided. When the linoleum has dried a suitable polish should be applied. The frequency of application of polish may vary between one and six weeks depending on the amount of traffic. Linoleum has now been largely superseded by cushion flooring.

Rubber flooring may be made from natural or synthetic rubber in varying thicknesses and is fixed with adhesive. Oils, fats and grease may prove harmful, but it has good wearing, resilience and sound absorption qualities. It is best maintained by cleaning with a damp cloth and ordinary household soap. Drying with a soft cloth will produce a natural polish.

Mastic asphalt and *pitchmastic* both provide dustless, jointless and impervious floors, but are liable to soften if in prolonged contact with fats, greases and oils. They are maintained by washing with warm water and suitable detergents.

Cement rubber-latex is normally maintained with wax polish. *Magnesium oxychloride* or *magnesite* is available in various colours to mottled or grained finishes and requires adequate protection against damp. Maintenance is carried out by scrubbing with warm water and the occasional use of mild household soap.

Timber finishes to solid floors are now quite common, but special care is needed in their construction to prevent damp penetration and in extreme cases outbreaks of dry rot. Boarding may be nailed to timber fillets which are either embedded in or resting on the upper surface of the concrete slab or screed. The fillets must be pressure-impregnated with a suitable preservative and the upper surface of the concrete or screed effectively waterproofed. The underside of the boards should be treated as an additional precaution. Excess water in washing such floors should be avoided.

Hardwood strip flooring provides an attractive finish and requires sealing and polishing. Another alternative finish is wood blocks fixed with a suitable adhesive to a screed. The most common defects are unevenness resulting from unequal wear which is normally cured by planing or sanding, loose blocks often caused by shrinkage or expansion with subsequent loss of key and cured by resetting the blocks in adhesive, and dry rot caused by damp penetration and requiring removal of the infected blocks, treatment of adjoining flooring and renewal with suitably treated blocks on a substantial bed of bitumen or other appropriate material. It is advisable to leave an expansion joint between the edges of the block flooring and adjoining walls, often with a cork strip under the skirting, to allow room for movement if the blocks expand on absorbing moisture from the atmosphere.[24]

Cork is available as tiles, mainly 300 mm square and from 5 mm upwards in thickness, or as carpet 1.80 m wide in various thicknesses. A damp-proof membrane is needed when laid on a solid sub-floor and an underlay of hardboard may be provided on softwood flooring. Fixing is by adhesive and sealing with solvent-based seals is important to prevent dirt absorption. Water-based emulsion polishes may be used over these seals.

Carpets are becoming a popular finish in offices, educational buildings and shops, where comfort and colour is an important consideration. This approach has been assisted by new carpet manufacturing processes and materials. Carpet tiles are particularly popular made up of 400, 450, 500 or 600 mm squares usually made from man-made fibres on a suitable backing making them dimensionally stable and hard wearing, but they do result in higher maintenance and replacement costs than most other types of finish. For instance, in the Royal Berkshire Hospital, carpets accounted for 17 per cent of the floor area but 43 per cent of total floor maintenance costs excluding cleaning.[25] Wharton[26] has described how carpets are particularly useful at the entrance to shops to create a pleasant atmosphere and filter off dirt to protect more brightly coloured and delicate floor coverings further inside the shop. He advocated cleaning by vacuum cleaner daily and shampooing only when necessary. Care is needed to select the right fibre, colour, weave and backing for the particular situation.

Floor seals are needed with wood, wood composition and cork surfaced floors in particular for protection, hygiene and decoration. A seal is a semi-permanent material which protects the floors from dirt, absorption, stains and foreign matter. Most floor finishes are porous to some degree and the filling of the pores extends the life of the floor. There are various categories of floor seal, and a knowledge of these seals and consideration of the prevailing conditions are needed in making a choice.

Where a floor may be subject to spillage of chemicals, a chemically-resistant floor seal should be used. Water-based seals are used on thermo-plastic, vinyl and rubber finishes. When re-sealing, compatibility of seals is important. In locations such as hospitals, where long life is essential to reduce to a minimum interference with the use of the floor, a very durable seal is desirable. All seals should be maintained with a floor wax to prolong their life, but if waxing is a problem then a very durable seal should be used.

Staircases

The main defects found in timber staircases are worn nosings, creaking treads, cracked balusters and handrails, and loose newel posts. On uncarpeted stairs nosings may become worn and a suitable remedy is to cut out a length of tread and nosing about 50 mm wide, insert a replacement piece and nail it to the old tread and riser.

Creaking treads often result from the lack of angle blocks between treads and risers and the insertion of two or three angle blocks to each step should provide a cure. In the case of wide stairs, in excess of 900 mm, the insertion of a rough carriage, usually about 100 × 75 mm, centrally under the flight may be required. Cracked balusters and handrails are normally repaired by splicing and joining by screws or wood dowels. Loose newel posts can generally be stiffened by fixing angle brackets at their feet.

Every year there are about 200 000 accidents on stairs in the home. BRE site surveys[27] found many defects: newels and top nosings were dangerously insecure; flights were not rigidly fixed; handrails were not sanded smooth; handrail brackets presented sharp obstructions; wooden spacers between handrail and apron lining obstructed passage of the hand; and mouldings on the apron lining were fixed so that they wedged the hand between moulding and handrail.

Roofs

Pitched Roof Timbers

The design and construction of timber roofs are controlled by the Building Regulations and Supporting Documents but weaknesses still occur. In some cases there is extravagant use of timber coupled with haphazard nailing which is neither effective nor economical. The design of a roof is influenced by the clear span, type of building, covering material, and situation and shape of building.

Roof timbers may be affected by wet rot resulting from leaks in the roof covering or condensation, normally involving the replacement of the defective sections of timber, and splicing old to new where necessary. In cases where the rot has not penetrated too deeply into the timber, it is possible to treat the affected timber and to strengthen it with timber or steel members bolted to it. Timbers should also be closely examined with the aid of a powerful torch or hand lamp for possible insect attack. The most vulnerable parts of the roof structure are those under gutters or partly buried in masonry.

With softwood roofs not more than 10 years old, the most likely form of attack is *Ernobius mollis* (bark borer) which can be effectively dealt with by cutting away all timber to which bark is adhering to a depth of about 6 mm. Where a modern softwood roof has been damaged by fire and soaked with water, look for dry rot, particularly in wall plates. With older softwood roofs the most common form of damage results from the common furniture beetle. Damage by house longhorn beetle is mainly confined to a fairly clearly defined area of north-west Surrey. Old hardwood roofs suffer many forms of insect attack particularly under gutters, at ridges, and where timbers are built into masonry and this can make necessary the expensive treatment and replacement of timbers.

Where a roof has sagged or, through insufficient ties, has forced walls out of plumb, it is not feasible to force the structure back into its original position. In extreme cases reconstruction will be necessary. Further movement can be prevented by inserting wood or steel ties between the ridge and ceiling beams or joists, between wall plates or feet of rafters, or under purlins. Roofs of semi-detached or terraced houses sometimes show a hump over party walls and this may be caused by inadequately sized timbers, insufficiently strutted purlins, excessive shrinkage in purlins, or raising of upper raking surface of a party wall above upper face of rafters with tiling or slating battens bent over the wall. Movement may occur at the eaves if rafters are not suitably birdsmouthed over wall plates and adequately nailed to ceiling joists. Inadequate bracing of trussed rafter roofs may be identifiable by verge tiles oversailing at gables.[28] Dampness and staining of ceilings may be caused by sarking felt not fitting closely around the soil and vent pipe, or not being properly lapped or dressed out over gutter and barge boards.[29]

The risk of condensation causing damage to roof timbers increases with improved loft insulation. The risk can be reduced by closing gaps in the ceiling to prevent the ingress of water vapour. Ventilation of the roof space is vital and constitutes the usual method of removing moisture from the roof. This can be achieved by forming ventilation openings at the eaves in the fascia or soffit boards, or at both eaves and ridge. Roofs of more than 15° pitch should be provided with ventilation openings equivalent to a continuous opening of not less than 10 mm wide.[30]

Pitched Roof Coverings

Often the first indication of trouble is a damp patch on a ceiling or in the top corner of a wall. Localised leakage may occur as a result of defective flashings, cement fillets which have shrunk or broken away from adjoining surfaces, choked or defective gutters or slipped or broken slates or tiles. Defective flashings need redressing, raking out the brick joint and rewedging and repointing the flashing. Zinc flashings may perish and become pitted in industrial atmospheres and are best replaced with lead, copper or other suitable flashings. Defective cement fillets should be replaced with metal flashings.

Choked gutters need cleaning and checking to ensure that they are satisfactory. Eaves gutters may need resetting to falls and rejointing, and defective lengths replaced. Coating all internal surfaces of parapet and valley gutters with bituminous composition may extend their useful lives. Check on the adequacy of tilting fillets and cover at junctions of gutter coverings with adjoining slating or tiling, and ensure that horizontal damp-proof courses are provided under copings to parapet walls.

Slipped or broken plain tiles are fairly readily replaced as the tiles are usually only nailed on every fourth or fifth course, and adjoining tiles can be lifted sufficiently to permit a replacement tile to be hooked over a tiling batten. The replacement of slates is more difficult as it entails removing the defective slate by cutting off the nail heads with a slater's ripper and fixing the new slate at its tail with a copper clip or tack bent over the head of the slate in the course below.

Rain and snow may penetrate a pitched roof because the slates or tiles are laid to too flat a pitch without increasing their lap. The problem is aggravated on exposed sites and the use of a flatter bellcast at eaves to improve appearance creates a vulnerable condition at the point of greatest rainwater runoff. Plain tiles for instance should never be laid to a flatter pitch than 40°. In severe cases it is necessary to strip the roof covering and to replace it with one suited to the particular roof pitch, as for example to replace plain tiles with single lap interlocking tiles on a 30° pitch roof. Where verge tilting is absent, edge slates and tiles are more vulnerable to frost action.

The life of slates or tiles is dependent upon a number of factors including the physical properties of constituent materials and method of manufacture, climatic conditions, degree of pollution and method of fixing. Poorer quality slates may have a life of up to 70 years while some of the poorer machine made clay tiles may be restricted to 40 years on account of their laminar structure which is susceptible to freezing conditions. Concrete tiles may have a longer life but their colour is often bleached over a comparatively short period. Galvanised nails are unlikely to last the life of the slate or tile and are a poor investment. Acceptable nails are aluminium alloy, copper, stainless steel and silicon bronze.

When large numbers of slates or tiles are defective it is generally more satisfactory to strip and renew rather than to carry out extensive patching. With older buildings problems sometimes arise through manufacturers ceasing to produce certain single lap tiles. One local authority faced with this problem stripped off the clay pantiles from a pair of houses and recovered with concrete interlocking tiles. All sound pantiles were taken into stock for repairing the remaining houses. In older houses, sarking felt is rarely provided under the tiles or slates and so rain or snow penetrating the roof covering has access direct into the roof space with the most unfortunate results. In extreme cases it is necessary to strip the tiles or slates and battens in order to nail a layer of felt to the upper side of the rafters. Plate 15 shows slates which are so badly laminated that they require replacing.

Problems can arise with other coverings to pitched roofs ranging from the failure of aluminium sheeting laid over polyurethane foam, because of differential movement, to the splitting of battens and rafters under thatch, where the whole of the old thatch had not been stripped on a previous re-thatching, or the absence of galvanised steel or PVC mesh has allowed birds or vermin to remove the straw or reeds. In addition, thatched roofs require periodic inspection and overhaul and are subject to insect attack and have a high fire risk.

Norfolk reed is the most durable thatch with a life in excess of 60 years, compared with combed wheatreed at 30 to 40 years and long straw at 15 to 25 years. All require re-ridging at intervals of 10 to 15 years. The life expectancy periods drop the further west of the British Isles the property is situated, as the warmer climate with higher humidity encourages the breeding of fungi, although this can be overcome by periodic chemical treatment.

Flat Roof Construction

A flat roof may be the only practicable form of roof for many large buildings or those of complicated shape and can be a more economical proposition

Plate 15 Laminated roof slating

than a pitched roof. It does unfortunately constitute a common area for premature failure in modern building. Most flat roof failures could have been avoided if the design principles now outlined had been adhered to. The technical options are described in BRE Digest 312[31].

Movement. Continuous coverings on flat roofs are much more susceptible to the effect of movement than the small units on pitched roofs. All forms of roof construction are liable to thermal movement and deflection under load, and in the case of a timber roof can subject the roof finish to considerable strain. Timber construction is subject to moisture movement, concrete to drying shrinkage and new brickwork to expansion. Hence only those roof finishes which are able to withstand some movement should be used on the more flexible types of roof, and asphalt, which is ill suited to accept movement, should not be used on timber roofs. Upstands should not be less than 150 mm high and if movement between the vertical and horizontal sections is possible, a separate metal or semi-rigid asbestos bitumen sheet flashing should be provided to cap the top of the turned-up edge of the covering. Adequate expansion joints should also be provided, consisting of an upstand not less than 150 mm high with a metal capping.

Falls. The retention of water on most forms of roof covering is undesirable and constitutes a common cause of failure. This often results from ponding caused by inadequate falls. In the past the normally accepted minimum finished fall has been 1 in 80, but after making allowance for building inaccuracies and structural deflection, BS 6229[32] recommends a fall of 1 in 40.

Insulation. The standards of insulation prescribed in The Building Regulations[33] should be regarded as minima. Care must be taken to keep all insulating materials dry as they cease to be effective and may deteriorate when wet.

Solar protection. With higher standards of insulation, solar reflective treatment is a necessity for asphalt and bitumen-felt roofs. White asbestos tiles or white spar chippings are useful, while concrete tiles may be justified if the roof is accessible and likely to receive regular use. Upstands which cannot be treated with chippings should receive an applied reflective coating, such as metal foil on a felt backing. Moreover, solar reflective finishes prolong the life of the roof covering.

Condensation. The humidity and vapour pressure are normally higher inside an occupied building than outside. Water vapour will usually penetrate most of the internal surfaces of a building and when the outside temperature is lower, it may condense at some point in the roof structure, often on the surface of insulation, and this is known as interstitial condensation.[34] Where the insulation is immediately under the roof covering it must be placed on a vapour barrier, often consisting of bitumen-felt laid and lapped in hot bitumen and turned past the edges of the insulation to meet the roof covering. With timber roofs it is necessary to cross-ventilate the spaces between joists[35] and where the roof extends over a cavity wall the cavity should be sealed at the top to prevent moist air entering the roof. Where cavity barriers restrict the crossflow of air in a roof structure, cowl type ventilators penetrating through the roof surface are needed.[36]

Flat Roof Design

The Bituminous Roofing Council[37] has defined the following three principal types of flat roof.

(1) *Cold roof.* In this type of construction the thermal insulation material is placed below the roof deck, normally at ceiling level. Heat loss through the ceiling is thus restricted, keeping the cavity, roof deck and covering at low temperature during winter. If condensation problems are to be avoided in cold roofs, adequate provision must be made for efficient ventilation of the roof space. It is important to provide a sufficient open area on each side of the roof cavity to ensure a free, unobstructed path for ventilation purposes.

(2) *Warm roof.* In the construction of warm roofs the thermal insulant is placed immediately below the weatherproof covering and on top of the roof deck and vapour barrier. The deck and cavity are thus maintained at warm temperatures during the winter. The thermal insulant is secured to the

deck by bonding or mechanical fasteners, while the weatherproof covering is bonded to the top surface of the insulation. With this type of construction, high levels of thermal insulation are more easily achievable and more positive condensation control is possibly by the selective use of high efficiency insulation material and vapour barriers.

(3) *Inverted roof (protected membrane)*. In this type of construction, the thermal insulation material is placed on top of the weatherproof covering so that the complete roof construction, including roof covering, is kept at warm temperatures during winter and at moderate temperatures during summer. With inverted roofs, the important requirement is that the thermal insulation material has low moisture absorption and water vapour transmission characteristics. An important advantage is that the insulation protects the weatherproof covering from extremes of temperature and differential movements within the roof structure are thus reduced to a minimum. Furthermore, the weatherproof covering is protected from various forms of damage, but it must be able to withstand continually wet conditions and is not immediately accessible for inspection and repair.

Vapour barriers. The Property Services Agency[38] has emphasised that all insulants must be laid on and protected by an efficient and properly laid vapour barrier. Failure to do this in warm roof construction is likely to result in moisture vapour from the building affecting the insulation, reducing its thermal efficiency and starting up cyclic interstitial condensation. Feedback reports to PSA have indicated that far too many roof defects are due to ineffective vapour barriers. High-performance felts are most likely to achieve the required level of performance, and these include polyester bitumen, bitumen polymer and pitch polymer. Vapour barriers should be fully bonded to the substrate (deck, topping or screed) with hot bitumen, and be provided with end and side laps with a minimum width of 100 mm.

In cold roof construction, it is more difficult to ensure that the vapour barrier will remain effective by always being above dewpoint temperature because: (1) it is very difficult to construct an efficient vapour barrier at ceiling level; and (2) the weatherproof covering, which is the first real vapour barrier, is during the winter at a temperature well below dewpoint for the internal conditions, and good roof cavity insulation is essential.[37]

Elimination of Faults in Flat Roofs

Types of fault. The decision as to whether remedial action should take the form of isolated repairs or a new covering depends on whether the fault is:

1. A system fault—where the conditions which produce the fault are inherent in the roofing system and can occur in any part of the general roof area; or

2. A detail fault—relating to a particular construction detail, such as perimeter, gutter or upstand.[39]

A DES design note[39] describes how on felted roofs, certain visual symptoms are frequently taken to indicate system failure, whereas in fact they represent detail faults. Two of the most common instances are splitting of the membrane at internal gutter positions or changes of deck/roof surface levels, and felt blistering on general roof surfaces.

Falls. The provision of falls as a design feature of flat roofs is recommended by BS 6229.[32] However, the creation of effective falls on most existing dead flat roofs may not be feasible, because:

    1. eliminating all flat areas or level intersections on any but the simplest plan forms is very complicated; and

    2. the resultant increases in height towards the apex creates considerable reconstruction problems at upstands to rooflights, clerestories and related features.

Rainwater outlets. Most roofs covered by a DES survey[39] drained through outlets situated within the building perimeter. These outlets are often either positioned incorrectly or too few in number to drain the roof effectively, in addition to sometimes being installed at too high a level in relation to the roof surface. Additional outlets should be provided to serve inadequately drained areas. These may not be sufficiently close to an existing surface water drain, so that horizontal runs are required if the disruptive construction of a new underfloor branch drain is to be avoided. Such runs are best positioned in the roof void or encased along the ceiling/wall junction.

Protection against solar degradation. The surfaces of most flat roof membranes require protection from solar gain, radiation loss and ultraviolet rays, if excessive exposure is not to reduce the optimum life of the covering. The main types of protection are paints, metal and mineral facings, chippings and overlay insulation slabs. However, paints and self-finished felts have not proved satisfactory in the longer term.

The overlay or 'inverted roof' solution is a relatively new technique on lightweight roofs. Apart from the structural support implications, careful attention must be paid to stability under conditions of severe wind loading, and valuable guidance is given in BRE Digest 295.[40]

Chippings have good thermal storage properties and thus raise the surface temperature of the membrane on cold nights, they offer protection against ultraviolet radiation and minimise blistering. Controlled tests by BRE and industry research bodies have shown that chippings subjected to foot traffic do not puncture the membrane and cause leaks as is commonly supposed.[39]

Vapour barriers. A sound vapour barrier is essential where a high level of moisture is likely to accumulate in the roof structure during the winter and may be retained at too high a level during the summer, with a consequent cumulative build-up. BS 6229[32] includes calculations which establish whether the retained moisture is likely to exceed tolerable levels.

Bonding of first membrane layer. About half of the system failures on the roofs covered by the DES survey[39] resulted from over-stressing of the membrane at joints of insulation boards which were subject to excessive thermal movement. This problem can be avoided by:

    1. specifying a more stable insulant; and/or
    2. spreading any stresses which may occur over a larger membrane area by the use of a partially bonded multi-layer felt system, and/or
    3. using a high-performance membrane.

A combination of any two of these measures will provide a satisfactory solution.

Investigation and diagnosis. A DES design note [39] identifies a prerequisite of effective remedial work on defective flat roofs as an initial careful and systematic investigation, leading to accurate diagnosis of faults and their causes. The aims of a thorough investigation should be to establish:

    1. the origin and extent of the leak(s);
    2. whether they are due to a detail fault, when a local repair will provide the remedy, or to a fault symptomatic of a system problem, in which case refelting or replacement will be required; and
    3. the degree of deterioration, since in the case of a system fault this will determine the extent and cost of the required remedial action.

Remedial solutions. The DES design note[39] recommends a comprehensive fault investigation procedure assisted by reference to behaviour profiles whereby the user can determine whether the defective roof is suffering from a fault system and, if so, recovering or renewal will provide the only satisfactory and long-lasting solution. Once that necessity has been accepted, the next step is to prepare a remedial specification appropriate to the type of roof and its current stage of deterioration. Detailed tables prepared by the DES provide invaluable guidelines covering a wide range of conditions.

    BMCIS (BMI) Design/Performance Data—Building Owners' Reports: 2. Flat Roofs (1986) has illustrated the use of a fault checklist to provide a broad guide to some of the main faults found in flat roofs, which are listed as follows.

    1. blisters;
    2. ponding;
    3. dents, punctures and rips;
    4. splits, tears and cracks;
    5. crazing;
    6. ripple, creak or sag;
    7. lifting of lap joints;
    8. chemical damage;
    9. water ingress at pipes, vents and rainwater outlets;
10. splits at joints in metal eaves trim;

11. felt skirtings coming loose;
12. embrittlement of asphalt;
13. surface corrosion and pitting of zinc and lead; and
14. lifting of sheets.

Bitumen-felt Roof Covering

The early felts were based wholly on organic fibres which were very strong when new but dimensionally unstable, and could rot when moisture eventually penetrated the bitumen coating. The introduction in the early 1950s of bitumen-felt based on asbestos fibre provided a more stable alternative, but not complete freedom from the risk of rotting. The development of polyester-based felt in the mid 1970s virtually eliminated dimensional instability and the risk of rotting. Subsequent modifications to coatings and increases in base weights have further improved performance.

The bitumen outer coating of bitumen roofing felt, if exposed to the weather, is gradually attacked by solar radiation. This deterioration can be postponed if the uppermost layer of felt, which is normally laid in three layers, has a surfacing or mineral aggregate, preferably white, partially embedded in a coat of bitumen dressing compound. Roofs over habitable rooms should be laid to this or a higher standard.

Bitumen roofing felt is generally unable to withstand more than a slight amount of stretching without splitting or tearing apart. This defect may be remedied by patching with a strip of felt reinforced with hessian bedded in bitumen. The next most common cause of failure results from differential movement at skirtings to parapets and at other peripheral weatherings. Sometimes blisters develop between the layers of felt as a result of insufficient pressure being applied when rolling a layer of felt into hot bitumen bonding compound or the entrapping of moisture between two layers of felt. They do not often lead to leakage and no remedial action is usually necessary.

Upstands and skirtings should be integral with the surface felt and be formed by turning up the second and top layers against abutments to a minimum height of 150 mm. The felt should be turned up over an angle fillet at the base of the upstand to prevent the felt cracking at the bend or becoming damaged as a result of lack of support. The angle fillet should be securely fixed to the roof to prevent distortion. Skirtings and upstands should ideally be masked by a metal or semi-rigid asbestos/bitumen sheet (SRABS) flashing.

Small holes in bitumen felt roofing can be sealed with a patch of felt bedded in bitumen. More extensive repairs may make necessary the removal of an entire sheet of felt by heating and softening a lapped joint and bedding a new layer of felt. Where general deterioration of the felt has occurred without fracture, a top dressing of hot bitumen and stone chippings may suffice. A survey of maintenance work on hospital buildings[25] showed that felt roofs were replaced on average at 16 year intervals, while asphalt lasted 28 years and tiled roofs 59 years. Maintenance costs of felt roofs were also high; they covered 45 per cent of the total hospital roof area but accounted

Plate 16 Split roofing felt to weather kerb

Plate 17 Defective work around rainwater outlet to bitumen felt roof

for 70 per cent of the total maintenance and replacement cost of all roofs.[25] Regular inspections are advisable to identify and repair small leaks before they accumulate into more widespread defects, involving higher expenditure, increased disturbance and possible damage to the roof structure. Plate 16 shows split roofing felt to a weather kerb through which rainwater can penetrate the roof, and plate 17 illustrates unsatisfactory waterproofing work around a rainwater outlet on a chapel roof.

Asphalt Roof Covering

Two layers of asphalt are always necessary on flat roofs, with a finished thickness of 20 mm and joints staggered with a minimum lap of 150 mm between layers. Weak spots may occur if asphalt is reduced in thickness to obtain an even finished surface over raised parts, such as welted joints in flashings. Where a roof is likely to take considerable traffic, the asphalt is best finished with asbestos cement or concrete tiles, preferably with solar reflective properties. It should have a life of 40 years.

A survey of 130 mastic asphalt covered flat roofs to Crown buildings showed a 28 per cent failure rate, resulting either from splitting and cracking of the asphalt due to movement of the substrate and the absence of an isolating membrane, or peripheral cracking due to differential movement between a roof deck and a non-integral parapet wall to which an asphalt skirting was fixed without any provision for movement. Slight hollows in a roof result in ponding and this may cause crazing of the asphalt but is unlikely to lead to water penetration. Cracked and blistered areas should be heated, cut out and made good with new asphalt without delay.[41] The new asphalt should be carefully bonded to the old by stepping the edges of the existing asphalt. Minor surface crazing is likely to result from over trowelling.

Oxidation occurs when asphalt is exposed to light and heat, water-soluble products are formed, slow hardening occurs and cracks may form following a rapid decrease of temperature. The Local Government Operational Research Unit identified five 'states' through which an asphalt roof passes as it deteriorates, namely

(1) no visible defects
(2) not leaking but slightly worn and wrinkled
(3) some leaks due to blistering and slight cracks and crazing
(4) broken blisters, extensive cracking and leaking
(5) asphalt broken up, perished and off key.[25]

In practice the deterioration processes are not uniform and at any point in time different parts of an asphalt roof will be in different states. In order to specify the condition of a roof it is generally necessary to measure or estimate the percentage in each of these states.

Problems have arisen through the application of white paint to the asphalt to reduce absorption of solar heat. The shrinkage of the relatively tough paint film is sufficient to pull the asphalt with it and cause cracking of the asphalt with consequent loss of watertightness. Dampness in ceilings below asphalt roofs may result from interstitial condensation rather than moisture penetration through the asphalt.

Excellent guidance on the maintenance and re-roofing of mastic asphalt and built-up roofing and advice on design aspects is contained in the Tarmac Guide.[42]

Polymer Roofing

Polymeric materials include polyisobutylene, butyl rubber, PVC and chloro-sulphonated polyethylene in single layer systems, stuck to the substrate with special adhesives and the joints between the sheets solvent or heat welded. They are more flexible than bitumen but require a high standard of installation and problems have occurred through inadequately bonded joints and mechanical damage causing splits.[43]

Metal Flat Roof Coverings

Lead is susceptible to two forms of corrosion—in slightly acid conditions and under alkaline conditions, as with lime or cement mortars. Acidic conditions may occur through the discharge of rainwater on to the lead from pitched roofs containing algae, moss or lichen, but the corrosion process is generally very slow. A protective coat of bitumen made up of one coat of hot bitumen or two thick coats of bituminous paint in vulnerable locations should provide adequate protection. In general the resistance of lead to atmospheric corrosion is very high, following the formation of a protective film of basic lead carbonate or sulphate and may have a life in excess of 100 years, if of adequate thickness. Lead has low strength and is very heavy in weight, but is ductile, flexible, and easily cut and shaped. Cracks in lead sheet occurring through movement of boards beneath are usually rectified by soldering. In course of time lead flashings tend to work loose and curl up from the bottom edge, and they then require rewedging, repointing and redressing. Sagging parapet gutters need stripping and replacement of bearers and boarding. Rippling and splitting of the lead may occur in the absence of rolls and drips or where there are sharp arrises. Creep is only likely to be a problem on pronounced slopes. Defective lead roofing is illustrated in plate 18 with poorly constructed lead rolls.

Copper, like lead, forms a very effective protective film. Furthermore it is tough and durable, readily cut and bent, is of light weight, does not creep and acquires a pleasant protective green film patina. Copper is resistant to alkalis but rainwater with an acid content dripping from algae covered roofs or cedarwood shingles has been known to cause perforation of copper roofing within 40 years, and protection with bitumen is advisable in these situations. The majority of repairs to copper roofs are concerned with redressing rolls, seams and welts and repointing flashings.

Plate 18 Defective sheet lead roofing

Zinc has an average life of 20 to 40 years which is much shorter than that of lead or copper. Like lead, zinc is liable to corrode in both alkaline and acidic conditions. It is advisable to give at least one coating of bitumen to zinc embedded in plasters and mortars. Cracks can be repaired with bitumen and a surface dressing will prolong the life of a zinc roof.

Sound Insulation

A full understanding of the science of sound transmission requires an extensive and highly technical study. It is becoming increasingly important as people become more noise-conscious. Insulation is required against sound generated in two different ways:

. (1) A source such as a radio may produce sound waves in air which in their turn produce vibrations in a party wall or floor (airborne source).

(2) A wall or floor separating two dwellings may vibrate by the direct impact of a solid object such as footsteps (impact source).

A radio or television in a ground floor room of one house can frequently be heard in the bedroom of a neighbouring house. Vibrations induced in the party wall at ground floor level are transmitted up the party wall and then pass into the upper floor rooms. This is known as *flanking transmission*. It can also occur when the rooms adjoin, either horizontally or vertically, and often provides an additional path to direct passage through the common wall or floor.[44,45,46]

In standard tests (BS 2750)[47] the insulation of airborne sound is measured in each of sixteen one-third-octave bands, the centre frequencies of which range from 100 to 3150 Hz. In order to determine whether a satisfactory standard has been achieved results are expressed in terms of *weighted standardised level difference*, calculated in accordance with BS 5821,[48] when checking the performance of existing walls; this is the difference in decibels (dB) between the energy levels in the rooms corrected to allow for a standard amount of absorption representative of normal furnished conditions. For example, the sound reduction of the brick or block components of an external wall is in the range of 45 to 50 dB (decibels), whereas the reduction with closed single windows is about 20 to 25 dB. For party walls between dwellings a one-brick wall plastered on both sides is considered acceptable, although the provision of wall linings on battens or studs can improve sound insulation. Floating floors, as described earlier, enable an acceptable sound reduction factor to be achieved between flats. Alternative constructional forms for use in party-floors are illustrated in BRE Digest 266.[46]

The object of sound insulation is almost entirely one of reflecting energy back into the source room; the role of absorption is limited to supplementing reflection at high frequencies in some types of wall or floor,[44] and is of particular benefit to the occupants of the room in which they are used. The amplitude of wall vibrations is inversely proportional to the mass of the wall, and so the amplitude of the sound waves radiated into the receiving room is inversely proportional to the mass of the wall. By doubling the mass of the wall transmission is reduced to a quarter.[44]

Where a dividing element is made up of parts each of different sound resistance, then however small the parts of lower resistance, they will reduce the overall insulation to a value lower than that of the most highly sound-resisting part. This is particularly significant with voids, gaps and cracks. The subjective sound insulation between two rooms is influenced considerably by the acoustics in the rooms, which are largely dependent on the amount of the reverberation—how long it takes a sound to die away after the source has stopped. This is very much affected by the amount of sound-absorbing material such as curtains and furniture and the nature of wall, floor and ceiling finishings. For instance, a sound-absorbent ceiling of materials like acoustic fibreboard and perforated facings of metal, plaster, hardboard and asbestos board backed with mineral wool, in a typing pool or workshop, will reduce the noise level in the room by preventing sound reflection. The relative cost and sound insulation performance of many forms of wall and floor construction are detailed in *Sound Insulation in Buildings*.[49]

Good sound insulation stems from adequate weight, airtightness and uniform resistance to sound. The following methods can be used to insulate existing buildings.

Windows provide the most common method for entry of outside noise into buildings. The sound insulation of a single window can be improved by making it airtight with phosphor-bronze draught-excluding strips or seals of plastics, rubber or felt and by fitting heavy glass; but for good insulation double windows with heavy glass are essential. The air space should be at least 150 mm wide and preferably 200 mm wide, with the head, reveals and sill lined with a sound-absorbent material such as acoustic fibreboard. Double windows should be as airtight as possible and it may be necessary to install mechanical ventilation. The Government grants scheme administered by the British Airports Authority for noise protection of existing dwellings in the vicinity of London's Heathrow Airport provided for the payment of grants for rooms fitted with a double window with a minimum width of air space related to the glass thickness used, as follows:

3 mm glass—not less than 200 mm cavity
4 mm glass—not less than 150 mm cavity
6 mm glass—not less than 100 mm cavity

In addition, an approved ventilator had to be installed with an air delivery capacity of not less than 1.84 m^3/min and a sound reduction of at least 49 dB at 500 Hz.[50]

It is rarely economical or even practicable to improve the insulation of the *walls* of an existing building. Reducing direct transmission by insulating a party wall is generally rendered valueless as by-passing by flanking transmission leaves the net insulation almost unchanged. The blocking up of airbricks and chimneys to improve sound insulation can cause condensation. The sound insulation of brick or block walls initially left fair faced, for reasons of design or cost, can be improved by plastering. If there is leakage through underfloor voids and gaps, these should be sealed.

Partitions can be improved by adding a heavy lining, such as 20 mm plasterboard on a framework, giving a wide air space, over a sound-absorbent material such as mineral wool quilt, with all joints between boards and around edges effectively sealed. Demountable partitions need to be well sealed around the edges.

To make an existing *wood-joist floor* more resistant to airborne noise it is necessary to increase the weight of the floor. One method is to lay heavy pugging such as 50 mm dry sand weighing about 90 kg/m^2 between the joists, provided the ceiling and joists are strong enough to take the extra weight, otherwise the pugging should be laid on pugging boards fixed between the joists, as illustrated in *Building Technology*.[51] Impact noise on a suspended floor can be reduced by covering the floor with rubber on a sponge rubber underlay, felt backed PVC, cork tiles or carpet on a suitable underlay. Another alternative is to provide a *floating floor* such as 20 mm tongued and grooved boarding on 50 mm square battens laid on 13 mm mineral wool resilient quilt. The resilient layer must be turned up at all edges which abut walls, partitions and other parts of the structure. Partitions should be built off the structural floor so that the floating raft is self-contained within each room. Doors will need altering and thresholds or ramps provided at door openings. Noisy machines should be isolated from floors by resilient mountings.

Noise often emanates from *plumbing* systems. WCs are particularly noisy but fittings with a double-syphonic trap and a close-coupled cistern are less noisy. When a WC is sited next to a main room or bedroom, the dividing partition should have a sound reduction value equal to a 100 mm brick wall plastered both sides. The WC door should be as heavy as possible, preferably a solid core flush door, with draught-excluding strips. Ball valves to cisterns and water storage tanks should ideally be of the Skevington BRE pattern to reduce noise. Internal stack pipes should be enclosed in ducts made with a material weighing at least 15 kg/m^2, such as 25 mm chipboard. Heating and hot water pipes should be fixed with pipe clips that give some clearance for expansion or contraction, with sleeves or resilient packing where pipes pass through the structure and clearance over pipes where they pass between floorboards and joists.

Thermal Insulation

There is an increasing demand for improved comfort coupled with a reduction in heat losses from buildings. This has been accentuated by the escalation in heating fuel costs. The economic level of insulation depends upon the cost of providing the additional insulation as compared with the reductions in the cost of heating plant and fuel over a period of time. The rate at which heat is transferred through an element of a building is termed the thermal transmittance or *U*-value. The lower the *U*-value, the better the insulation and the lower the heat loss. The total heat loss through the building fabric can be found by multiplying the *U*-values and areas of the externally exposed parts of the building, and then multiplying the result by the temperature difference between inside and outside.[52] *U*-values are expressed in W/m^2 K (watts per square metre for 1° Celcius difference between internal and external temperatures). *U*-values for new dwellings in England and Wales are controlled by the Building Regulations 1985 and are 0.35 W/m^2 K for roofs and 0.6 W/m^2 K for exposed walls and floors. These standards are based on windows and rooflights not exceeding 12 per cent of the areas of external walls with a *U*-value of 5.7. The procedures for implementation of the Building Regulation requirements are detailed in *Building Technology*.[51]

When selecting materials to achieve the required *U*-values, the comparison should include installation and maintenance costs. Where air spaces are incorporated they should not be less than 20 mm wide. Valuable data on thermal insulating materials is provided in *Thermal Insulation of Buildings*.[53]

With ground floors, concrete slabs on hardcore have generally been considered to give adequate thermal insulation without additional treatment, but the heat loss through the floor can be reduced significantly by inserting a layer of material of high thermal resistance between the waterproof membrane and the screed. When incorporating under-floor heating systems it is advisable to include a rot-proof insulating layer of mineral fibre or expanded plastics over the damp-proof membrane.

Suspended wooden ground floors should be provided with additional insulation, either in the form of a continuous layer of semi-rigid or flexible material, such as paper-backed insulation quilt (paper side downwards) over the joists, with butted longitudinal joints to allow the floor to breathe, or alternatively a semi-rigid material, such as expanded polystyrene boards fixed between the joists.[54]

External *cavity walls* with a brick outer leaf and an inner leaf of aerated concrete insulating blocks can combine strength with good thermal insulation without increasing costs significantly. The thermal insulation value of a cavity wall can be further increased by filling the cavity with a UF foam produced from urea formaldehyde resin, under suitable conditions and taking adequate precautions. The foam is injected into the cavities of existing buildings by drilling 19 mm diameter holes mainly through mortar joints in the outer leaf at approximately 1 m staggered centres. The holes are subsequently made good with mortar colour-matched to the existing. Building Research Establishment investigations have indicated some cases of damp penetration across the filled cavity where the outer leaf had a high porosity and exposure to large amounts of driving rain, and there have been cases of fumes penetrating the building. Other insulants are glass and rock fibre slabs, expanded polystyrene boards, rock fibre tufts, polyurethane granules, expanded polystyrene (EPS) loose fills, glass-fibre loose fill and foamed polyurethane. BRE Digest 236[55] describes the principal methods used in cavity insulation and the problems that can arise in their installation.

Framed structures with sheeting rails may be lined internally with rigid insulating boards or slabs fixed to the rails with a T-grid or clip system. Alternatively, insulating quilts or boards can be hung behind a rigid sheet lining which may not itself have high thermal insulating properties.

Double glazed windows can reduce the heat loss through windows by about 50 per cent and eliminate the 'cold zone' near windows. They also reduce the risk of condensation on windows but the cost cannot always be justified in the British Isles. In a typical semi-detached house, only about one-tenth of the heat loss from the house will be saved by double glazing. It is usually more economical to improve the U-values of the walls and roof, and to stop draughts. With existing single glazed windows there are two possible ways of improving thermal insulation. The first method is to fix a second line of glazing with wood or plastic faced beads, although they are unlikely to remain airtight indefinitely. The seals may disintegrate under movement and shrinkage, allowing water vapour to enter the airspace and to condense on the inside of the outer glazing. The wood exposed to the air space should be painted or varnished to prevent evaporation of moisture from the timber into the air space and breather holes should be provided at the rate of one 6 mm diameter hole for 0.5 m^2 of window, with the holes plugged with glass fibre or nylon to exclude dust and insects. The second method is to fix secondary windows to the existing frames or to ancillary frames. There are proprietary secondary windows available in aluminium or PVC, which may be either hinged or sliding to give access for cleaning. The casement types usually have a compressible strip to provide a seal against the existing window or the ancillary frame. The glazing joints of double glazing units, glazing compound and beads should be checked on repainting. With sealed *in situ* systems it may be necessary to remove the glass on the room side, clean it and reseal it at intervals of two years or more depending on the effectiveness of the seal.[56]

Pitched roofs can have a rot-proof thermal insulating quilt fixed over the ceiling joists or, alternatively, a loose fill material or quilt may be spread between the ceiling joists. The thickness of insulation should be at least 100 mm to reduce heat loss effectively. The insulation is omitted under cisterns in the roof space but all pipes and the sides and tops of cisterns need lagging. Top floor ceilings can be formed of aluminium foil-backed plaster-board, rigid polyurethane, expanded polystyrene/plasterboard laminate or other insulating board to reduce heat loss.

Concrete flat roofs can have the thermal insulation either over or under the structural units. Timber flat roofs having thermal insulation at ceiling level require a vapour check below the insulation and ventilation of the roof space. Where the insulation is placed over the roof deck immediately below the roof covering, a vapour barrier is needed between the insulation and the deck to prevent interstitial condensation in the insulation.

A high standard of thermal insulation can be obtained in a house by using an external cavity wall of 102.5 mm brick and 100 mm aerated concrete blocks with plasterboard dry lining and 50 mm foam fill or other acceptable insulant in the cavity, double glazed windows and a roof with a vapour barrier above the ceiling and 100 mm glass-fibre mat. On an average house the annual energy savings are likely to amount to about 30 per cent of the cost of the provision of the insulation work. Costs of energy are likely to continue to rise, making lower *U*-values economical. Double glazing of windows cannot usually be justified on economic grounds unless a very high standard of heating is required, an expensive form of energy is used, or a high percentage of the wall area is glazed. Improvements can be made in the energy utilisation of most buildings consequent upon effective energy auditing and management.[57]

Vibration

Developments in machinery, road and rail traffic and aircraft are increasing vibration and noise to an extent that they may become objectionable to people and interfere with laboratory work and some trade processes. Wind-generated vibration also needs considering when designing structures exceeding four storeys in height. Fears are also expressed that buildings may be damaged by vibration, but investigations by the Building Research Establishment show that the risk of damage to normal buildings is extremely rare, even when the level of vibration is considered objectionable or even intolerable by the occupants of the building. BRE Digest 278[58] describes how the response of buildings to a vibration source is governed by various factors, such as the relationship between the natural frequencies of the building, the damping of the resonances of the building or elements, the stiffness of the building or elements, the magnitude of the forces acting on the building, and the interaction of the building or elements within the vibration source. Cracks in plaster brickwork and glass should not be attributed to the effects of vibration until other possible causes have been eliminated. For example, most cracks in plaster ceilings result from movement of the plaster itself or of deflections of the timber joists, and such cracking often occurs in areas known to be free from external sources of vibration. Some of the vibration factors such as damping values and stiffness of the building will probably require specialist advice.[58]

The occurrence of repetitive loading, such as that caused by machinery, rarely creates a structural problem, unless the frequency coincides with a natural frequency of some element of the building. However, the effect on occupants may be unacceptable well before any structural damage occurs. People's perceptions of levels of vibration will vary. Under certain conditions the human body can detect amplitudes as small as one micron. BRE have shown that human tolerance is dictated not only by scientific but also by psychological factors.[58]

References

1 *BRE Digest 287*. Specifying structural timber (1984)
2 British Standards Institution. *BS 3452: 1962 Copper/chrome waterborne wood preservatives and their application*
3 British Standards Institution. *BS 5056: 1974 Copper naphthenate wood preservatives*
4 *BRE Digest 201*. Wood preservatives: application methods (1984)
5 BRE Princes Risborough Laboratory. *Methods of Applying Wood Preservatives* (1974)
6 British Standards Institution. *BS 5268: Code of practice for the structural use of timber. Part 2: 1984 Permissible stress design, materials and workmanship*
7 BRE Princes Risborough Laboratory. *Decay of Timber and its Prevention* (1976)
8 *BRE Digest 299*. Dry rot: its recognition and control (1985)
9 British Standards Institution. *BS 5707: Solutions of wood preservatives in organic solvents. Part 1: 1979 Specification of solutions for general purpose applications, including timber that is to be painted. Part 3: 1980 Methods of treatment*
10 British Standards Institution. *BS 4072: 1974 Wood preservation by means of water-borne copper/chrome/arsenic compositions*
11 British Standards Institution. *BS 913: 1973 Wood preservation by means of pressure creosoting*
12 British Standards Institution. *BS 5589: 1978 Code of practice for preservation of timber*
13 *BRE Digest 18*. Design of timber floors to prevent decay (1975)
14 *BRE Digest 307*. Identifying damage by wood-boring insects (1986)
15 *BRE Defect Action Sheet 47*. Suspended timber floors: notching and drilling of joists (1984)

16   *BRE Defect Action Sheet 58*. Suspended timber floors: joist hangers in masonry walls—installation (1984)

17   British Standards Institution. *BS 5669: 1979 Specification of wood chipboard and methods of test for particle board*

18   *BRE Defect Action Sheet 32*. Suspended timber floors: chipboard floorings—storage and installation (1983)

19   *BRE Digest 239*. The use of chipboard (1980)

20   *BRE Digest 54*. Damp-proofing solid floors (1971)

21   *BRE Digest 313*. Mini-piling for low rise buildings (1986)

22   *BRE Digest 79*. Clay tile flooring (1976)

23   *BRE Digest 33*. Sheet and tile flooring made from thermoplastic binders (1971)

24   British Standards Institution. *CP 209 Care and maintenance of wood surfaces. Part 1: 1963 Wooden flooring*

25   Local Government Operational Research Unit/Royal Institute of Public Administration. Report C146. *Aids to Management in Hospital Building Maintenance*. HMSO (1972)

26   A. Wharton. Free estimates cost you money. *Building Maintenance* (July/August 1973)

27   *BRE Defect Action Sheet 54*. Stairways: safety of users—installation (1984)

28   *BRE Defect Action Sheet 24*. Pitched roofs: trussed rafters bracing and binders—installation (1983)

29   *BRE Defect Action Sheet 10*. Pitched roofs: sarking felt underlay—watertightness (1982)

30   *BRE Digest 270*. Condensation in insulated domestic roofs (1983)

31   *BRE Digest 312*. Flat roof design: the technical options (1986)

32   British Standards Institution. *BS 6229: 1982 Code of practice for flat roofs with continuously supported coverings*

33   *The Building Regulations 1985*: SI 1985 Nr 1065. HMSO (1985)

34   *BRE Digest 180*. Condensation in roofs (1978)

35   *BRE Information Paper 35/79*. Moisture in a timber-based flat roof of cold deck construction (1979)

36   *BRE Digest 218*. Cavity barriers and ventilation in flat and low pitched roofs (1978)

37   Bituminous Roofing Council. *Information Sheet 1: Flat roof design and construction: types of flat roof* (1983)

38   DOE, Property Services Agency. *Flat Roofs Technical Guide: Vol. 1. Design*. HMSO (1981)

39   Department of Education and Science, Architects and Building Group. *Design Note 46: Maintenance and Renewal in Educational Buildings: Flat Roofs: Criteria and methods of assessment, repair and replacement* (1985)

40   *BRE Digest 295*. Stability under wind load of loose-laid external roof insulation boards (1985)

41   *BRE Digest 144*. Asphalt and built-up felt roofings: durability (1972)

42   Tarmac. *Flat Roofing: A Guide to Good Practice* (1982)

43 W. H. Ransom. *Building Failures: diagnosis and advoidance*. Spon (1981)
44 *BRE Digest 143*. Sound insulation: basic principles (1976)
45 *BRE Digest 252*. Sound insulation of party walls (1981)
46 *BRE Digest 266*. Sound insulation of party floors (1982)
47 British Standards Institution. *BS 2750 Methods of measurement of sound insulation in buildings and of building elements. Part 4: 1980 Field measurements of airborne sound insulation between rooms. Part 7: 1980 Field measurements of impact sound insulation of floors*
48 British Standards Institution. *BS 5821 British Standard method for rating the sound insulation in buildings and building elements. Part 1: 1984 Method of rating the airborne insulation in buildings and of interior building elements. Part 2: 1984 Method for rating the impact sound insulation*
49 DOE. *Sound Insulation in Buildings*. HMSO (1971)
50 *BRE Digests 128* and *129*. Insulation against external noise (1971)
51 I. H. Seeley. *Building Technology*. Macmillan (1986)
52 *BRE Digest 108*. Standard *U*-values (1984)
53 DOE. *Thermal Insulation of Buildings*. HMSO (1971)
54 *BRE Digest 145*. Heat losses through ground floors (1984)
55 *BRE Digest 236*. Cavity insulation (1984)
56 *BRE Digest 140*. Double glazing and double windows (1980)
57 E. D. Mills (Ed.). *Building Maintenance and Preservation*. Butterworths (1980)
58 *BRE Digest 278*. Vibrations: building and human response (1983)

5 BUILDING MAINTENANCE PROBLEMS AND THEIR SOLUTION—IV

Joinery, Corrosion of Metals, Plastics, Plasterwork, External Renderings, Internal Finishings, Decorations and Glazing

Joinery

General Defects

Good-quality joinery should be free from cracks, large or loose knots and rough or raised grain. The softer, lighter spring grain has a tendency to greater shrinkage with the possibility of the darker, harder autumn grain being raised on the surface. Careful selection and proper conversion of the timber will assist in minimising this defect and a good standard of rubbing down with glasspaper by the painter will also help. External joinery which contains too much sapwood may deteriorate and shrink and cause fracture of the paint film. In extreme cases, rotting may occur beneath the paint film. It is now well established that external window joinery made from present-day supplies of redwood (*Pinus sylvestris*) is liable to decay within a few years, because of the presence of a large proportion of sapwood in the timber. Sapwood offers little resistance to wood-destroying fungi, and soon decays if it remains wet. This has resulted in the widespread adoption of preservative treatment of redwood joinery timber since the mid 1970s. On the other hand, Western red cedar and Douglas fir with no sapwood have sufficient natural resistance to decay without preservative treatment.[1]

Water causes most of the trouble in wood—if timber with the correct moisture content were installed in a building and if the moisture content were kept at the right level, most of the problems of timber maintenance would be removed. Moisture penetration causes swelling, distortion, failures of surface finishes, and introduces risks of complete failures through decay caused by wood-destroying fungi.[2] Timber should be seasoned to the correct moisture content, be protected from moisture on the site, with joints properly sealed; its exposed end grain—which absorbs moisture hundreds of times faster than side grain—should be kept to an absolute minimum, and exposed horizontal surfaces sloped to throw off rainwater. Condensation is another cause of moisture gaining access to timber.

Windows

In recent years there has been a substantial increase in the number of instances of decay in wood windows in comparatively new houses. Decay occurs both in opening lights and in frames permanently in contact with brickwork or blockwork. It occurs most frequently in ground floor windows and in the lower parts of the members concerned, such as the lower rail of an opening light, the bottoms of jambs and mullions, and the sill itself, often at or near a joint.[3]

Types of decay.   In old buildings, decay in window joinery may be part of a widespread attack of dry rot fungus—*Serpula lacrymans*. Most decay in window woodwork exposed to the weather is of the wet rot type (both brown rot and white rot) which will not spread to other timber in the building. The rot is almost certain to be of this variety where the decay is confined to relatively small localised pockets detectable only from outside the building or if no actual fungus growth can be found. The brown rots cause dark discoloration of the wood in its early stages and cracks along and across the grain occur later. The white rots bleach the timber and it finishes up in a stringy condition. The basic causes of decay are the low natural resistance of sapwood and the presence in the wood of sufficient moisture to permit the growth of wood-destroying fungi, usually in excess of 20 per cent.[4]

Plate 19 shows a badly decayed wood sill and mullion to a window in a modern college gymnasium, while plate 20 illustrates a wood window sill to a

Plate 19 Rotting wood window sill and mullion

Plate 20 Badly decayed wood window frame

commercial building which has completely disintegrated as a result of gross neglect.

Entry of moisture.   Decay of existing joinery involves in the main preventing access of moisture to the wood and the following factors accentuate the problem:

(1) flat surfaces on horizontal rails which do not effectively shed water;

(2) failure to seal joints and exposed end grain, resulting in capillary entry of water;

(3) use of animal and casein glues which fail under damp conditions;

(4) failure to cover joinery in transit and on site, placing too much reliance on pink shop primers which are of variable quality, soon become weak or powdery and often allow some moisture to penetrate;[3]

(5) failure to prime rebates or careless puttying leading to putty failure;

(6) failure to provide an effective seal under wood glazing beads;

(7) overstressing of joints in opening lights, causing joints to open and putty to come away from glass;

(8) poor paintwork maintenance leading to excessive swelling and shrinkage with consequent opening of joints and putty lines, which new paint fails to seal;

(9) excessive condensation, particularly in bathrooms and kitchens, where temperatures and humidities are often high, and aggravated by defective back putties.[4]

Remedial measures.   Early indications of conditions conducive to decay are given by putty failures, the waterlogged condition of wood beneath defective paint or discoloration of paintwork near joints. Discoloration is caused by the growth of fungi through the paintwork from underlying moist sapwood. Open or strained joints and failure of paint over back putties may give rise to moisture penetration. Extensive swelling and jamming of opening lights indicate that moisture has gained excess. Decay may first show as depressions in the surface of the wood or there may be wrinkling, discoloration or loss of paint. In these instances, the underlying wood should be probed with a bluntly pointed tool such as a small screwdriver to assess the extent of the decay.

Where there are indications of water penetration but no decay is detected, remedial measures should be undertaken during a dry period of the year. Paintwork should be stripped in suspect areas and extending about 100 mm around them, and loose or cracked putty removed. Horizontal wood surfaces should be cut to form a slope for drainage. After an adequate drying out period, follow with generous applications of wood preservative worked well into the joints of the woodwork. Strained joints should be strengthened by metal brackets. Putty can then be renewed and open joints carefully sealed.

The easiest way to preserve sound but wet wood against decay is to drill and insert pellets at joints and other vulnerable positions. These pellets contain a fungicide based on boron compounds which dissolve in the moisture of the timber, diffuse into the wood and will stop any decay. An alternative method is to inject wood preservative at the joints, using specially designed plastic injectors.[5]

Regular maintenance inspections ensure that decay is diagnosed at an early stage when minor repairs will suffice. Relatively inexpensive repairs will prolong the life of a window by several years adopting the following procedure:

(1) strip paint and cut out decayed wood in late spring or early summer and allow to dry;

(2) inject preservative, often of organic solvent type, and allow to dry;

(3) apply priming coat of paint;

(4) fill holes using a suitable proprietary hard filler—a large hole may be filled with a piece of wood, preservative-treated and primed after shaping, and sealed with filler;

(5) seal whole of repair area with primer;

(6) apply undercoats and finishing coat of paint.

Where the woodwork is in an advanced stage of decay, the whole part of an affected window may have to be replaced. The sectional sizes of timber scantlings for windows have become smaller in recent years. It is bad practice to use flimsy sections particularly in areas subject to severe weather conditions.[3] Opening lights should neither be very loose nor very tight, but should fit snugly and fasten fairly tightly. An alternative approach is to use weathersealed joints, encompassing bold detailing with relatively open joints between the sash and frame backed up by a weatherseal joint in the rebate of the frame at the rear of the sash. Avoidance of excessive exposed end grain in wood windows and frames is essential.

Wood sliding sashes are liable to decay at sills, bottoms of jambs and angle joints of sashes which may need bracketing on fracture. Sash fasteners sometimes break and need replacement, but the most common defect of all is broken sash cords. Sash cords may be tested by pulling out the cord as far as it will extend and then releasing it—a rotted cord will break under the strain. Where one defective cord is found, it is advisable to renew both using best flax line or gunmetal chains for heavy sashes. Dampness below windows may be due to one of several causes—absence of water bar between wood and stone sills, defective joints between wood sills and tile or brick subsills, or insufficient weathering or checkthroating, or lack of throat on overhang of sill.

A decayed portion of a wood sill can normally be cut out and replaced. In cases of limited decay or a missing water bar, a piece of sheet lead or other suitable waterproof material can be dressed over the sill and subsill. A decayed stone subsill can be cut back to a sound face, hacked to provide a key, and made good with cement and fine stone aggregate. Throatings and grooves can be readily formed where required and inadequate ones enlarged. Open joints between wood sills and subsills in other materials should be raked out and filled with a suitable sealer. Where excessive condensation runs down the inside face of a window, a small hollowed wood moulding may be fixed to the inside of the sill to form a channel, in which condensed water may collect and evaporate or discharge through small weep holes drilled through the bottom rail of the sash.

BRE Digest 262[6] emphasises the importance of ensuring that all opening joints on windows are close fitting or the provision of separate weather-stripping seals to reduce energy waste and prevent draughts, although some ventilation is essential for the comfort of the occupants. The digest also describes how to withstand three types of loading-forces applied to fasten or unfasten and move an opening light, moving a jammed window, and persons on ladders supporting themselves on windows—collectively referred to as resistance to abuse.

Doors

Doors are sometimes a high-cost maintenance feature and could desirably receive more attention from designers. External doors are more likely to be protected from the weather than windows by porches or balconies, but if subject to rain-splash may receive more wetting. Garage doors in particular are seldom protected by overhangs, and their large size and outward opening makes them more liable to mechanical damage which can strain joints and cause cracking to the paintwork.

Water penetration.   Rainwater does not usually penetrate into flush doors unless openings are cut into them, as for glazed areas. It is essential that flush doors used externally shall be of exterior quality and be fitted together with suitable glues. On the other hand, panelled doors resemble windows in that their numerous joints provide many possible points of entry for moisture. A wide bottom rail is normal practice, and an equally wide middle rail, usually 180 to 200 mm wide, is now required to accommodate a British Standard letter box. The amount of dimensional change in such a wide rail resulting from seasonal variations in atmospheric humidity is considerable. If the joints between horizontal rails and vertical stiles are well glued there is a risk that the rails will split when shrinkage occurs during a dry summer. In practice, only limited amounts of glue are normally used so that swelling and shrinkage take place, and the protective paint film over the joint is broken, with the unsealed joint acting as a moisture reservoir. Risk of water penetration into the lower rail is increased if the interface between the weather moulding and the rail is not adequately sealed. In like manner, external glazing bars should be undersealed with putty or mastic. Use of unsuitable glues which are not resistant to moisture or an insufficient application of glue can lead to early joint failure, accentuated by the increasing tendency in small modern houses to fit doors to open outwards. BRE recommend weather and boil proof (WBP) glues, primarily the phenol formaldehyde (PF) and resorcinol formaldehyde (RF) types.[7] The low natural resistance to fungal attack of the softwoods commonly used in door manufacture is a further factor contributing to deterioration following moisture penetration.

Framed doors.   The majority of doors used in local authority or speculative houses are mass-produced and some are poorly designed and constructed. Some types of framed door with dowelled joints are liable to open up with shrinkage, permitting the entry of moisture and possible decay. Dowels should be made from impermeable and durable timbers or be treated with a water-repellent preservative prior to assembly, provided the treatment is compatible with the glue.[4] Panels of external doors should not be of plywood unless it is of good exterior quality and glued with water-insoluble adhesives in accordance with BS 6566,[8] otherwise the panels may be subject to blisters, cracking and peeling.

In older houses, framed doors often contain 12 mm thick panels set into grooves in stiles and rails with mouldings planted or worked solid around the edges of the panels. Where panels crack, the split usually follows a glued joint and this can be repaired by soaking off the old glue and rejointing. A panel mould parallel with the split is removed and the joint closed by levering the panel with a chisel. Where damage is more extensive the panel may have to be removed. A panel should not be glued into the grooves and nails or pins fixing panel mouldings should not penetrate the panel, as they could prevent natural shrinkage and cause the panel to split.

Door frames and thresholds.   External doors need the bottom rail rebated over a galvanised steel water bar and to have a throated weatherboard fixed to the bottom rail, to prevent rainwater penetrating under the door, particularly when it is in an exposed position. On occasions, weatherboards are too narrow to shed the water satisfactorily over the timber threshold, and thresholds are sometimes machined to weather to an excessively thin front face which may subsequently curl up and form a channel which collects water and may even discharge it over the threshold into the building. It may be necessary to replace the threshold with a wider one and possibly to replace the water bar. Door frames set within 75 mm of the outer face of walls in exposed positions need protection by door heads. The bottoms of external door frames may decay owing to lack of painting. Defective lengths can be cut out and new sections spliced on with the joints put together in white lead or thick paint to make them watertight, replacing defective thresholds at the same time. Flexible sealants need to be applied between frames and the adjoining masonry to secure a watertight joint. A suitable sealant such as oil-based mastic, butyl rubber or acrylic (solvent type) should be used, although some may not be compatible with wood stains or preservatives. Ideally a groove or rebate should take a 10 × 10 mm section of sealant with adequate back-up material, such as foamed polyethylene to control the depth of sealant. In the absence of a joint gap, corner fillets may be used which are prone to failure.[9]

Joinery defects.   Doors which shrink unduly should be taken off their hinges and a strip fixed to the hanging stile. Other defects include damaged arrises (external angles) and indentations. Slightly damaged arrises may be rectified with a plane and chisel. Where the damage is more extensive the arris could be rounded. Shallow indentations can be removed by hot water treatment and finished with a smoothing plane. In more severe cases, the dent can be drilled out with a centre bit and filled with a treated wood plug or piece of new wood, followed by redecoration. On redecoration, defective panels can be hidden by covering the door faces with plywood or hardboard. Warped doors can be adjusted by easing (planing the shutting stile). A survey by Sinnott[10] of 183 new dwellings erected by 92 builders in the West Midlands revealed a generally low standard of joinery work, mainly resulting from poor workmanship on site and careless handling and storage of materials rather than the quality of materials used, although the finish of timber and timber products supplied to the builder is often poor. The contrast between manufactured fitments and work done on site was markedly to the disadvantage of the site work.

Ironmongery

The provision of good-quality ironmongery reduces failures and hence maintenance costs. Bronze metal antique finish (BMA) is often polished which results in exposure of the brass and subsequent damage to surrounding paintwork. Anodised aluminium is generally preferable with stainless steel for high-class work. Aluminium ironmongery is best fixed with aluminium screws as steel screws are apt to rust.

Corrosion of Metals

The corrosion of metals in buildings may result in one or more serious defects, namely:

(1) the structural soundness of the metal may be reduced;
(2) it may cause distortion or cracking of some other building material in which the metal is bedded;
(3) failure of the metal may result in water entering the building;
(4) unsightly surfaces may be produced.

Dissimilar metals in contact can result in bimetallic corrosion. Where they are unavoidably in contact, they should be insulated from each other by impervious non-conducting materials, such as bitumastic coatings to prevent the electro-chemical process occurring.

Low-alloy steels or *weathering steels* have been used extensively in the United States and are now being used increasingly in the United Kingdom. When permitted to rust under suitable conditions, corrosion soon ceases after the formation of a protective rust coating.

Ordinary steels, including mild steel and low steel alloys, require suitable protective treatment to prevent corrosion. Preventing corrosion of steelwork is necessary not only to protect the steel but also other adjoining materials. For example, corrosion of steel embedded in concrete or masonry may fracture the encasing material. For protective purposes, it is necessary to exclude water, to apply sacrificial metals to protect the steel components or provide a protective environment. Light stains from corroded steelwork on masonry or concrete surfaces can be removed with oxalic acid solution, while very severe staining will require a sealing coat followed by a pigmented masonry finish.[11] Corroded steel can cause degradation of certain timbers and pronounced blue-black staining (iron stains), which can be overcome by galvanised coatings.[12]

Ungalvanised steel windows, mainly found in houses built before 1950, suffered from severe rusting, distortion of the windows and cracking of the glass. It is generally advisable to replace this type of window.

Heavy steel sections carry millscale and, if these are permitted to weather, rusting is likely to occur under the scale. Paint adheres well to scale but the scale itself may flake off taking the paint with it. Painting over rust, although not so serious as painting over millscale, is undesirable as the paint is likely to deteriorate.

Stainless steel rarely requires maintenance and if it was costed over the probable life of the building could prove economical in a variety of situations, possibly including cavity wall ties.

Iron is liable to severe corrosion under normal climatic conditions. The rusting of iron requires the presence of water, oxygen and carbon dioxide (or other acid) and is accelerated by increased acid concentration. Exposed ironwork requires protective treatment often in the form of corrosion resistant paint.

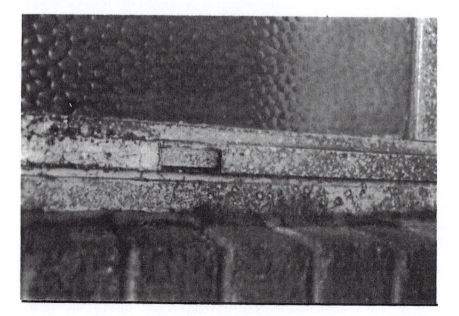

Plate 21 Badly pitted aluminium window frame

Corrosion of *non-ferrous metals*—lead, copper, zinc and aluminium—was considered in chapter 4 under flat roof coverings. Aluminium window frames are now used increasingly and opinions vary as to the form that maintenance might take. Some believe that no maintenance is required with anodised aluminium in only mildly corrosive environments. In moderately corrosive situations, provision should be made for washing the window frames at fairly frequent intervals to prevent accumulation of dirt and subsequent corrosion by pitting beneath it. One arrangement is for the frames to be washed over when the windows are cleaned, but separate cleaning is much more effective. Plate 21 shows a badly pitted anodised aluminium window to a hospital.

Plastics

BRE Digest 69[13] describes how sunlight, particularly in the form of ultraviolet radiation can cause colour changes in plastics, while the mechanical and electrical properties are significantly impaired by a rise in temperature. Most plastics are impervious to water. In many cases mechanical action is likely to be the ultimate cause of failure of plastic components. for example, impact may produce indentation or brittle fracture, while continuously applied loading can cause collapse or rupture.

Unplasticised PVC can be used in a variety of building components including soil and vent systems, underground drains and fittings, rainwater goods, wall cladding, skirtings and architraves, and window frames. Lack of rigidity in this material has resulted in frames being of rather thick section unless reinforced by timber or steel, and in the latter case perfect seals are required at joints of the plastic frame to protect the reinforcement.[14] The most common colour for uPVC window frames is white as it is the most durable and is likely to have a life in excess of 20 years. The main problem will be high thermal movement, as a temperature range of 45°C can cause a movement of up to 4.5 mm in a length of 2 m, and joints must be able to accommodate this movement and also remain watertight.

There are now many uPVC windows and doors on the market normally having profiles of cellular structure with a minimum wall thickness of 3 mm, with welded T-joints and corners and the profiles reinforced with steel or aluminium channel where necessary. They can be single or double glazed usually fixed by snap-in glazing beads. UPVC patio doors are also being used extensively incorporating toughened glass to BS 6262[15] and adequate weather seals.

A number of local authorities were becoming very concerned at the continual window repainting and replacement programmes in the mid 1980s, and set up fabrication units producing uPVC windows often at around 100 windows per week. Some believed this would result in maintenance-free windows but only time will tell whether their assumption is correct. As previously outlined, there are inherent defects in this material.

BRE Digest 224[16] described how foamed and expanded plastics were useful building materials on account of their good thermal insulating and acoustic properties combined with low cost and weight and ease of handling. Typical examples are extruded expanded polystyrene, expanded polystyrene beads and granules, foamed phenol–formaldehyde resin, foamed polyurethane and foamed polyurethane granules.

Plasterwork

The characteristics of the principal plaster finishes are given in table 5.1. Some of the more common plastering defects with their main causes and remedies are listed in table 5.2

Other Plastering Defects

On occasions problems of flaking and bulging of the final plaster coat have occurred within six to nine months of plaster being applied to clinker or breeze block walls and partitions. The usual cause is the greater drying shrinkage of the blockwork and undercoat plaster relative to the finishing coat. Concrete blockwork tends to shrink as it dries out, particularly if the blocks have not been matured for a sufficiently long period, or if they are wet when laid. Plaster undercoats, particularly of cement, lime and sand, tend to shrink with the blockwork. Little drying shrinkage occurs in finishing coats of gypsum plaster and hence these may break away from the undercoats and blockwork as the latter shrinks, resulting in hollowness and flaking. If the finish is thoroughly keyed to the undercoat, cracking only may occur. Defects of this kind normally arise within a year of plastering.

In repairing the plasterwork, the application of a new finishing coat to the existing undercoat is usually satisfactory, provided it is left for a sufficient period of time. The undercoat must be thoroughly cleaned down. If the suction of the undercoat is very high, a finish of well soaked lime putty and sand with a light gauging of gypsum plaster usually gives good results. Alternatively, a PVAC emulsion bonding agent can be applied to the undercoat prior to applying the finishing coat.

Plaster joints between structural columns and internal walls may crack owing to differential movement between column and wall. This defect can be avoided by forming a groove in the plaster at the juncture, or by making a cut in the plaster and masking it with a cover strip fixed at one edge only. External angles of plastered walls are liable to damage and the provision of a metal or plastic angle piece will give protection. The insertion of galvanised metal scrim containing an angle bead into the plaster at external angles is probably the ideal arrangement.

Where plaster extends across different backgrounds, the fixing of expanded metal across the junctions will reduce the risk of cracking due to differential thermal expansion and drying shrinkage. Thus in the case of the columns previously described, the columns could be bridged by expanded metal over building paper or polythene sheeting fixed to the wall on either side, to isolate the plaster from any movement of the column. Where cracks can be anticipated, as at the junction of wall and ceiling, it is advisable to

ble 5.1 Characteristics of plaster finishes

Plaster	Surface hardness and resistance to impact damage	Other surface characteristics	Restrictions on early decoration	Shrinkage or expansion	Remarks
me plasters **ne**	Weak and very easily indented	Open-textured (depending on sand), absorbs condensation	Initially only suitable for permeable finishes that are unaffected by alkali	Shrink on drying, but shrinkage is reduced by addition of fine sand	Slow hardening. Only apply on dry undercoats
uged lime	Resistance to damage increases with proportion of gypsum plaster	Similar to above, but smoother finishes obtainable		Shrinkage is restrained by the gypsum content provided that over-trowelling is avoided	Only apply on dry undercoats
ypsum plasters **ass D (Keenes)**	Very hard and resistant to damage	Very level and smooth. Particularly suitable for low-angle lighting conditions	None, except on undercoats containing cement or lime, or unless lime is added to the finishing coat	Expand during setting. Subsequent movements usually small, but too rapid drying can lead to delayed expansion	Set slowly and so allow ample time for finishing to a smooth surface. Should not be allowed to dry too quickly
ass C **nhydrous)**	Hard and resistant to damage	Slightly less smooth than Class D			
ass B **emihydrate)**	Sufficiently hard and resistant for most normal purposes, but weakened by additions of lime	Sufficiently smooth and level for most purposes		Expand during setting, though extremely slightly with board finish plasters. Subsequent movements are small	Set quickly, should be allowed to dry as soon as possible
ghtweight	Surface hardness similar to Class B plasters. Ease of indentation varies with the type of lightweight undercoat, but resilience tends to prevent serious damage	Sufficiently smooth and level for most purposes	None, but the higher water content of lightweight undercoats makes these somewhat slower to dry than sanded gypsum plaster undercoats	Expand during setting. Subsequent movements usually small and easily restrained by background	Maximum fire resistance. Lightweight plaster surfaces warm up more quickly than others and so help to prevent temporary condensation
ement/lime/sand **0–¹/₄:3**	Very strong and hard	Wood float finish	Initially only suitable for permeable finishes that are unaffected by alkali	Shrink on drying, but surface cracking can be minimised by avoiding over-working	Suitable for damp conditions
1:6	Strong and hard	Wood float finish			
2:9	Moderate	Wood float finish			
ngle-coat finishes **oard finish gypsum** **lasters** **Class B)**	Surface hardness similar to Class B above, but resistance depends on background	On suitable backgrounds, similar to Class B above	None; finish dries very quickly	Extremely small expansion on setting. Subsequent movements small	
hin-wall finishes **ased on gypsum**	Softer than board finishes	Smooth and level on sufficiently level backgrounds	Dries very quickly. No restrictions when dry	Extremely small expansion on setting. Subsequent movements small	
ased on organic inders	Moderately hard. Resistance depends on background	Matt surface, closely following the level of the background	Dries very quickly. No restrictions when dry	The very thin coats are restrained by the background	
ypsum projection lasters	Properties generally intermediate between Class B and Class C gypsum plasters				

ource: BRE Digest 213[17].

Table 5.2 Some plastering defects and their causes and remedies

Defect	Cause	Remedy
CRACKING		
Fine hair cracks on the finished plaster	Use of loamy sand (if work is in gypsum plaster throughout). Excess lime in final coat. Applying final coat before initial shrinkage of undercoats is complete (if undercoats are based on cement or lime)	Filling of fine hair cracks is extremely difficult; it is often better to apply wallpaper—or a lining paper or polyethylene sheeting—if a painted finish is required
Clearly defined cracks following a definite line, particularly with plaster on building boards	Poor treatment of joints. Shrinkage or warping of timber studding or joists. Settlement or thermal movement	Cut out and fill. In some cases cracks are liable to reappear and repairs should be postponed for as long as possible. Reinforce plaster with jute scrim or metal mesh, or mask cut at joint by cover strip fixed to one side
LOSS OF ADHESION		
Loss of adhesion of final coat in work of two or more coats or single coat work on building board*	On gypsum undercoats—a strong final coat over a very weak undercoat. On cement or cement:lime based undercoats—applying final coat while undercoat still 'green' and/or inadequate mechanical key. Dirty surfaces. *Wrong type of board—such as plasterboard intended for direct decoration. *Use of lime in plaster. Thermal expansion may cause stresses between plaster and background	If the undercoat is too weak the only cure is to strip and replaster. If the undercoat is sound strip the final coat, allow the undercoat to thoroughly dry, roughen surface, remove dust with damp brush and replaster. *Strip and replaster. Check suitability of board for plastering. Treat the surface with one or two coats of a suitable PVAC emulsion type of bonding agent if there is any doubt about its providing a satisfactory bond
QUICK SETTING PLASTER		
Plaster sets too quickly	Plaster very fresh from works and still hot. Dirty mixing water or partially set plaster from a previous gauging. Unsuitable sand—normally the setting time of sanded plaster decreases as the volume of sand is increased. The set will speed up if the plaster has been kept too long or if bags have become damp	Open tops of bags and allow plaster to cool. Use clean water and keep mixing platform, buckets and tools clean. Test the sand for grading and cleanliness. As a temporary cure reduce the proportion of sand. Store in dry conditions and do not store for more than 2 months—checked from the date stamped on the bag

Defect	Cause	Remedy
SLOW SETTING PLASTER Plaster sets too slowly	An unsuitable sand can extend the setting time of a sanded plaster to 12 hours, whereas the setting time with a standard sand is about 3 hours	Test the sand and change if necessary. With sanded mixes the setting time will increase as the proportion of sand is decreased
DRY OUT Plaster surface soft and powdery with very fine cracks	Plaster drying before setting. Gypsum plaster requires as much water in setting as is driven off from the gypsum in manufacture	Use the correct grade of plaster applied to the correct thickness. On work already plastered the only remedy is to strip and replaster
EFFLORESCENCE Soluble salts on plaster face	Soluble salts brought forward from the background to which the plaster has been applied, to the face of the plaster as the building dries out	Dry brush the surface carefully and repeatedly as the salts appear. The salts should be swept up and thrown away. Good ventilation will hasten the drying process. Decoration is best delayed until the structure is thoroughly dry; if thus is not practicable use a porous paint suitable for early decoration
SEALED IN WATER Moisture trapped in new plaster	Much of the water used in construction can be retained in the structure for a considerable time. Wet plaster should not be sealed with impermeable finishes	Use permeable paints to allow the moisture to evaporate and facilitate proper drying out. If impermeable paints are to be used the walls must be allowed to dry out thoroughly
MOULD GROWTH	The growth starts in minute windborne spores which alight on and develop in the newly applied coating. The spores will only develop if dampness is present	Once the construction has dried out the growth will stop. Any existing mould and decoration should be scraped off the surface of the plasterwork. When dry the affected area should be treated with a fungicidal wash keeping the work dry and ventilated
FLAKING AND PEELING OF FINAL COAT	Persistent moisture penetration through the background	Strip defective plaster and provide positive barrier to dampness or substitute cement-based mixes

Defect	Cause	Remedy
IRREGULARITY OF SURFACE TEXTURE	Uneven trowelling or marked differences in suction of backgrounds often due to the mortar having a very different suction from that of the bricks or blocks	Lay or point in a mortar similar in character to the bricks or blocks, or apply an additional coat of plaster
POPPING OR BLOWING	Particles in the background or in the plaster, lime or sand which expand after the plaster coat has set	Small holes may be filled by brush treatment with thick slurry of plastic paint or quick setting hemihydrate plaster. Larger holes filled by normal patching techniques, matching surrounding material in porosity and colour
RECURRENT SURFACE DAMPNESS	Deliquescent salts attract moisture from the air. They can result from the use of unwashed sea sand, and be carried from the background into the plaster by, say, condensation in an unlined flue	Strip the plaster and provide an impervious barrier
RUST STAINING	Application of unsuitable plaster to metal lathing or plaster in contact with corrodible ferrous metal in persistently damp conditions	Strip the plaster and re-plaster with metal lathing plaster or mixes containing Portland cement and/or lime. Paint ends of galvanised metal lathing or metal scrim. Metallic conduit or channelling need treating to prevent rust and positioning sufficiently deep below the surface to prevent cracking of plaster
SOFTNESS OR CHALKINESS	Excessive suction of the undercoat, undue thinness of the final coat, working past the setting point, or exposure of final coat to excessive heat or draughts during setting	Adequate wetting, use of a bonding agent, proper application of a final coat of adequate thickness, or use of a special type of plaster

Note: Plastering defects may be due to causes other than the use of faulty plastering materials or techniques. For instance, moisture penetrating an external wall may cause blistering, efflorescence, flaking or complete disruption of the plaster—or a plastered ceiling may crack because the construction permits excessive deflection. No plaster repair can be expected to remedy these problems, and it is essential to determine the real cause of the problem before spending money on remedial work.

make a straight cut through the entire thickness of the plaster along the line of the junction. Plaster scrims can also be used to prevent unsightly irregular cracking. Small movements between walls and ceilings can be covered by plasterboard cove to pleasing effect.[17]

Premixed lightweight plaster or thin-wall plasters are best suited for plastering on dense concrete. Surfaces should not be plastered until the concrete has dried out sufficiently. Bonding agents must be used in accordance with the manufacturer's instructions. Where no bonding agent is used, the surface of the concrete should be pre-wetted immediately before plastering.

When it is necessary to plaster painted brickwork, good adhesion is required. This may be obtained by hacking the surface over at least one-half of the area and raking out the joints or by fixing wire mesh or fabric about 6 mm from the surface of the brickwork. The choice of plaster is important—suitable plasters being retarded hemihydrate gypsum plasters or a proprietary concrete bonding plaster. Where used in conjunction with metal mesh, the undercoat should contain a small proportion of lime.

External Renderings

External renderings can be used to prevent rain penetration through porous brick walls, since the rendering not only fills the hair cracks in the mortar joints, but a textured or dashed finish tends to throw off water from the face. Nevertheless, renderings even when incorporating waterproofers are not entirely waterproof and good workmanship and suitable mixes are essential if cracking is to be prevented.

Plain renderings should ideally be porous and weak, so that the contraction occurring during setting and drying out will be widely distributed and not cause large cracks which are characteristic of strong mortars. A porous rendering will absorb some rainwater but will not readily transmit it to the background and it will dry out during the next fine spell. In conditions of severe exposure, the first coat at least of the rendering should be fairly impervious. In exceptionally severe conditions where the wall is likely to be subject to hard frost or continuous driving rain, pebbledash or roughcast is preferable to plain rendering as both shed water well and do not crack excessively with a rich mix. Hence the selection of mix is influenced by the type of background, degree of exposure and decorative considerations. Mixes suitable for rendering are detailed in table 5.3, of which the most suitable are cement:lime:sand mixes II, III and IV.

The *background* must be thoroughly cleaned and any fungi algae destroyed with a suitable fungicide and removed. Hollows should be dubbed out using a mix that is stronger and coarser than the first undercoat but weaker than the background. Where the thickness exceeds 16 mm it should be applied in two or more coats. Backgrounds should be dampened to reduce excessive suction.

The strength of the background should be adequate to restrain shrinkage or thermal movements of the rendering. The degree of porosity and suction

Table 5.3 Mixes suitable for rendering

Mix type	Cement : lime : sand	Cement : ready-mixed lime : sand		Cement : sand (using plasticiser)	Masonry cement : sand
		Ready-mixed lime : sand	Cement : ready-mixed material		
I	1 : $\frac{1}{4}$: 3	1 : 12	1 : 3	—	—
II	1 : $\frac{1}{2}$: 4–4$\frac{1}{2}$	1 : 8–9	1 : 4–4$\frac{1}{2}$	1 : 3–4	1 : 2$\frac{1}{2}$–3$\frac{1}{2}$
III	1 : 1 : 5–6	1 : 6	1 : 5–6	1 : 5–6	1 : 4–5
IV	1 : 2 : 8–9	1 : 4$\frac{1}{2}$	1 : 8–9	1 : 7–8	1 : 5$\frac{1}{2}$–6$\frac{1}{2}$

Note: In special circumstances, for example where soluble salts in the background are likely to cause problems, mixes based on sulphate-resisting Portland cement may be employed.

Source: BRE Digest 196[18].

Table 5.4 Recommended mixes for external renderings in relation to background materials, exposure conditions and finish required (the type of mix shown in *italics* is to be preferred)

Background material	Type of finish	First and subsequent undercoats — Exposure			Final coat — Exposure		
		Severe	Moderate	Sheltered	Severe	Moderate	Sheltered
(1) Dense, strong, smooth	Wood float	II or III	*II or III*	*II or III*	III	*III or IV*	*III or IV*
	Scraped or textured	II or III	*II or III*	*II or III*	III	*III or IV*	*II or IV*
	Roughcast	I or II	*I or II*	*I or II*	II	II	II
	Drydash	I or II	*I or II*	*I or II*	II	II	II
(2) Moderately strong, porous	Wood float	II or III	*III or IV*	*III or IV*	III	*III or IV*	*III or IV*
	Scraped or textured	III	*III or IV*	*III or IV*	III	*III or IV*	*III or IV*
	Roughcast	II	II	II	as undercoats		
	Drydash	II	II	II			
(3) Moderately weak, porous*	Wood float	III	*III or IV*	*III or IV*	as undercoats		
	Scraped or textured	III	*III or IV*	*III or IV*			
	Drydash	III	III	III			
(4) No fines concrete †	Wood float	II or III	*II, III or IV*	*II, III or IV*	II or III	*III or IV*	*III or IV*
	Scraped or textured	II or III	*II, III or IV*	*II, III or IV*	III	*III or IV*	*III or IV*
	Roughcast	I or II	*I or II*	*I or II*	II	II	II
	Drydash	I or II	*I or II*	*I or II*	II	II	II
(5) Woodwool slabs*‡	Wood float	III or IV	*III or IV*	*III or IV*	IV	*IV*	*IV*
	Scraped or textured	III or IV	*III or IV*	*III or IV*	IV	*IV*	*IV*
(6) Metal lathing	Wood float	I, II or III	*I, II or III*	*I, II or III*	II or III	*II or III*	*II or III*
	Scraped or textured	I, II or III	*I, II or III*	*I, II or III*	III	III	III
	Roughcast	I or II	*I or II*	*I or II*	II	II	II
	Drydash	I or II	*I or II*	*I or II*	II	II	II

*Finishes such as roughcast and drydash require strong mixes and hence are not advisable on weak backgrounds.

†If proprietary lightweight aggregates are used, it may be desirable to use the mix weaker than the recommended type.

‡Three-coat work is recommended, the first undercoat being thrown on like a spatterdash coat.

Source: BRE Digest 196[18].

affect the adhesion of the rendering and influence the choice of the type of mix and method of application. If the background does not afford sufficient key, it may be necessary to fix metal mesh, roughen the surface or apply a spatter coat or bonding medium.[18]

Choice of undercoats is very much affected by the type of background as shown in table 5.4. There are six generally recognised background types:

(1) Dense, strong and smooth materials—very dense bricks, blocks or concrete with low suction and smooth surfaces, requiring a mechanical key such as bush-hammering for concrete or a bonding coat of spatterdash.

(2) Moderately strong and porous materials—most bricks and blocks are in this category and joints are raked out to provide a key.

(3) Moderately weak and porous materials such as lightweight concrete or bricks of low strength. The rendering must be weaker than the background otherwise shrinkage could result in failure of the background surface.

(4) No fines concrete with many large voids provides a good mechanical key.

(5) Woodwool slabs afford a good mechanical key but are weak and should be completely dry when rendered.

(6) Metal lathing is best plastered on both sides. Where fixed to battens, self-firring nails or distance pieces should be used so that the first coat can be forced through the lathing to encase it.

Some backgrounds contain appreciable amounts of salts, particularly sulphates, which can in the presence of moisture attack cement in mortars and renderings. A good rendering will normally prevent damp penetration but in the case of rising damp metal lathing should be fixed to impregnated timber battens, bitumen-impregnated fibre lathing used or, better still, a damp-proof membrane inserted.

Undercoats should be between 8 mm and 16 mm thick, with each succeeding coat thinner and weaker than the preceding and lightly scratched to provide a key for the next unless it is machine applied. Finishing coats normally vary from 5 mm upwards in thickness. In conditions of severe exposure three-coat work is recommended and two-coat work for moderate conditions. The undercoat must be allowed to dry out thoroughly and for the worst shrinkage to take place before the next coat is applied. Renderings require adequate curing (at least three days).

Finishing coats may take various forms, as listed in table 5.4, and are now described.

(1) Pebbledash or drydash is a rough finish of exposed pebbles or crushed stone, graded from about 6 to 13 mm, thrown on to and sometimes lightly pressed into a fresh applied coat of mortar. A pebbledash finish on an undercoat with a spatterdash coat beneath is particularly suitable for walls exposed for long periods to driving rain and wind, being very durable and generally free from defects. It must be applied to fairly strong backing.

(2) Roughcast or wet dash is a rough finish thrown on as a wet mix. The aggregate in the finishing coat consists of sand and crushed stone or gravel

graded from about 6 to 13 mm. A roughcast finish on an undercoat with a spatterdash coat beneath is very suitable for severe conditions.

(3) Plaincoat or smooth floated finishes are subject to surface crazing and this is accentuated with mixes that are rich in cement or which use fine sands. Best results are obtained with wood, felt, cork or other suitably faced floats.

(4) Textured or scraped finishes are obtained by the use of different tools on finishing coats. Textured and scraped finishes are suitable for all conditions and are less susceptible to crazing than smooth finishes.

(5) Machine applied finishing coats, such as *tyrolean* vary in texture with the materials used and the type of machine, and they are suitable for all conditions. Proprietary materials are normally supplied ready for mixing but in the absence of a special undercoat a 1:1:6 cement—lime—sand would be suitable. Some mixes are self-coloured.

A coloured final coat can be obtained by any of the following methods.

(1) Using a selected sand or other aggregate.
(2) Using white or coloured cement.
(3) Mixing in suitable pigments.
(4) Using proprietary finishing materials.

One of the worst problems with external renderings is shrinkage cracking. This can be reduced by using weaker mixes containing coarse sand with a low clay or loam cement. Rainwater trapped behind a crazed rendering may penetrate inwards when drying out.

Spalling and lack of adhesion of external renderings may result from poor workmanship, frost action and/or sulphate attack. Sulphate concentrations may occur in chimneys or where a cement and sand mortar has been used for bedding or repointing, where the sulphates tend to crystallise, forcing off the render. BRE[19] have identified the main causes of cracking and detachment of renderings and subsequent rain penetration, as inadequate bond or key to the wall, continuous rendering over zones where relative movement occurs in the background, the rendering is stronger than the background or preceeding coats, is too weak to exclude rainwater adequately or too rich or too wet to avoid cracking. Loose renderings should be cut away to a firm edge, undercutting it a little. It is advisable to cut away from about 100 mm around any movement cracks, wash out the gap and fill any cracks in the backing masonry. If structural movement is continuing or the masonry surface is friable, then it is best covered with expanded metal fixed with non-ferrous nails. Plate 22 shows how rapid deterioration of the structure can follow the failure of a protective rendering.

External Wall Tiling

Some extensive failures of external wall tiling have occurred in recent years and defective tiling on tall buildings can cause serious problems. The wall

Plate 22 Loss of rendering and disintegrating walling

tiles can be fixed with a cement:sand bed or with adhesives and each is considered.

When using a *cement:sand bed*, the tiles should preferably have a good undercut key in their back surface. The background should also have an adequate key by using keyed bricks, or deeply raked brick joints, or by grit blasting or bush hammering concrete or by using some other suitable method. Bedding and render coats should incorporate the coarser sands and thick coats should contain galvanised reinforcement adequately tied back to the structure but spaced out from the background to provide a key. Ample soft joints should be provided in accordance with the manufacturer's instructions; passing through the thickness of tile, bed and render, and coinciding where possible with any discontinuities in the background.[20]

When fixing tiles with *adhesives*, all surfaces must be clean, dry and free of loose material. At least 14 days should elapse before fixing tiles on to new rendering and protection from rain is necessary during fixing and from 7 to 14 days afterwards. The adhesive must form a solid bed and with thin-bed adhesives this effect is obtained by applying an even layer of adhesive to the wall and pressing each tile into position with a twisting action and gentle beating immediately after the application of adhesive. With thick-bed applications the normal method is to 'butter' adhesive on the back of each tile before placing and beating. Flexible movement joints sealed with a flexible sealant must be provided both horizontally and vertically at 3 to 4.5 m centres. Structural joints should extend through to the surface of the tiling. After fixing, joints should be left open for a few days to permit adequate drying out of the adhesive before grouting.[21]

Internal Finishings

With the wide variety of surfacing materials and decorative finishes now available, selection poses problems. Appearance, durability, acoustics, ease of cleaning and cost are all important. They are also subject to the changes of fashion and widely differing forms of usage. A choice has often to be made between a long life surface finish and shorter life coating or decoration with the consequent need for more frequent maintenance. The relative merits of plastered solid partitions and 'dry' plasterboard and similar materials require careful consideration, including regard to the inconvenience and cost of replacing the more vulnerable material.

Ceramic wall tiles provide a very popular finish for bathrooms, toilets and kitchens. They should be of good quality and be laid with straight joints in both directions. Most tiles are now fixed with adhesives, a method which permits them to be applied to hardboard and plasterboard surfaces, and makes the replacement of cracked or loose tiles easier. After each section of tiling is fixed it should be washed down and the joints suitably grouted.

Attractive and durable finishes to walls and fitments can be obtained by using decorative laminates veneered to asbestos, chipboard or plywood. This finish is hardwearing and offers good resistance to heat and liquids. Defective wall plaster can be replaced with new plaster or with a dry lining of plasterboard fixed to impregnated timber or plasterboard battens.[22]

Decorations

The importance of decorations is evident from the fact that the maintenance costs of buildings are approximately equally divided between decoration, services and structure. Decorations enhance the appearance of buildings and in many cases also protect materials which would otherwise deteriorate. In general the protective function lasts longer than the deterioration in appearance, although there are design details which sometimes make it difficult to renew the paint coating adequately and hence fail to ensure continued protection. Typical examples are the difficulty of painting behind downpipes fixed close to walls and securing an adequate thickness of paint on angular or sharp edges of joinery.

Paintwork Qualities

The quality of paintwork can be assessed in various ways and the following indicate some of the more important desirable features of good paintwork:

(1) attractive, bright, evenly coloured and smooth finish, with uniform gloss, sheen or texture;

(2) absence of paint on glazing and other adjoining unpainted surfaces;

(3) satisfactory colour scheme;

(4) absence of rust marks;

(5) absence of resin exuding from knots;

(6) adequate body to paint coating, entirely obliterating any background colours;

(7) freedom from cracks, blisters and other defects;

(8) ability to protect substrates and withstand regular cleaning.[23]

Defects in Paintwork

Paintwork may become defective for many reasons and the more important ones are described later in this chapter. Defects can occur through poor application of the paint. The most common examples are ridges in the paint film caused by brush marks, sagging due to downward movement of the paint film and wrinkling during drying resulting from the paint being applied too thickly.

The Local Government Operational Research Unit has defined 'states' through which an element passes as it deteriorates from a new condition. With emulsion paint the states were defined as

(1) no visible defects,

(2) soiled,

(3) badly soiled but paint film intact,

(4) slightly crazed and flaking,

(5) extensively crazed, flaking or peeling.[24]

In practice, deterioration processes are not uniform and at any point in time different parts of the decoration are likely to be in different states, and the situation is further complicated by variations in assessment by different technical personnel.

Painting Cycles

There are three main reasons for redecoration:

(1) part of a policy of preventive or planned maintenance;

(2) part of a policy of redecoration for reasons other than maintenance; that is, for the benefit of users who may enjoy an occasional change of colour scheme;

(3) redecorating carried out because the protective coat has deteriorated.

Apart from the frequency of painting it is important to use a suitable paint. It has been estimated that damage to British industrial buildings through corrosion and neglect costs around £4 million per day and that improved surface protection could save at least half of this amount. The difference in capital cost between an appropriate surface protection and a cheap paint is insignificant against the labour and disruption costs involved in having to, say, paint one year earlier. Furthermore, the cheaper paint may not be giving adequate protection over the shorter life span with the possibility of consequent damage to the element concerned.

Painting costs depend on a variety of factors including the amount of paintwork, frequency of painting, painting specification, accessibility, availability of labour and current costs of labour, materials and overheads. The cheapest and most effective painting policy aims at obtaining a balance between excessive painting costs and insufficient painting, leading to the eventual replacement of expensive building elements. In general, annual painting costs increase with more frequent applications but annual replacement costs are reduced.[25] However, a government costs in use study of twenty-four Crown office buildings[26] found that office painting cycles were not rigorously followed, owing to differences in climatic conditions, degree of exposure, amount of public use, and opinion as to proper standards. Buildings are also decorated piecemeal to avoid undue disturbance to the occupants, and there is some deferment of redecorations if an alterations scheme is proposed.

In times of financial stringency, painting is often deferred, supposedly to save money, but it results in much higher costs in the long term.[27] The costly consequences of lack of painting are illustrated in plates 23, 24 and 25. Plate

Plate 23 Breakdown of paintwork and woodwork to timber facade

Plate 24 Loss of paint and decaying timber to pilaster base

23 shows the progressive decay of the entire timber facade to a modern hospital extension, plate 24 a rotting base to an attractive pilaster to the entrance to a well-designed bank, and plate 25 illustrates the rotting weatherboarding to a higher educational building, about 15 years old, which permitted the entry of rainwater, into the lecture rooms and offices beneath and the boarding has since had to be replaced at high cost.

Investigations into factory painting highlighted the problem of scheduling the work to avoid interference with production. This usually entails a concentration of work during the annual shut-down period, and this is assisted by the use of modern painting systems and techniques, including one coat spray on paints. These are often based on high-build alkyd resin blended with special pigments to provide adequate obliterating power and film build in one coat and can be applied by airless or conventional spray equipment.

The recommended intervals between paint cycles vary enormously with different conditions from possibly 5 years in relatively mild areas to 3 years in very aggressive locations. The cycles may not however be based on complete repainting. A sound initial paint base may be followed by spot-priming as necessary and one finishing coat, as preparation is more costly than paint application. One industrialist estimated the cost of

Plate 25 Breakdown of paintwork and decaying timber to weatherboarding

repainting a large factory chimney every 2 years at £3000, while full paint treatment every 5 years combined with disruption of production could amount to £30,000. Table 5.5 shows possible painting cycle costs on steelwork.

Table 5.5 Painting cycle costs on steelwork

Exposure time	1 year	3 years	5 years	6 years
Condition	excellent	good	patchy	corroded
Treatment required	no cleaning, 1 coat gloss paint	local scraping, wire brush, spot prime, 2 coats gloss finish	25% chip, scrape, wire brush, prime overall, 2 coats gloss finish	100% chip, scrape, wire brush, prime overall, 2 coats gloss finish
Cost comparison/ treatment	1.00	2.80	5.11	5.91
Labour/materials ratio	1.8 to 1	3.8 to 1	3.15 to 1	3.8 to 1

Painting Processes

There have been many developments in both paint production and application in recent years. Although some of the new paints may be less tolerant of poorly prepared surfaces or bad weather than older materials, nevertheless they possess important advantages and can be used beneficially under certain conditions. Careful preparation of the surface and skilful application are major factors determining the performance of any paint.[27] Considerable guidance in the selection of painting systems is given in BS 6150.[23]

Repainting Woodwork

Attention should be concentrated on the vulnerable, weathered areas of sills and lower rails. Paintwork which is chalked or dirty but otherwise sound needs thorough washing before repainting, preferably with a detergent solution or a proprietary cleaner. Once the dirt is loosened it should be removed with copious quantities of clean water, the surfaces rubbed down with wet or dry abrasive paper and allowed to dry. All cracks should be carefully sealed by using hard oil-based stoppers, which do not shrink on setting like putty, applying any necessary preservative treatment, and priming any bare wood before filling. Any knots or resinous wood should be sealed with shellac knotting or leafing aluminium primer to prevent resin exuding through the paint. Any bare wood should be primed to fill the pores, stop the suction of the wood and form a base for the undercoat. Primers may be low-lead alkyd based or aluminium and are best applied by brush, paying particular attention to corners, nail holes, joints and end grain. Bring primed areas of woodwork when dry forward with undercoating, applying filler if necessary, within 48 hours of priming. A compatible finishing coat should be applied after the undercoating is dry by lightly rubbing down and cleaning the surface of any loose matter, all in accordance with the manufacturer's instructions.

The moisture content of joinery should not exceed 18 per cent for exterior joinery and 12 per cent for interior work. Exterior painting should not be undertaken on damp surfaces and should preferably be carried out between mid-April and mid-September.[23]

If the existing paint is soft, very chalky or eroded, cracked, blistered or peeling, or shows any adhesion weakness, it should be completely removed. Complete removal is also desirable if the paint has been affected by mould growth or by bleeding through of stains or preservatives, or if there is already an excessive number of coats. Decayed timber should be cut out and replaced; both old and new timber should be treated with preservative, particularly end grain. *In situ* pressure injection of preservative is possible where existing joinery is at risk of decay. The blowlamp provides the quickest and most effective way of removing paint from wood, but where it cannot be used, as against glass, paint removers of the organic solvent type (BS 3761) are recommended if care is taken to remove all traces of paint remover. The bare wood should be rubbed smooth with abrasive paper and the painting process continued as for new work—knotting, priming, stopping, undercoat(s) and finishing coat. Some manufacturers recommend a four-coat system consisting of a primer, undercoat and two finishing coats. Conventional alkyd resin paints form the major type of exterior finish on woodwork, although their elasticity is lost on ageing. Claims were being made in the mid 1980s for paints with greater permeability and flexibility, but usually accompanied by reduced gloss (breathing or microporous paints).[28]

New or bare exterior woodwork should be primed as soon as possible to protect it from the weather, but damp surfaces are best left unpainted until dry. Where timber has been left exposed to the weather, any loose grain must be removed by thorough sanding, scraping or planing. On one site the prolonged period during which primed timber remained exposed to the weather (up to 12 months) was considered a contributory factor in the early breakdown of external paintwork on new dwellings. Paint must not be applied on top of creosote, bitumen or tarry materials as they are likely to bleed through the new paint and discolour it. One approach is to scrape the surface well and to seal it with two coats of aluminium wood primer or with one or two coats of good-quality shellac knotting before applying the new undercoat. Another alternative is to apply one coat of thinned knotting followed by one coat of aluminium primer.

Table 5.6 details painting defects on woodwork and possible causes.

Clear *varnish* can give an attractive finish to exposed external woodwork where a natural effect is required. At least four initial coats are essential. For maintenance thoroughly clean all surfaces, strip any brittle or flaking areas with a solvent remover and wash down with clean white spirit, stop cracks, holes and open joints, and brush two coats of varnish on the dry surfaces. Varnishes become brittle by weathering and ultra-violet rays can bleach or degrade the underlying wood surface.

Exterior wood stains, mainly high solid, are being used increasingly as alternatives to paint. They are less durable but their renewal is easier and cheaper. They do not, however, avoid the need for preservative treatment, hide knots or imperfections, or protect putty, so they require bead glazing. By permitting a greater range in moisture content, especially where dark stains are exposed to sunlight, they allow greater dimensional changes and sometimes splitting of the timber.[28]

Repainting Metalwork

Repainting metalwork should not be delayed beyond the appearance of the first traces of rust. This avoids the more costly work later of removing all rust and paint. The old paint surfaces can be rubbed down and finished with one or two suitable coats. Any very small patches of rust can be removed and touched in with an inhibitive primer. Complete removal of the paint followed by suitable surface preparation is necessary if rust covers more than 0.5 per cent of the area. Old paint is often difficult to remove. Solvent or alkaline paint strippers are reasonably effective provided their residues are removed with white spirit or large quantities of water. Alternatively a

Table 5.6 Painting defects on woodwork

Painting defect	Possible causes		
	Preparation of wood	*Application*	*Exposure*
Blistering	Damp or unseasoned wood; liquid or vapour beneath coating. Knots not properly treated	Paint poorly applied	Excessive heat
Peeling Poor adhesion Flaking	Damp or unseasoned wood. Surfaces not properly cleaned, powdery or friable	Paint poorly applied, particularly the priming coat, or omission of it or use of unsuitable primer. Too long between coats.	
Irregular cracks	Damp or unseasoned wood. Paste or size left on wood	Hard drying paint applied over soft coatings	Excessive heat
Chalking Powdering	Early coats in system may have failed to satisfy porosity of substrate	Could be incorrect or unsuitable formulation	Lengthy exposure, possibly severe
Insufficiently opaque ('grinning') Colour uneven	Surface not cleaned	Paint poorly applied. Undercoat too thin or uneven	
Resin coming through	Knots not properly treated. Very resinous knots not cut out	Too much resin for knotting to seal	Resin softened by heat
Delayed drying Uneven drying	Surface not properly cleaned. Residues of paint removers. Painting on creosote without sealing	Finish applied before undercoat completely dry	Damp, cold or frost

Table 5.6 (cont'd)

Painting defect	Possible causes		
	Preparation of wood	Application	Exposure
Discoloration	Painting on creosote or bitumen without sealing. Surface not properly cleaned. Knots not properly treated. Stains from various species of timber, either as solid, ply or veneer. Pigment dyestuffs, generally reds or maroons, melt when burned off, are absorbed into the wood and discolour the paint system, unless sealed	Possible use externally of colour intended for internal use	Lengthy exposure to bright sunlight. Chemical attack on yellow, where direct daylight is excluded
Poor gloss	Alkaline materials left on wood	Paint poorly applied. Still, cold air in unheated rooms	Damp, fog or frost; lengthy exposure
Bloom—hazy or white appearance on gloss		Gas or paraffin heating during drying	Condensation
'Sinkage'—patchy low gloss		Unsuitable undercoat. Finish applied before undercoat sufficiently dry	
'Sheeriness'—uneven gloss		Paint not properly stirred. Failure to maintain a wet edge. Vigorous brushing of matt paint in dark colours	
'Wrinkling'—loss of gloss	Can also be caused by excessive retention of preservatives or their solvents, owing to insufficient drying out	Paint dries too quickly. Skin-drying of paint film. Mixing resin-based paints with oil and other paints. Adding unsuitable solvents	Excessive heat

blowlamp, scrapers and wire brushes may be used. Angles, crevices, bolt heads and rivets must be cleaned with special care. More efficient cleaning processes include blast cleaning, flame cleaning and acid pickling.

Severely rusted metal window frames must have all putty and glazing removed to allow thorough cleaning by brushing and scraping. Two coats of zinc chromate, zinc phosphate or zinc-rich primer should be applied before reglazing and puttying. The putty should be bevelled to shed water from the joint and the exposed primer given undercoats and exterior quality finishing paint, even inside buildings.

Paint Maintenance Practice

The author carried out a paint maintenance survey on a national basis in 1983[27] and found that standards of painting maintenance in general were deteriorating significantly, creating serious problems. For example, it is generally recognised that the frequency of repainting is influenced by climatic conditions, atmospheric pollution, degree of exposure and condition of substrate. Most property managers aimed for the repainting of external gloss surfaces two to three years from the initial painting and at four to five year intervals thereafter. In practice the period between painting cycles had often been extended to 6 to 9 years. Local authorities suffered from substantial budget cuts and many private property owners afforded painting a low priority, as their primary concern was profits and productivity. These delays in painting resulted in expensive bills for joinery repairs and replacements preparatory to painting, which in badly neglected situations amounted to as much as six to fifteen times the cost of the painting.

In many cases the failure of components at an early stage in the life of the building, which had been attributed to paint failure, was in fact due to the poor initial quality of the substrates and the lack of necessary protective measures. Another major weakness was the failure to monitor and enforce the specified painting cycles in full repairing leases of properties. To prevent the decay of building components, it is essential that repainting takes place before the existing paint film begins to break down. Saving on painting and decoration is frequently false economy as it is merely storing up much greater and more costly problems for the future.

One of the greatest weaknesses in painting work was found to be the general lack of attention to surfaces to be painted and satisfactory application methods. All too common were failures to remove all loose and flaking paint, to burn off existing wood surfaces where paint had broken down, to clean adequately the surfaces to be painted, the omission of sealing to knots, the lack of a good coat of suitable primer on all bare surfaces, the failure to seal all cracks and holes with an appropriate stopping material, painting in damp conditions and on wet surfaces, the failure to lightly rub down between coats where appropriate, the mixing of incompatible paints, the omission of a specified undercoat or finishing coat and the excessive use of thinners. These faults, which demonstrate an overall decline in standards, drastically diminish the effectiveness of painting systems and result in premature paint failures.

Repainting Plastered Walls and Ceilings

Emulsion paints have superseded oil-bound distemper as a finish to plastered walls and ceilings with their wider range of finishes and better resistance to washing. They dry rapidly and two or more coats can be applied in one day. Emulsion paints can be applied to most surfaces but the manufacturer's advice should be sought for use externally, on wood or metal, or when condensation is present. Rollers, airless spraying and other modern techniques should be used as far as practicable, to reduce the labour component of maintenance work. Matt finishes reduce the reflection of light sources and minimise the effect of surface irregularities, but they are less suitable than gloss finishes when high wear resistance, ease of cleaning or maximum hygiene are desired.[29]

Smooth surfaces should be rubbed down to a matt surface prior to painting. If the surface is soft or powdery or has been decorated previously, all loose material must be removed by washing and brushing off, while soft size-bound distemper or ceiling distemper must be washed off completely. Dirty surfaces must be thoroughly cleaned. Before applying emulsion paint over wallpaper it is advisable to start with a small area, in case the paper contains colours which bleed into the paint. Emulsion paint is normally applied in two coats with a wall brush, flat distemper brush up to 150 mm wide, 100 mm varnish brush, roller or spraygun, and should be thinned only with clean water.

Surfaces that have previously been painted with emulsion paint may be redecorated with other materials provided that the emulsion paint is sound and adhering well. It should be rubbed down wet before redecorating and then allowed to dry before applying oil paint. A coat of size may be needed after rubbing down and before hanging wallpaper. Emulsion paint which is not adhering satisfactorily may be removed by scraping or wire brushing.

Surface defects most likely to affect painting will be those caused by dampness from direct rain penetration, rising damp or condensation, and the only really satisfactory remedy is to stop the damp penetration. A compromise solution is to use a permeable paint which will allow salts or moisture to escape. Slight efflorescence occurring after a first decoration can usually be cured by applying an alkali-resistant primer. Where extensive making good of plaster is necessary, this may result in areas which are alkaline or of different absorption from the remainder, and it is best to coat the whole wall with alkali-resistant primer, even when using emulsion paint, for safety and uniformity of finish. Tarry or sooty stains on chimney walls cannot be sealed by paints although lead foil is sometimes satisfactory. If the staining is extensive the plaster may have to be removed and recurrence can be prevented by installing impervious flue linings. Persistently damp conditions encourage mould growth and improved ventilation is often a better remedy than fungicidal paints. A comprehensive list of defects on painted walls with their causes and cures is given in BRE Digest 198.[30]

Painting of Masonry and Asbestos Cement

It is advisable to use alkali-resistant paints for painting masonry and asbestos cement. A less satisfactory alternative is to apply one or two coats of alkali-resistant primer followed by a normal paint system such as oil gloss paint. With asbestos cement, porous paints are preferable as they allow the material to breathe with less risk of damage. Suitable paints include emulsion paints based on alkali-resistant polymers. Alkali-resistant paints, such as those based on chlorinated rubber, are also suitable for use on brick, stone and concrete walls. Asbestos cement should not be painted with an impervious finish on one side only, since differential carbonation may cause warping or cracking; hence back-painting with the same paint or a cheaper impermeable one is necessary.[29]

Pattern Staining

Pattern staining consists of dark and light patterns which appear on plaster surfaces. The pattern on a lath and plaster ceiling forms virtually a complete replica in light and shade of the lathing and joists. It may also occur on the soffits of hollow tile floors, on ceilings of wallboard or plasterboard nailed to joists, on hollow block partitions, on frame walls and on wall linings or coverings where these are fixed to battens.

Some dirt or dust from the air is bound to accumulate on walls and ceilings, and with pattern staining it stands out clearly and spoils the appearance of the wall or ceiling long before the general darkening of the surface would make redecoration necessary. It is caused by the cooler surfaces receiving a greater amount of dust. Wood laths are poor conductors of heat and the heat flows more slowly through the lath and plaster than through plaster alone, and so the areas under the laths collect more dust; similarly with upper floor and ceiling joists.

Various methods can be used to remedy this defect.

(1) To make the surface warmer than the air by choosing a form of heating that will keep the wall or ceiling surface warm. Radiant heating systems are preferable to convection systems.

(2) To maintain a uniform temperature over the whole surface of the wall or ceiling by (a) selecting a suitable type of plaster; (b) adding insulation locally at points where heat flow is high, or (c) adding insulation over the whole structure to obtain a general reduction in heat flow.

Insulating materials which have been used with success include 20 mm of slag wool, 25 mm glass silk and 15 mm insulating board, each on brown paper.

Glazing

Prior to glazing, rebates should be cleaned and primed. Glass should be cut to allow a small clearance at all edges and then be back-puttied, by laying putty along the entire rebates and bedding the glass solidly, sprigged for timber rebates and pegged for metal rebates, and neatly front puttied, taking care to ensure that putty does not appear above the sight lines.[22] The putty is designed to prevent the passage of air, dust and moisture past the glass and consists of linseed oil putty to BS 544 for wooden frames and metal casement putty for metal and non-absorbent hardwood frames. As linseed oil putty sets, some absorption takes place through the primer into the wood frame, and the putty shrinks.

The putty requires protection with a minimum of two coats of paint as soon as it has set sufficiently to receive it. The undercoat can usually be applied within 7 to 14 days after glazing depending on atmospheric conditions and size of fillet, and the final gloss coat of paint should be applied within 28 days of glazing and extend about 2 mm beyond the inside edge of the face putty to prevent rain eroding the putty edge. Aluminium frames should be treated with zinc chromate primer to ensure effective adhesion of the putty. When reglazing, all old glass and putties must be hacked out and the rebates thoroughly cleaned prior to inserting the new glass. The majority of glass used is of ordinary glazing quality (OQ) to BS 952. For panes exceeding 1 m^2 clear sheet glass should be at least 4 mm thick.

Considerable wilful damage to glass in large panes and vulnerable positions, such as in doors to lock-up garages, has occurred on housing estates. In some cases it would be better to use transparent plastics instead of glass to reduce the risk of breakage.

Double Glazing

With the rising cost of heating fuels and the desire to reduce condensation, there is an increasing demand for the double glazing of existing windows. One common arrangement is to fix a second line of glazing with wood or plastic face beads to existing frames. As described in chapter 4, no matter how well the glazing seals are made, the cavities cannot be expected to remain airtight indefinitely. Another approach is to fix separate secondary windows.[31]

References

1 BRE Princes Risborough Laboratory. *Technical Note 24: Preservative treatments for external softwood joinery timber* (1982)
2 BRE Princes Risborough Laboratory. *Technical Note 29: Ensuring good service life for window joinery* (1974)
3 *BRE Digest 73.* Prevention of decay in external joinery (1978)
4 *BRE Digest 304.* Preventing decay in external joinery (1985)
5 A. Oliver. Can paint call a halt to development of rot? *Chartered Surveyor Weekly* (2 August 1984)
6 *BRE Digest 262.* Selection of windows by performance (1982)
7 *BRE Digest 175.* Choice of glues for wood (1975)
8 British Standards Institution. *BS 6566: Plywood. Part 7: 1985 Specification for classification of resistance to fungal decay and wood borer attack. Part 8: 1985 Specification for bond performance of veneer plywood*
9 *BRE Defect Action Sheet 69.* External walls: joints with windows and doors—application of sealants (1985)
10 R. Sinnott. *DOE Construction 8: Quality of surface finish in new homes.* HMSO (December 1973)
11 E. D. Mills (Ed.). *Building Maintenance and Preservation.* Butterworths (1980)
12 *BRE Digest 301.* Corrosion of metals by wood (1985)
13 *BRE Digest 69.* Durability and application of plastics (1977)
14 E. J. Gibson (Ed.). *Developments in Building Maintenance—1.* Applied Science Publishers (1979)
15 British Standards Institution. *BS 6262: 1982 Code of practice for glazing for buildings*
16 *BRE Digest 224.* Cellular plastics for building (1979)
17 *BRE Digest 213.* Choosing specifications for plastering (1978)
18 *BRE Digest 196.* External rendered finishes (1976)
19 *BRE Defect Action Sheet 38.* External walls: rendering—application (1983)
20 *DOE Construction 4: External wall tiling with cement:sand bedding.* HMSO (December 1972)
21 *DOE Construction 5: External wall tiling with adhesives.* HMSO (March 1973)
22 I. H. Seeley. *Building Technology.* Macmillan (1986)
23 British Standards Institution. *BS 6150: 1982 British Standard code of practice for painting of buildings*
24 Local Government Operational Research Unit. *Report C144.* Hospital building maintenance—can decision making be improved? HMSO (1972)
25 Local Government Operational Research Unit. *Report D2.* How often should you paint? HMSO (1970)
26 DOE. *Costs in Use: A Study of 24 Crown Office Buildings.* HMSO (1971)

27 I. H. Seeley. *Blight on Britain's Buildings: A Survey of Paint and Maintenance Practice*. Paintmakers Association (1984)
28 *BRE Digest 261*. Painting woodwork (1982)
29 *BRE Digest 197*. Painting walls. Part 1: Choice of paint (1982)
30 *BRE Digest 198*. Painting walls. Part 2: Failures and remedies (1984)
31 *BRE Digest 140*. Double glazing and double windows (1980)

6 BUILDING MAINTENANCE PROBLEMS AND THEIR SOLUTION—V

Plumbing, Heating and Hot Water Supply, Air Conditioning, Electrical Installations, Gas Installations, Lifts, Refuse Collection from Flats, Drainage, Safety, Security, Fire Resisting Construction and Fire Precautions, Cleaning, Pest Infestation and Repair of Flood Damage

Plumbing

Building services are costly maintenance items and their lives are usually much less than those of the buildings which accommodate them. Hence particular care should be taken in the selection, design and installation of these services, to ensure that maintenance can be carried out easily, quickly and economically. All services should be readily accessible with adequate access and working space provided. This becomes all the more important with tall buildings, as was evidenced by the lack of initial consideration in the United Nations Building in New York, with unfortunate consequences.

General Faults

The composition of water varies as between different areas of the country and some waters tend to corrode certain metals when used alone or in combination with other metals. It is preferable to keep to the same metal throughout a water supply system wherever practicable. There have been cases of premature failure of galvanised steel pipes laid in clay soils caused by anaerobic bacteria. Suitable protective measures include a reinforced bitumen coating or surrounding the pipes with 225 mm of sand or gravel. In 1974 the Department of the Environment warned of the dangers of cheap lead-based solder in the joints of copper water pipes polluting drinking water, but there was little evidence of any action being taken in respect of the thousands of dwellings involved in the mid 1980s.

Pipes need to be securely fixed at intervals not exceeding 1.2 m on horizontal runs and 1.8 m on vertical runs to give adequate support. All joints should be made in accordance with best practice. Many failures of capillary joints on copper pipes have resulted from insufficient preparation—grease preventing adherence of the solder. Plastics pipes of polythene or unplasticised PVC (uPVC) are now being used for cold water supplies and should comply with BS 3284 for high density polythene, BS 1972 for low density polythene and BS 3505 for uPVC, which are stronger and more rigid than polythene.

Concern has been expressed at the high incidence of trouble with ball valves and leaking taps and it has been suggested that a study of the number and cause of failures would assist in determining the design improvements that are most urgently needed. Most overflow pipes discharge in con-spicuous positions so that the defect is soon noticed and remedial action taken. On occasions however overflow pipes discharge internally into fittings or hose pipes are taken from external overflows to take any overflow water into the nearest gully. In high-rise buildings, there is a possibility of winds carrying away the overflowing water so that it goes undetected. Overflow pipes must always be of a larger diameter than inlet pipes to avoid water overflowing into the building if the valve becomes stuck in the fully open position.

A ball valve may fail to close for one of several reasons—perforated float, eroded seating, defective washer or the presence of grit or lime deposit. Copper floats, especially when soldered, may become corroded resulting in breaking away or perforation of the float; this will not occur with plastics floats. High velocity discharge of water from a ball valve may erode the seating and cause leaks. The remedy is to install a new seating, preferably of nylon, or to reseat the valve with a special tool. Worn washers require replacement and the seating should be inspected at the same time, as it may be the cause of the trouble. With hard water, corrosion coupled with lime deposition may cause the valve piston to stick in the open position; where this is repeated, the valve should be periodically dismantled and cleaned, and the piston greased. Grit may be a similar source of trouble, particularly in newly built houses where the water system has not been thoroughly flushed.

Sticking of the valve in the closed position may occur with unoccupied houses where dirt and lime have dried out on the working parts. This can be remedied by moving the float up and down a few times or, better still, by dismantling the valve and cleaning it thoroughly. The splashing of ball valves can often be reduced by fitting a silencing tube or drown pipe to the valve provided this is permitted by the Water Authority, as there could be a risk of back-siphonage. A number of these problems can be avoided by using a quieter type of ball valve, such as the diaphragm variety.[1]

Leaking taps waste water and are a nuisance, particularly in baths, where they cause stains. Taps leak when the washer becomes worn or when metal seatings are used and become eroded. Black synthetic rubber washers are suitable for either cold or hot taps and have a long life. It is important that all stop valves shall be in good working order.

Pipes and fittings need to be accessible after installation for purposes of examination, repair, replacement and operation. Some cold water tanks in flat-roofed buildings are placed so close to the ceiling, to gain height, that it is impossible to adjust the ball valve let alone change it. Some service cores become so congested that it is not always possible to reach important stopcocks and valves.

Exposed pipework provides maximum accessibility but is often resisted on aesthetic grounds, although it may not be too objectionable if fixed to follow skirtings, architraves and similar features. Concealment in cupboards provides accessibility and eliminates problems of appearance. Pipework installed in ducts in solid floors which require breaking up to reach a leaking joint cannot be regarded as reasonably accessible. Pipes below solid floors should preferably be laid in sinkings in the floors covered with access panels or be encased behind skirtings set forward from the wall face. Vertical pipes may be fixed in recesses in walls faced with removable panels.

An investigation into service installations in high-rise flats in Singapore found that all too frequently there was inadequate space around service installations to satisfactorily maintain pipes and associated equipment and insufficient control valves, so that large parts of the system needed to be shut down for maintenance purposes.

Frost Precautions

Extensive damage to plumbing systems can occur in severe winters unless adequate precautions against frost are taken. Surveys of frost damage have shown that most frost damage is avoidable and that the greatest source of damage is outside waterclosets and washhouses, followed by internal plumbing and rising mains. Essential frost precautions are as follows:

(1) Run pipes in safe places wherever possible giving external pipes at least 750 mm cover and siting them in ducts where they approach the ground floor. Internal pipes should preferably be located on inside walls with appliances grouped to keep pipe runs as short as possible.

(2) Fix pipes to proper falls to permit emptying when the building is left unoccupied and unheated, with emptying cocks at all low points.

(3) Protect all pipes in vulnerable places by suitable lagging.

Unheated out-buildings such as WCs and garages are particularly vulnerable, and any pipes and tanks installed in them should be well lagged and stopcocks fitted to cut off the supply in the event of a burst, while outdoor standpipes require adequate lagging with a waterproof covering. Unheated bathrooms can also be a source of trouble and wherever practicable some heating should be provided by means of an adjoining domestic boiler or hot water cylinder, radiator or heated towel rail. As far as possible, pipes and tanks should not be located in the roof space, but where this does occur, the storage cistern should be placed immediately above the hot water cylinder and be suitably insulated on all sides except the bottom, often with expanded polystyrene panels. All pipes in the roof space, except warning or overflow pipes, should be lagged or fixed below the ceiling installation. Pipes fixed tightly to external walls and under boarded ground floors are liable to frost damage unless protected.

Occupants of buildings should be advised of any frost precautions that they need to take and the following recommendations are applicable to occupiers of dwellings.

(1) Know location of main stopcock and be sure that it operates easily and closes tightly.

(2) Know location of all pipes and provide protection in draughty places.

(3) Pipes must be lagged where the temperature cannot be kept above freezing point.

(4) Before leaving a house unoccupied in the winter months, turn off the water supply preferably at the main and drain off the whole system. This may not be necessary where a heating system continues to operate on time control and the dwelling is unoccupied for a short period only.

(5) To reduce the likelihood of WC taps freezing when a dwelling is left unoccupied and unheated, dissolve a handful of salt in half a litre of hot water and pour into the pan.

(6) Attend to any dripping taps or leaking ball valves at once before they cause trouble. The water authority may rewasher cold water taps free of charge.

(7) Ensure that windows near WC cisterns and in other vulnerable locations are kept closed and are draught-proof at nights when heavy frost is anticipated.

(8) Where no frost precautions have been taken and the dwelling has been unoccupied during a spell of severe frost, refrain from switching on the immersion heater or lighting the boiler until sure that the system is completely free of ice.

(9) Where there is severe leakage from burst pipes, switch off the electricity supply at the main.

Storage Tanks

Many failures of galvanised steel tanks have occurred where they have been used in association with copper pipes. Certain types of water are capable of dissolving minute particles of copper from the pipes of a hot water system. When the water comes into contact with the galvanising of the tank, some of the copper is deposited on the zinc and an equivalent amount of zinc is dissolved. Electrolytic action between the copper and zinc causes rapid attack on the zinc coating and ultimate perforation. Rusting of the unprotected steel then takes place and before long the tank begins to leak.

In general a hot water system is best constructed throughout of copper, unless previous experience in the district shows it to be safe to use these materials together. If there is any doubt about a galvanised steel cold water tank, it is best to paint the internal surfaces with a non-toxic bituminous composition, which will form a protective coating and so reduce the electrolytic action. Other alternatives are to use tanks of plastics or glass fibre or to fit a sacrificial anode consisting of an aluminium block earthed to the steel tank which may extend its life. The tank should be adequately lagged at top and sides.

The BRE[2] have found cases where trussed rafter members have been distorted because the tank bearers were incorrectly sized and positioned, pipe joints disturbed and chipboard tank platforms have become wetted by condensation and collapsed. Moisture damage has also been caused by overflow pipes having too small a bore, sagging or having inadequate slope.[3]

Plumbing Noises

The avoidance of plumbing noises is often a matter of good planning and design and the choice of quiet appliances. Sanitary accommodation should ideally be separated from living accommodation but this may prove difficult in blocks of flats and the conversion of existing properties to separate dwelling units. The dividing wall between a bathroom or WC and a bedroom should not be less than 75 mm concrete block, plastered both sides. The door to a bathroom or WC should be as heavy as possible (solid core if flush doors) and well fitting.

In existing dwellings where bathrooms or WCs are located over main rooms with wood joist floors, it is advisable to introduce as much pugging as the ceiling will take and possibly to reconstruct the floor as a floating floor to give maximum insulation. In the latter case a flexible rubber joint will be required between the WC and the soil stack. The flushing cistern should be isolated from the wall by resilient pads of thick cork or rubber.

A WC flushing cistern generates noise in two ways—(1) from the flushing mechanism, and (2) from the ball valve. The quietest form of flushing mechanism is the siphonic type with a cistern of thick vitreous china or plastics, but this will increase cost. A ball valve of the Skevington/BRE type will reduce noise and wear.[4] The use of low-level flushing cisterns and insulation of internal stack pipes assist in reducing plumbing noises.

When water flowing in a pipe is suddenly stopped by the rapid closure of a valve or tap, the pressure causes a surge or wave which rebounds from the valve and passes back down the pipe. The loose washer plate or jumper on the valve oscillates giving a knocking sound known as *water hammer*. The resultant pressure in small rigid pipes may damage them. In cisterns fed through 12 mm pipes, the float actuating the ball valve should have a minimum diameter of 150 mm to avoid oscillating.

Sanitary Pipework

Most modern plumbing arrangements use the single stack system which is well described and illustrated in BRE Digest 249.[5] In single stack plumbing all forms of appliance discharge into multi-use discharge pipes or stack pipes, which also have air admittance valves. A bath waste must enter the stack above or at least 200 mm below the entry of a WC branch. Wash basins, baths and sink wastes should have 75 mm deep seals to guard against self-siphonage, and WCs must have a 50 mm deep seal. The maximum slope of a 32 mm wash basin waste depends on the length of waste pipe and where it exceeds 1.70 m in length it should be provided with a 32 mm diameter trap with a short 32 mm tail pipe discharging into a 40 or 50 mm branch pipe. Another alternative is to ventilate the branch pipe. The length and slope of bath and sink wastes are not critical but long wastes may become blocked by sediment settling out of the waste water and access for cleaning should be provided. Approved Document H1 of the Building Regulations 1985 recommends maximum lengths of 3 m for 40 mm branch pipes and 4 m for 50 mm branch pipes.

The bend at the foot of a stack should be of large radius (at least 200 mm at the centre line), or two 135° fittings should be used. Sanitary appliances on the lowest floor are best separately connected to the drainage system, particularly with high-rise buildings. The minimum vertical distance between the lowest branch connection and the drain invert should be 450 mm for three-storey houses with a 100 mm stack and two-storey houses with a 75 mm stack. Ample provision should be made for access, particularly at or near bends. Leaking pipe joints may result from cracked pipes, unsatisfactory joints or inadequate support.

Where appliances are widely spaced, as in a hospital, it may be impracticable or even impossible to use a single stack method. BRE Digest 81[6] describes how pipe systems in older hospitals often take tortuous routes with only limited access, stemming from lack of co-ordinated planning of different services, both with each other and with the main structure. Most stoppages occur where the pipework is complicated by knuckle bends, sharp offsets and $92\frac{1}{2}°$ junctions—clearance is expensive, can cause disruption of the hospital and may be a possible cause of cross infection. Pipework is sometimes blocked by disposable items, including plastic syringes and spatulas, which should be disposed of in ward incinerators. Access is often inadequate, badly sited and inconvenient in use. Planned maintenance is recommended, including regular cleansing of grease traps, and this requires adequate drawings showing the drainage layout and access points.

Appliances

All sanitary appliances should be durable, smooth, non-absorbent, non-corroding, largely self-cleansing, of simple design and construction, accessible, economical and of satisfactory appearance. Modern developments have seen the frequent replacement of wash basins by vanitory units, the fitting of waste disposers to sinks, and the increased provision of bidets and showers.

WC flushing cisterns should fill within 2 minutes and, to secure this, the size of the valve orifice must match the supply pressure. Hence with a very small head a 9 mm orifice may be necessary to give the required flow. Persistent slow filling may also be due to partial blockage of the valve by foreign matter which requires clearing. Inadequacy of flushing may result from the cistern not being filled to the prescribed water line. A perforated float partially filled with water needs replacing, otherwise careful bending upwards of the ball valve arm will remedy the defect. Corroded and ineffective old cast iron cisterns containing a cast iron bell lifted by a chain pull should be replaced with a modern cistern containing a piston actuated siphon, for more efficient and quieter operation.

Inadequate flushing may also be caused by an obstruction in the flush pipe, often of jointing material, or by the flush pipe entering the pan socket at an angle. The trouble may arise in the pan itself where, for instance, the flush of water is confined to one side of the pan. Pans with siphonic discharge are more positive in their action as the contents are pulled out by the suction created behind the trap. For maximum quietness and efficiency, a close-coupled double trap siphonic suite fitted with a Skevington/BRE ball valve is recommended.[4] Modern closets may be fitted with a macerator and pump connected to a small-bore drainage system discharging to a discharge stack.

Wash basins of vitreous china are preferable to those made from fireclay since they are stronger and less absorptive. Overflows are usually difficult to clean and dirty water rising in the overflow passage, when the waste plug is lifted, leaves behind an unpleasant scum which can produce unhealthy conditions. For this reason some housing authorities are specifying basins, baths and sinks without overflows.

Showers require a minimum head of 900 mm between the bottom of the cold water cistern and the shower rose to function satisfactorily. Where this head is not obtainable, the cold water cistern may be raised above the level of the roof or a pump may be installed to boost the pressure. Showers should be sited in rooms that are adequately ventilated to reduce condensation.

Heating and Hot Water Supply

Hot Water Supply

Defects in hot water supply systems may stem from one or more of the following factors:

(1) *Air locks.*   When there is air in the system and on heating the water the air is released and rises to the highest point. Ideally the pipes should rise towards a vent point at a slope of 1:120 and air then escapes through the vent. Where pipes contain dips or fall in the reverse direction, air becomes trapped and impedes the flow of water. Trapped air is released by draining and refilling or by blowing through the pipework.

(2) *Insufficient hot water.*   This can be caused by the inadequate size of boiler or hot water cylinder, excessive length of primary flow and return pipes, poor quality fuel, air locks, insufficient lagging of pipes and tanks, or possibly a combination of these defects.

(3) *Noises.*   A particularly troublesome noise is the knocking which may occur in primary flow and return pipes or in the boiler, resulting from expansion of water by freezing, furring or corrosion, and possibly involving descaling or renewal of pipes. Knocking in cold water systems is termed water hammer and usually results from faulty valves as described earlier in this chapter.

(4) *Poor flow.*   This can stem from air locks, insufficient head of water or air drawn into the system through a vent. The latter defect can be remedied by inserting a larger cold feed pipe or raising the storage cistern.

Back boilers for heating water are prone to leaks at about 15 to 20 year intervals and need checking, as do also immersion heaters and their thermostats. The thermostat is checked by switching off everything, setting the thermostat low and then letting the water heat up. When the meter stops, the thermostat is turned up and a check made to determine whether it has restarted. Feeling the water provides a crude temperature measurement.

Modernising Plumbing Systems

The modernising and conversion of old dwellings often includes the provision of or improvement of hot water services which will qualify for a grant when funds are available. It may at the same time or at some future date be decided to extend the system to include space heating, when it is advisable to install an indirect system to avoid radiators running cold when domestic water is drawn from the tank, a slower rate of circulation of heated water, and furring of the hottest parts of the system. In severe cases lime deposits may cause disruption of cleaning plates and gaskets on boilers and the blocking of pipes and bends near boilers.

Heating Systems

The principal task of the building services engineer is to create a comfortable and stimulating indoor climate. Building services have become a major item in the cost of new buildings and are also costly to maintain and operate, probably totalling nationally over £3000 million per annum, although there is a lack of systematic data on maintenance and energy costs. Both maintenance and energy costs for services are influenced by the capacity of the plant, the extent of use, and design aspects such as floor area, amount of glazing and quality of internal environment. The extent to which regular cycle maintenance or even periodic replacement of some components is justified by the avoidance of breakdowns, needs careful evaluation. Inefficient hot water heating systems may result from radiators or boilers which are of inadequate capacity.

A common problem with solid fuel independent domestic boilers has been down draught resulting from chimneys terminating below the ridge line. Rigid joints between appliance flue pipes and brick or block flues often fracture and are best remade with a more resilient material such as asbestos rope or cord. Ashes from solid fuel boilers are not always removed as frequently as they should be with resulting deterioration of firebars. Continual changes in the design of heating appliances and the multiplicity of small parts create problems in replacement. A reduction in the number and types of appliances and standardisation of parts would reduce maintenance costs.

The principal defects in boilers have been identified as noises resulting from inadequate water flow or ineffective design, air entrainment through leaks, and scale formation by fresh water inflow to replace water lost by air displacement or leakage. The operating efficiency of boilers has increased at the expense of the adequacy of the materials used to cope with more exacting conditions, requiring more frequent replacement, possibly at about 15 year intervals.

Old central heating systems often include cast iron radiators, which, although relatively inefficient, are uneconomic to replace, since it would take many years before the cost of removal and replacement could be recovered out of reductions in fuel costs. Where hot water and heating are combined in a single system, independent supply of hot water is required outside the heating season and control of priority is useful within it.

Radiators and pipework should be protected from chemical corrosion by an inhibitor. Most small-bore heating systems incorporate pumps, to secure quicker circulation, and these require occasional servicing in the same way as boilers. Self-priming cylinders can cause problems, and a separate header tank and indirect cylinder give better performance. One-pipe heating systems have the principal disadvantage that radiators along the system become progressively cooler. Micro-bore systems are being increasingly used as they occupy less space, but need careful design.[7]

Unvented Hot Water Supply Systems

As an alternative to the direct and indirect vented hot water supply systems, Schedule 1 to the Building Regulations 1985[8] (paragraph G3) covers unvented hot water supply systems, which have been used extensively in Germany, the United States and other countries. It requires that if hot water is stored and the storage system does not incorporate a vent pipe to the atmosphere, there shall be adequate precautions to: (a) prevent the temperature of the stored water at any time exceeding 100° c; and (b) ensure that the hot water discharged from safety devices is safely conveyed to where it is visible but will cause no danger to persons in or about the building.

A typical unvented hot water system based on German practice embraces the following arrangements. All cold water outlets are supplied direct from the mains through a check valve, a meter and, if mains pressure is high, a pressure-reducing valve. The hot water supply is directly fed, and a second check valve is fitted in the feed pipe. Because this check valve prevents water expanding into the cold feed pipe when a charge of water is heated, it is necessary to provide a pressure relief valve and a drain to allow expansion water to leak away. Similarly a temperature-operated energy cut-out is normally used to disconnect the supply of energy. This ensures there will be no explosion if a thermostat failure results in uncontrolled heating of the water.[9]

Approved Document G3 of the Building Regulations contains the following recommendations with regard to such systems. Any unvented hot water storage system should be in the form of a proprietary Unit or Package which is the subject of a current British Board of Agrément (BBA) certificate. A unit or package comprises a system incorporating a minimum of two temperature-activated devices operating in sequence: a non re-setting thermal cut-out to BS 3955 and a temperature relief valve to BS 6283, in addition to any thermostatic control which is fitted to maintain the temperature of the water. It will also include any necessary operating devices to prevent backflow, control working pressure, relieve excess pressure and accommodate expansion fitted on the unit by the manufacturer.

Other Aspects

Schedule 1 to the Building Regulations 1985[8] (paragraph L5) requires hot water storage vessels to have adequate thermal insulation, and also hot water pipes, unless they are intended to contribute to the heating of a part of the building which is insulated or they give rise to no significant heat loss. For example, insulation material should have a thermal conductivity of not greater than 0.07 W/m K and a thickness equal to the outside diameter of the pipe up to a maximum of 50 mm. Heat pumps, using outdoor ambient air as their heat source and supplying hot water as the heating media have energy and cost-saving potential as described in BRE Digest 253.[10]

BRE Digest 205[11] describes the practical aspects of using flat plate solar collectors for augmenting the heating of domestic water in the UK, while Digest 254[12] analyses the reliability and performance of solar collecting systems and gives guidance on the methods of checking system operation, in view of the failures that have occurred in practice.

Air Conditioning

Air conditioning comprises filtration, heating, ventilation, cooling and dehumidification by mechanical means, to provide a comfortable environment for occupants and acceptable conditions for specialised activities, such as computer suites, hospital clean rooms and sophisticated manufacturing processes. Air conditioning is becoming more common as some of the accompanying advantages are more generally recognised—more uniform temperature range, healthier environment, more alert staff, reduced cleaning and redecoration costs, and less external noise. The design of air conditioning equipment is also changing and packaged equipment can be located in corridors, above false ceilings, on the roof or even in the conditioned space itself. Furthermore, the equipment is being designed to permit longer periods between maintenance visits.

For instance, roller bearings have replaced sleeve bearings in most fan drives—they are cheaper in first cost, reduce noise and do not require attention more than once a year and with small plants may be sealed for life. Refrigeration compressors are often in sealed hermetic units to be replaced in the event of plant failure. A low cost sealed unit of limited life is a better buy than a longer life slow turning open drive machine which can be serviced.[13]

With filters the rate of media replacement depends on the extent of pollution in the environment. There are two main types of filter—the throwaway and the cleanable. As a rough guide filter cells are changed or cleaned once a month in heavily polluted locations, once every two months in built-up areas and city centres and once every three months in rural or suburban areas. The life of the cleanable filter is about $1\frac{1}{2}$ to 2 years. Modern filter developments for the more complex installations include rolltype and electrostatic filters. In practice throwaway filters have proved the most acceptable as they can be regularly maintained by relatively unskilled persons. Probably the most common cause of air conditioning plant failure is blocked filters, resulting in reduced air flow which can be followed by freezing of refrigeration plant and reheater batteries, and breakdown of fan bearings causing downflow of cold air with consequent occupancy discomfort. Condenser coils need cleaning each spring by spraying with acid solution and make-up water often requires chemical treatment.[13]

A checklist should be prepared of the items to be covered on a maintenance visit. Resident staff can deal with matters within their capacity and the remainder given to a reputable service company. Gosling[13] drew attention to the shortage of qualified service personnel. The .dangers inherent in air conditioning systems which rely on wet cooling towers were highlighted following the outbreak of legionnaires' disease at Stafford district general hospital in 1985, and also the need for a comprehensive operating and service manual.

There is no need to hold the temperature constant in an air conditioned building—it is better kept within an acceptable range for comfort. Since heat inputs in summer are intermittent in nature, energy can be saved by running plant continuously at a fairly low rate, allowing heat to pass into storage at peak periods and to be emitted at other times. The plant should be monitored for efficient working and appropriate action taken, such as shutting down coolers before the air handling equipment.

A DOE subcommittee[14] recommended that manufacturers of air conditioning equipment could assist maintenance by improving designs and in some cases by reducing the complexity of the control system. Building owners need to understand the value of properly implemented preventive maintenance. At the other extreme some very sophisticated plant is being used in prestige office buildings in countries with very hot, humid climates. A typical example is the computerised air conditioning plant installed in the main Standard Chartered Bank building in Singapore.

Some recent developments in air conditioning engineering comprise:

(1) heat recovery: extracting and using heat from vitiated room air, that would otherwise be dissipated, in the external air; and

(2) the concept of thermal balance, involving full consideration of thermal and other properties of a building's structure and fabric.

Electrical Installations

Modern electrical installations generally incorporate ring circuits; a recognised provision being one circuit for each 100 m^2 of floor area. Each ring can carry an unlimited number of 13 amp socket outlets. The ring system is cheaper and more convenient than the earlier arrangement in which each socket outlet required an individual sub-circuit with its own cable run from a separate fuseway. In a ring circuit a number of outlets is served by a loop of cable-run which forms one sub-circuit; each pair of conductors forming the loop starts from and returns to a single terminal on the fuse distribution board, with each outlet receiving current from two directions. Spur connections may be taken from a ring circuit to serve outlying socket-outlets; only two socket-outlets or one fixed appliance is fed from each spur, and not more than one-half of all points are fed by spurs. Cookers, immersion heaters and central heating boiler controls should be wired on separate circuits. Cartridge fuses for ring circuit plugs are standardised at two ratings—3 amp (blue) when the appliance has a rating of not more than 720 W and 13 amp (brown) with appliances of 720 W to 3 kW rating. White meter or economy 7 tariff is the most economical method of using off-peak electricity.

Socket-outlets should be provided on a generous scale to give ample facilities for the use of electrical appliances without the need for trailing flexes and multiple adaptors. The following represents a minimum desirable level of provision: kitchen—4; living area—4; dining area—4; each bedroom—3; hall and landing—1; garage if integral with house—1; store or workroom—1; although a larger provision is desirable. Where two or more outlets are provided in a room they should wherever practicable be positioned on different walls and preferably be switch-controlled, so that each appliance can be isolated from the supply before the plug is removed. Switch-controlled double socket-outlets are preferable in most cases to reduce or eliminate at very little extra cost the use of adaptors and consequent possible overload. The siting of outlets in the floor for aesthetic reasons is not very practicable as it can result in their being covered by carpets or, worse still, water on the floor may seep into the electrical system.

Cables or conduits should be located well below surface finishes to avoid surface cracks. Metal conduits or channelling and fixing nails require galvanising or other suitable treatment to prevent rust staining through wet finishes. Some timber preservatives can attack plastics insulation and sheathing on electric cables. High standards of workmanship are essential for both efficiency and safety. Electrical installations deteriorate as a result of ageing of the insulation material and cumulative mechanical damage, hence installations should be inspected and tested in accordance with the IEE Wiring Regulations at least once every five years. BRE[15] have identified cases of damage to cable insulation leading to risk of short circuit and fire, with the normal heat dissipation impeded by thermal insulation and other heat sources, with consequent cable overheating or through incorrect choice of cable type or size. If repairs or alterations are required to electrical wiring built in and concealed within the carcass or structure of a building, especially in domestic properties, it becomes necessary to break open the wall or floor surface to gain access. There is a need for detailed wiring and location diagrams to minimise the work of disturbance. The design of components, such as hollow detachable skirtings, to accommodate electric wiring would also be beneficial.

Wiring normally lasts about 30 years and aged circuits, installed before the early nineteen fifties can be identified by their round-pin plugs and rubber-sheathed cable, which is probably perished, and a shortage of socket outlets. It is advisable to open up sockets and switches, particularly on damp walls, to check the cable type and to inspect for rust.

Lighting

More than half the electrical energy supplied to commercial buildings in England and Wales is used for lighting. With the emphasis on energy conservation it is important to obtain the best results from the energy consumed. The main function of a lighting installation is to convert electrical energy into useful light of suitable colour, in the required places and transmitted from the best directions. Sodium discharge, mercury fluorescent and white or natural fluorescent tubes may often provide economical substitutes for ordinary tungsten (GLS) lamps. Continual progress is also being made in the design and production of low-energy lamps.

The design of luminaires (lighting fittings) is often a compromise between projecting the maximum light on to a working plane and achieving a balance of brightness on other room surfaces to promote visual comfort. The decorations of a room can have a significant effect on the illuminance—the ceiling ideally being white in colour, whereas white walls can be trying on the eyes, particularly in small rooms. Glazed areas must be cleaned regularly to secure a significant daylight contribution and the relative economics merit evaluation.

The illuminance from a lighting installation decreases from the first day of use. All relevant factors require analysis to produce the optimum maintenance pattern. The design and installation of luminaires should aim for easy maintenance, as the cheapest equipment may be false economy in the long term. The blackening of ceilings above electric lamps can be reduced substantially by fitting light shades to diffuse the heat. There is a fire risk from plastics light diffusers and it is advisable to use extended aluminium or other non-combustible material. The continual redesigning of luminaires by manufacturers causes maintenance problems in matching existing installations. Finally, very small installations with no resident maintenance staff are relatively more expensive to maintain than larger installations.

Planned lighting maintenance (PLM) involves regular cleaning and replacement of all the lamps in an installation. The frequency will depend on the burning hours per year of the lighting. As a guide, tubes that are used 100 hours per week will need changing every 12 months, while those burning 25 hours per week will only need changing every 4 years. Many factories use weekends or factory shut-downs for maintenance, so as to cause minimum interruption of work. Lighting circuits in older properties may need replacing, while in modern dwellings, more attention is now paid to the provision of external lighting and more sophisticated switching devices and lighting fittings.

Much of the wasted energy for lighting results from traditional switching arrangements, with a single or multiple on/off switch/panel positioned by the entrance to the space. BRE Digest 272[16] discusses automatic lighting controls and gives guidance on the types of control best suited for particular types of installation. It shows how energy savings can be predicted and compared with traditional manual switching and suggests alternatives to the traditional approach.

Gas Installations

The Gas Safety Regulations 1972 created a number of offences with a maximum penalty of £400 for using gas appliances which are unsafe. There are four main hazards which may be encountered in the use of gas:

(1) inadequate supply of oxygen;
(2) inadequate flue for the escape of products of combustion;
(3) escape of gas; and
(4) inadequate protection of hot components, flames or utensils.

Natural gas is not poisonous, but the build-up of it in a room would cause oxygen starvation. The main danger is one of explosions, such as those which occurred in the winter of 1976–77. Subsequently the King Committee was established to investigate the matter. The main conclusion of the Committee was that more action should be taken to identify gas leaks and deal with them quickly, and to engender greater awareness by the public of the dangers and to secure their active co-operation.

Smells of leaking gas must be reported immediately to the local office of British Gas because of the possible dangers to occupants, and all basic precautions taken, such as extinguishing naked lights, not operating electric switches, opening doors and windows, checking gas taps and the pilot light and turning off the supply.

Normal wear and tear on all gas appliances reduces their efficiency and safety. Vital dimensions and adjustments change and soot builds up in unwanted places. All appliances, therefore, require periodic servicing as well as repairs when they cease to function satisfactorily. The Gas Safety Regulations require installation and maintenance work to be carried out by competent persons. It is likely that only employees of British Gas, members of the Confederation of Registered Gas Installers, or persons with acceptable qualifications or experience would be recognised as competent.

The products of combustion must be discharged through suitable flue pipes or blocks into the open air. Dampness in the roof spaces of houses has, for instance, resulted from flue pipes from gas-fired water heaters terminating in unventilated roof spaces, whereas they should have been carried up to the ridge and fitted with a ridge terminal. Both flues and flue outlets must be of suitable size for the appliances served and have sound joints.

A check of gas heater efficiency can be undertaken by checking the rating from the maker's plate, running the appliance on full for a few minutes and taking the meter readings, before and after running. Metered consumption divided by the time taken should match the rating. However, the problem is not usually that the appliance is inefficient but that it is of an inappropriate type for the anticipated use.

Lifts

Lifts are provided with different types of drive (traction and hydraulic) operating to different speeds (single, double and variable) and various forms of control (automatic push-button, down collective, selective collective, and group). Consideration must be given to the capacity of the lift car and the lift speed, as well as the floors to be served.

A limited number of firms produce lifts to similar criteria but their parts are not interchangeable. The best policy is to invite tenders for the maintenance contract at the same time as tendering for the new installation, so that a decision can be taken on the two items jointly, as this work cannot readily be undertaken by direct maintenance staff. The best guide to lift maintenance costs from the known costs of existing installations is on the basis of the number of passenger-floors (the product of the passenger capacity and the number of floors served by each lift).

The ease of maintenance is largely dependent upon the types of finishes adopted in the lift and lift shaft, as well as the degree of accessibility to plant and equipment. One of the most practical lift doors is that made of plastic laminate recessed into a stainless frame and faced on both sides. In luxury flats, stainless steel is often used for aesthetic reasons although any scratches are easily visible. Spray-painted metal surfaces provide another alternative but they are vulnerable to chips and scratches. Regular cleaning is necessary in all cases.

Lift floors are usually covered with PVC sheets, although these are subject to damage from hot tobacco ash. Waste receptacles and cigarette urns should be located close to lifts to reduce possible damage and maintenance problems. Costly studded rubber sheets are often used for lift flooring in blocks of luxury flats, and these are more difficult and more costly to maintain. Ceilings to lifts should be kept simple, and normally incorporate fluorescent lighting behind polycarbonate panels.

Lift control panels should be made as vandal-proof as possible. Plastics control buttons may be burnt or prised out and so dislocate the lift service, and are best replaced with metal anti-vandal buttons. Lift doors or the control buttons may be damaged by being wedged open by persons delivering goods to flats, and this is best resolved by the provision of stop buttons, although even these are subject to abuse on occasions.

The pit floors to lift shafts should have a smooth finish for ease of cleaning. Lift plant rooms require adequate space both to house and maintain the plant and equipment.

After installation, a lift must be regularly inspected and properly maintained to ensure satisfactory operation. The responsibility for the maintenance of lifts in high-rise buildings is normally given to lift manufacturers, because of the technical complexities and the fact that it is the most economical arrangement.

In a comprehensive maintenance contract, the preventive maintenance programme comprises the maintenance of all machinery, car, hoistway and pit. Normally, the maintenance schedule is prepared on a weekly basis, whereas inspection and testing of safety equipment should be done at least once a year to ensure the safe operation of the lift. Contracts usually provide for routine lubrication and checking in accordance with an agreed schedule. However, maintenance does not only involve periodic lubrication and inspection, but also covers the detection and anticipation of possible breakdowns. If a lift is operating below optimum efficiency, the lift contractor will be called upon to carry out the necessary repairs or replacement of parts. The cleaning of the lift pit is included in the lift maintenance contract, whereas cleaning of the lift car will be done by the normal cleaning staff.

A lift maintenance engineer must be able not only to plan and supervise the maintenance work, but to be on call at all times. He has a very heavy responsibility as an unsafe lift could put human life in jeopardy.

Refuse Collection from Flats

Most blocks of flats incorporate refuse chutes which terminate in refuse storey chambers with the refuse frequently falling into containers with a capacity of around 1 m^3, and a caretaker replaces the containers as they are filled, usually on a rotating turntable. Larger containers holding up to 9 m^3 are sometimes used in conjunction with a special vehicle equipped to lift and carry them. Furthermore, machines for reducing the volume of refuse have been developed in recent years to make better use of the space available in storage chambers. These compress refuse into rigid containers or disposable sacks, or burn it in incinerators designed to ensure that the flue gases are free of dust and noxious gases.[17]

Incinerators reduce refuse to about one-tenth of its original volume. On the other hand they add to the general level of atmospheric pollution, and they need skilled maintenance and an expensive chimney. Compactors do not reduce the volume of refuse as much as incinerators; nor do they increase the rate of decomposition significantly. Hence at least a weekly collection is necessary, but compactors do not cause pollution, are cheaper than incinerators and can be readily installed in the majority of existing buildings.[17]

Refuse chutes should be designed in accordance with BS 5906[18] and be constructed of non-combustible, fire-resistant and moisture-proof walls, and often either 450 or 600 mm internal diameter. Adequate access is needed for inspection, cleaning and clearing of chutes after misuse and there must be ample provision for ventilation. Refuse is deposited into chutes through hoppers located on private balconies or in naturally or mechanically ventilated public lobbies with self-closing fire doors, opening off open spaces. The maintenance programme should include regular inspections and cleansing of the chute, chamber and associated fittings to ensure a satisfactory standard of hygiene.

Drainage

Drains can cause trouble in a number of different ways. Loads from foundations of buildings or vehicles, or ground movement below drains can cause fracture of pipe joints or, in severe cases, fracture of the pipes themselves. Rigid cement mortar joints used extensively in the past with clay pipes are particularly vulnerable, and very old drains may have clay joints and be bedded on bare earth. Drains may also become choked through the deposition of silt or objects such as brushes and rags, particularly where the pipes are laid to flat gradients with restricted flows, and there may be no provision for access at changes of direction or gradient. Intercepting traps, now rarely installed, are another cause of blockage.

BRE[19] have shown how leaking flexibly jointed clay drains may be caused by damaged pipes, unsuitable bedding or fill, or badly made joints. Blocked drains are cleared by rodding, water jetting or winching from manholes and inspection chambers, and chemical cleaning is sometimes used for industrial drains. Defective pipes require replacing and leaking joints cutting out and making good. Drains can be repaired by pressure grouting using either cement-based grout, which is strong and rigid, or a plastic gel, which fills cracks and cavities and remains flexible. If leaking drains are suspected, they should be tested by water, air or smoke.

Inspection chambers and manholes should be inspected periodically to check that they are in sound condition, particularly the benching and rendering, and that the drains are running freely. These chambers are liable to cause blockages, either because of their construction or because of disintegration resulting from age or chemical attack.[20] Gully traps require cleansing; the frequency being determined by local conditions, and checking to see that they retain an acceptable seal and are soundly jointed to the drain. If the gully is loose or the ground around it is sodden, further investigation is required.

Cesspools frequently have crumbling rendering, leak and permit foul discharges into the surrounding ground and sometimes into nearby water-courses and even wells. The usual remedy is to waterproof the interior surfaces of the cesspool with asphalt, waterproofed cement mortar or a bitumen-based application such as synthaprufe. Septic tanks may on occasions require similar remedial treatment. Metal covers to manholes and other chambers may rust and require an application of bituminous paint and bedding in grease to prevent the escape of gases. Cast iron covers cracked by vehicles need replacing with heavier covers or possibly suitable steel covers.

Safety

Modern building regulations and techniques provide effective safeguards and means of access during the construction process, but these are often removed on completion rendering subsequent maintenance unnecessarily costly or dangerous. Permanent provision should be made for access for maintenance purposes, particularly in multi-storey buildings, otherwise makeshift methods may be used which could lead to accidents.

With older buildings, maintenance operatives must be clearly instructed how to reach parts of the building which are difficult of access, and the instructions must be backed up with adequate supervision to ensure that the men do not take dangerous short cuts. They must also be made fully aware of the dangers and limitations of the materials and components with which they may come into contact. A large proportion of the accidents which occur result from falls, frequently from ladders, and these mishaps are often fatal or very serious. Safer conditions stem from the provision of permanent means of access, such as fixed ladders, roof walkways and handrails at places of high risk, coupled with the use of safety nets to provide protection against falls from high points during temporary work.

Accidents are seldom really accidental, but result from ignorance of, or failure to carry out, safety procedures, and from failure to enforce safety rules or carry out scheduled inspections; they are seldom due to sheer carelessness on the part of the victim. In practice, maintenance and safety are closely interrelated as the maintenance engineer is likely to be responsible for both ensuring safe working conditions and dealing with accidents. The appointment of a fully trained safety officer is desirable in factories to ensure the co-ordination of safety aspects by a single person—he may also carry out other functions such as fire prevention and security.

The *Factories Act 1961* requires that "all floors, steps, stairs, passages and gangways shall be of sound construction and properly maintained and shall, so far as is reasonably practicable, be kept free from any obstruction and from any substance likely to cause persons to slip." Furthermore, means of exit shall have handrails and openings shall be protected. The *Health and Safety at Work Act 1974* states that "It shall be the duty of every employer to ensure, as far as is reasonably practicable, the health, safety and welfare at work of his employees." It is also a statutory requirement to prevent areas around machines from becoming slippery or congested, and to provide effective and easy means for isolating electrically operated appliances.

Lovejoy[21] has highlighted many essential precautions to ensure safety in accessibility for maintenance work. For example, within a factory or storage warehouse, entrance doors should be protected by railings and barriers so that no one inadvertently enters a vehicular access. Sight glasses in doors, especially rubber doors used in warehouse division walls are essential, and their size and position must allow for all vision heights.

There are many buildings where the provision of a short flight of stairs or a fire lobby with too small a space between two sets of doors, has precluded the easy access required by users. There is often a conflict between the requirements of the user and the Fire Prevention Officer; the illegal propping open of fire doors is a typical example, which could be avoided by using pressure-pad operated doors.[21]

With services it should only be possible for trained operatives to gain access to vulnerable or dangerous areas. Hence, doors to lift motor rooms, transformer rooms and switchgear should be locked while in use, and when the plant is under maintenance, inspection or repair, these rooms should not be accessible to occupants of the building. Areas enclosing machinery or electrical appliances should be well lit and have adequate space for carrying out maintenance work safely. Where machinery may be in use while under maintenance, adequate safeguards and interlocks should be installed and used to prevent accidents. Where electrically operated overhead travelling cranes are in use, these should be isolated when access is required in the area of the crane rails.[21]

Where a manhole cover is lifted for access to a shaft, it should be placed in a secure position and the open hole fenced by means of a portable guard. Access into effluent tanks should not be made without prior testing of the air in the tank to determine whether there is any harmful gas present or an oxygen deficiency.

The provision of properly designed and constructed access to clean patent glazing and gutters will ensure that the hazardous activity of walking along valley gutters between two slopes of fragile roof sheeting, with the consequent risk of injury or even death, will be eliminated. Warning notices must be provided drawing attention to the fragile nature of roofing materials where applicable. Access for external window cleaning and internal wall and ceiling maintenance often includes the provision of appliances such as ladders, access towers, or suspended or slung scaffolds, and these must be soundly manufactured and properly used. Safe access must be provided to suspended cradles and suitable training given to persons using them.

Where a ladder is used to gain access to a platform, the head of the ladder should be at least 1 m above the level of the platform, unless an alternative handhold is provided. Ladders should not rest against eaves gutters, particularly those made of plastics, but should be supported from the wall by an effective ladder stay. The maximum height of a ladder should be 9 m. Step ladders and trestles should be regularly inspected to ensure that they are in good working order and that the ropes are sound and secure. The use of a roof ladder is strongly recommended on any angle of slope regardless of the condition of the roof covering. Erection and use of scaffold structures should comply with the requirements of BS 1139 (metal scaffolding).

In order to carry out inspection work in dangerous situations, it may be advisable to use safety belts or harnesses attached to anchorages by their lanyards. The use of longer fixed lines is also a possibility. In some cases it is possible to rig industrial safety nets beneath hazardous working areas.[21]

Security

The security of an external door depends primarily on the position, type and construction of the door and the strength and reliability of the hardware. The strength and fixing of the door frame and type and extent of glazing are also important.[22] The ideal construction is of solid wood, but this ideal has to be balanced against cost and appearance. If glass is fitted in a position where its breakage will give access to the lock or latch, it should be reinforced. Hinges should be of adequate length and positioned so that the pin is inside and the screws are concealed when the door is closed. Letter plates should conform to BS 2911 and be positioned at least 400 mm from the door locking device, and an internal cover plate offers additional security. All exterior doors should ideally be fitted with a mortice thief-resistant lock conforming to BS 3621, but doors less than 45 mm thick are weakened by such locks and should be fitted with an automatically deadlocking rim lock. BS 3621 requires locks complying with the standard to provide a minimum of 1000 effective differs. Other doors should be reinforced with securely fitted shoot bolts, top and bottom. The final exit door, which has to be locked from the outside, is obviously more vulnerable than those secured from the inside, and should be normally the one most overlooked by neighbours and passers-by. Safety chains provides an additional and effective security device. The glass can be strengthened by replacement with wired glass, or by polycarbonate sheeting. If a very high standard of security is required, as in a wages office, then bandit glass is recommended. Grilles or shutters can afford good protection, but will only be as strong as the frame to which they are attached, or the means by which they are attached to the surround of the window.[22] A variety of intruder alarms are available for use with windows.

Metal windows appear to offer slightly better security than wood; some manufacturers are offering security locks. A high standard of security for windows is only possible by the use of expensive glazing materials and techniques, and special fittings. A thief will normally try other means of gaining entry if it entails breaking glass. If panes of glass are to be small enough to prevent entry they must be not more than 0.05 m^2, but if this minimum is to be exceeded, then the larger the better. Some side-hung casements have external hinges from which the pin can be removed and the casement lifted out, in which case the pin should be made secure by welding or other means. Sliding sashes may need protecting with a locking device and louvres in windows should be checked to ensure that it is not possible to bend or remove the glass seating, clips or glazing beads and that the locking mechanism is adequate. The placing of small ventilating lights should be carefully considered to avoid external access to fasteners on larger opening lights. Plastic domed roof lights are often easily removable.

Balconies are a security risk and should be restricted to individual dwellings. Shelters and porches to doors should not obscure view or provide cover for a thief to work on the door. Careful attention should be paid to the siting of external pipework as it may provide convenient access to upper windows. Coin-operated gas and electricity meters are a security risk and where installed should be accessible only from inside the dwelling. Suitable fencing, such as chain link not less than 1.2 m high, around and between gardens will retard a thief. Lighting is a crime deterrent and a useful back-up aid to other security measures.

The annual cost of vandalism to buildings in England and Wales has been estimated at £30 million. There are probably two principal ways of combating it: (1) to discourage vandalism by avoiding features in the design which appear to attract damage and by using materials which are not easily disfigured; (2) to persist in the repair and replacement of damaged and defaced work.

Insurers often specify the types of locks, barriers and levels of alarm response according to the degree of risk. There are many types of intruder alarm system and they should all comply with BS 4737. Because it is difficult to find neat and secure routes for wiring, more equipment is now 'wirefree' and sends signals to the control panel by infra-red light, high-frequency sound or radio. Major alarm installations are wired through to the security services either; (1) by automatic dialling to the police; (2) to several telephone numbers including the police; or (3) to a commercially run central station, which monitors plant and keeps records of movements around the building. Film cameras and TV can be directed at vulnerable points and be activated by infra-red light or very low lighting levels. Telephones, fire alarms and intruder alarms can be linked.

Fire-resisting Construction and Fire Precautions

Internal Fire Spread (Surfaces)

The spread of fire over a surface can be restricted by provisions for the surface material to have low rates of surface spread of flame, and in some cases to restrict the rate of heat produced (paragraph B2 of Schedule 1 to the Building Regulations 1985).[8] The provisions in Approved Document B/2/3/4 prescribe class 1 materials for walls of dwellinghouses and these include plasterboard, and class 3 materials for ceilings, which include plywood and treated or painted hardboard.

Internal Fire Spread (Structure)

Premature failure of the structure can be prevented, and the spread of fire inside a building can be restricted, by provisions for elements of the structure to have a specified minimum period of fire resistance (paragraph B3(1) of Schedule 1 to the Building Regulations). The minimum periods of fire resistance for all loadbearing elements of the structures of dwelling-houses, including separating walls, are prescribed in table 1.1. of Approved Document B/2/3/4 as $\frac{1}{2}$ hour or 1 hour for basements of more than 50 m^2 floor area and houses of four storeys or more. Higher periods apply to some other buildings, ranging from 1 to 4 hours.

A BRE report[23] gives guidance on the notional periods of fire resistance for a wide range of constructions. For example, loadbearing timber framing members at least 44 mm wide and spaced at not more than 600 mm apart, with lining both sides of 12.5 mm plasterboard with all joints taped and filled; and 100 mm reinforced concrete wall with 25 mm minimum cover to reinforcement, both provide half-hour fire resistance. A solid masonry loadbearing wall, with or without plaster finish, at least 90 mm thick, will provide one hour fire resistance. For masonry cavity walls, the fire resistance may be taken as that for a single wall of the same construction, whichever leaf is exposed to fire.

In general, domestic loadbearing external walls of normal materials satisfying the conditions of strength, stability and weather resistance will usually provide sufficient resistance to fire. Methods of upgrading existing partitions are described in BRE Digest 230.[24]

Compartmentation

The spread of fire can also be restricted by provisions for subdividing the building into compartments of restricted floor area and cubic capacity, by means of compartment walls and compartment floors (paragraph B3(2) of Schedule 1 to the Building Regulations). This provision applies to flats and maisonettes.

Other forms of compartmentation apply in the case of a division between adjoining buildings by means of a separating wall, such as walls between terraced and semi-detached houses carried above roof level or suitably fire stopped or between a house and an attached garage (paragraphs B3(4) and (5) of Schedule 1 to the Building Regulations). The wall and any floor between a garage and a house are to have half-hour fire resistance and any opening in the wall is to be at least 100 mm above garage floor level and be fitted with a half-hour fire resisting door (Approved Document B/2/3/4).

Concealed Spaces and Fire Stopping

Hidden voids in the construction of a building provide a ready route for smoke and flame spread; this is particularly so in the case of voids above spaces in a building as, for example, above a suspended ceiling or in a roof space. As any spread is concealed it presents a greater danger than would a more obvious fire weakness in the building structure. Provision is therefore made in paragraph B3(3) of Schedule 1 to the Building Regulations to restrict the hidden spread of fire in concealed spaces by closing the edges of cavities, interrupting cavities which could form a pathway around a barrier to fire, and subdividing extensive cavities. It is also necessary to effectively seal the openings around all penetrating pipes, cables and other services.

Constructional techniques are illustrated in Appendixes G and H of Approved Document B/2/3/4. Ends of cavities should be effectively sealed to provide fire stops as described in BRE Digests 214 and 215.[25]

Stairways

An internal stairway in a house which has more than two storeys, excluding a basement, may need to be enclosed and protected to meet the provisions of Section 1 of the Mandatory rules for means of escape in case of fire (HMSO, 1985) in support of the requirements of B1 of Schedule 1 to the Building Regulations. BS 5588 gives recommendations for means of escape for flats and maisonettes, shops and offices.

Fire Hazards

BRE Digest 260[26] outlines the design principles for systems which provide safe escape routes from buildings, including the use of smoke doors to isolate lengths of corridor or stairway, ventilated lobbies, smoke ventilation and powered smoke extraction from large undivided buildings such as shopping malls and warehouses.

Sound management, adequate fire protection equipment and effective fire prevention systems will reduce the likelihood of a serious fire, but they cannot eliminate it. Precautions to limit the spread of smoke normally include the provision of smoke-stop doors in corridors, at entrances to staircases and lobbies and in other suitable locations. Staircases should be ventilated by opening windows and/or skylights. Automatic dampers should be provided at strategic points on conveyors and in air-conditioned ducts. In large single-storey premises roof ventilators assist smoke dispersal. Basements present a difficult ventilation problem and smoke outlets should be provided with fitted covers that can be removed or broken to permit the escape of smoke.[27] Lofts create a special hazard with their traditional use as storage areas for paper, cardboard boxes, timber and other combustible materials, and the surrounding structural timbers ensure a ready supply of fuel once a fire is started. In 1982 it was reported that well over half the domestic fires in the UK involved furniture and beds, with more than half the fatalities occurring in fires started by smokers' materials.[28] Many of the deaths were caused by the combined effects of smoke and toxic gases.[29]

Fires Emanating from Electrical Apparatus

Electrical appliances and installations account for nearly one-quarter of the known causes of fires, and approximately two-thirds of these start on the fixed part of the installation. The most likely initiating cause is a high tension resistance joint or a partially severed cable producing a local hot spot for which there is seldom complete electrical protection. Other contributory factors include leakage and short circuits as a result of old or damaged insulation, especially if coupled with over-fusing and overloading. Extensive temporary wiring of substandard quality and inadequate maintenance provide further causes of fires.

Faults on flexible leads are responsible for about one-third of the fires in the wire and cable category, and there is no doubt that flexes generally are a serious hazard. Hot spots can develop so easily if there is a taped joint, loose terminal connection or damaged insulation on a lead to a heating appliance. In industrial buildings, apparatus running unattended for long periods can introduce a serious fire risk. Much benefit would accrue were occupiers of buildings, and particularly maintenance personnel, to carry out simple megger tests and visual inspections of installations, especially of flexible loads, at least at the frequencies prescribed by the Institution of Electrical Engineers Regulations. This would result in many faults being rectified before they developed into breakdowns and fires.

Fire-fighting Equipment

The provision of adequate escape routes and fire alarms in the majority of buildings frequented by the public is required under various Acts of Parliament, but there is no statutory requirement for the installation of automatic fire detection or automatic extinguishing equipment whose main function is to restrict the damage to the building and its contents. Legislation in general provides for: (1) provision and, where necessary, the enclosure of escape routes in suitable materials; (2) aids to escape comprising fire alarms, hand extinguishers and emergency lighting; (3) display of escape instructions and training of staff. The legislation is framed to give local fire authorities a measure of flexibility in securing adequate escape arrangements and equipment at minimum cost and disturbance and to the mutual satisfaction of the persons responsible for the building and the fire authority.

Apart from the provision of adequate escape routes, legislation requires warning systems for buildings frequented by the public and adequate lighting of escape routes even in the event of failure of the public electricity supply. There are three recognised alarm systems—multi-zone, single zone and single point.

The *multi-zone* installation is suitable for complex industrial and commercial premises, including the larger hotels. It generally consists of a series of sounders (bells, klaxons and warblers) coupled with manual call points (break glass units) all powered from a central battery unit continuously charged from the a.c. mains. The more expensive systems incorporate detectors and automatic fire brigade call-out facilities. The *single-zone* is particularly suitable for smaller premises such as small hotels and guest houses. It consists of sounders and manual call points, and may be powered from mains or standby batteries. *Single point units* or self-contained alarms form the most economical way of protecting a small property.

Smoke detectors are available in several forms but for general fire protection in buildings one of two types are usually selected—(1) optical smoke detectors which are activated by the absorption or scattering of visible or near visible light by products of combustion, (2) ionisation chamber detectors in which combustion products entering the ionised chamber alter the conductivity to initiate the alarm circuit. The first category respond well to smoke particles and visible combustion products, while the second type are better suited to detect invisible gases from clear burning fires.

Portable fire extinguishers can be carried and operated by hand. Where all or most of the occupants of a room are women, the extinguishing equipment should desirably be made up of 4.5 litres water buckets and 3 kg dry powder or 7 kg carbon dioxide extinguishers. Extinguishers need regular maintenance and there should be adequate personnel with knowledge of and confidence in their use. Ideally extinguishers should be safe for the user, efficient and durable; they should discharge and reload rapidly, and be easy to maintain and of reasonable appearance. The extinguisher selected should be on the list of approved portable fire extinguishing appliances issued by the Fire Office's Committee.

Four classes of fire are recognised by BS 4547: *Class A* fires involving solid materials, usually of an organic nature in which combustion normally takes place with the formation of glowing embers. These are best extinguished by water type extinguishers. *Class B* fires involve liquids or liquefiable solids and are best extinguished by smothering to prevent oxygen from combining with the flammable liquid vapours or gas and this may be done by using a dry powder, carbon dioxide or a vaporising liquid extinguisher. *Class C* fires involve gases and are extinguished in the same way as Class B. *Class D* fires involve metals, and special powders have been developed for the extinguishing of various metal fires.

The simplest form of fire-fighting equipment is the *fire bucket*, usually of red-painted metal or plastics and of around 10 litres capacity, containing water or sand. They should be hung or stood on a shelf not more than 1 m above floor level in a prominent position and be covered to reduce evaporation of water. A typical provision is three buckets per 210 m^2 of floor area with not less than six buckets per floor. Hand pumps (stirrup pumps) for use with buckets of water are extremely useful.

Building controls normally rely on passive fire protection, and make little or no allowance for the beneficial action of sprinklers and other active measures. However, sprinklers reduce fire severity and hence the risk of large fire losses, and it is possible to make a case for relaxation of passive protection in favour of active protection.

Sprinklers are usually associated with commercial, industrial and storage buildings, where benefits include substantially reduced insurance premiums. There are wet systems for heated buildings, dry systems for unheated buildings, and combined wet and dry systems to provide flexibility of use in different climatic conditions. A deluge system is also available for special buildings to provide total coverage from all the sprinkler heads acting together. The coverage of sprinkler heads varies from 7.5 m^2 in high hazard buildings to 21 m^2 in low hazard. Some sprinkler heads are activated by the expansion of special liquids in glass bulbs, while others rely on the melting of solder, and those on deluge systems have open heads with the water flow controlled by special detectors.

Hoses of 19 or 25 mm diameter on reels, with adjacent stopcocks, placed about 1.5 m from the floor, can be used by untrained operatives and are normally located on escape routes and common walkways. Risers of either the wet or dry varieties, with a minimum diameter of 100 mm carrying water to each floor level of high-rise buildings, may be installed to permit the connection of fire brigade hoses.

Fire-fighting Arrangements

Clear and unambiguous fire instructions, clearly audible alarms and regular fire drills give the occupants of a building a feeling of security and a sense of involvement. They ensure that emergency equipment is tested at regular intervals and that fire exits are kept clear and in usable condition. Where there is insufficient staff available to regularly check and service fire equipment, a maintenance contract should be negotiated. In the majority of industrial premises fire fighting consists of nothing more than first aid fire fighting to hold the outbreak in check until the fire brigade arrives, and it is therefore advisable to have a number of staff trained to use the equipment and to be thoroughly familiar with all the buildings.

Whether or not a company wishes to set up a works fire brigade depends on the nature of the risk and the money available. In all cases there must be a foolproof method of ensuring that the alarm is given to the fire brigade as soon as the fire is discovered. Close liaison should be developed with the local fire service and opportunities provided for local firemen to visit the premises—this may one day prevent the loss of the buildings by fire and possibly loss of life or injuries to firemen through unknown hazards.

Cleaning

General Background

The cleaning industry in the United Kingdom was spending over £3000 million per annum in 1987 and yet its performance has been found not to compare favourably with the United States or Scandinavia.[30] Within 30 years after erection, or even less, cleaning costs may exceed the original cost of the building. Lack of co-operation between architects and maintenance and cleaning organisations all too often results in inadequate consideration of cleaning aspects at the design stage, and designers generally need to acquire greater knowledge of cleaning facilities, methods and equipment and to receive improved feedback of information on the performance of buildings.

The planning of even a single window in an inaccessible position can, over the years, generate cleaning costs many times in excess of the value of the window. Large areas of terrazzo and other flooring can be badly damaged by using the wrong cleaning materials. The effectiveness of maintenance and cleaning work can be increased by the architect supplying a maintenance manual which includes a full description of finishings, furnishings and fittings in the building. The manual should also contain the various manufacturers' recommendations for cleaning and maintenance. All cleaning and maintenance procedures should desirably be logged in the manual so that if the building changes hands the new occupier will see what has been done and what he should do to maintain the building to a reasonable standard.[31]

Method of Execution

The owner of a building often has the choice between carrying out cleaning work by direct labour or letting it out to a contractor. Cleaning contractors concentrate their energies on this class of work and inevitably develop efficient techniques and considerable know-how. They relieve the client of the problems involved in recruiting a workforce, and organising and equipping it. Taking flooring as an example—each type needs its own cleaning technique. Some of these needs for caution are fairly obvious but if overlooked can ruin a floor, such as using spirit-based cleaning materials on rubber or benzine-based materials on pitchmastic. Labour accounts for the greater part of the cost of cleaning and must be effectively used and supervised. With offices, specialist contractors are usually better suited to deal with cleaning of carpets, rugs, curtains and venetian blinds as well as telephone cleaning and disinfecting and cleaning of windows and roof lights. A costs in use study of offices by DOE[32] showed that direct labour cleaning costs considerably more than contract cleaning in the cases investigated.

When engaging a cleaning contractor, it is important to employ a reputable contractor, preferably on a three-year contract to provide incentive and continuity. There should be a penalty clause to deal with unsatisfactory work or non-performance in addition to provision for cancellation. The method of payment and notice for termination must be clearly prescribed, together with details of frequency, methods, equipment and materials, supervision and workforce.

The DOE survey[32] distinguished between daily cleaning of offices—floors, ash trays and waste baskets—and periodic cleaning involving a more thorough clean every 3, 6 or 12 months, when files are taken out and dusted or vacuum-cleaned, and it also includes the regular polishing of floors. CP 153: Part 1, specifies frequencies for the internal and external cleaning of windows, as listed in table 6.1.

In industrial buildings, cleaning of overhead pipework and steelwork should be carried out at regular intervals, otherwise dust will build up until draught and vibration cause it to fall and become both a nuisance and a danger.

Table 6.1 Frequency of window cleaning

Type of Building	Frequency
Shops	Weekly
Banks and business premises	Twice monthly
Offices and hotels	Monthly
Hospitals	Monthly
Factories (light industry)	Monthly
(heavy industry)	Every two months
Schools	Every two months
Domestic buildings	Monthly

Preventive Devices and Storage Space

Adequate preventive devices should be provided to eliminate difficult cleaning tasks. These include waste receptacles and cigarette urns, preferably wall-mounted to avoid creating an impediment to floor cleaning. Plastics liners for waste receptacles and garbage cans prevent rusting and need for cleansing, and also provide vermin and odour control.

Probably the most sadly neglected area is the adequate provision of storage facilities for cleaning materials and equipment. One example is the modern multi-storey teaching hospital in Nottingham where the cleaning closets are too small and are only provided on every other floor. Hence cleaning trolleys have to be left in corridors and cleaners have to transport equipment from one floor to another. Ideally storage closets should be at least 2.4 × 1.8 m to accommodate the equipment and materials for two cleaners. They should have good lighting, easily cleaned surfaces, a sink and hose pipe, ample shelving and fixings for hanging tools.

Windows

Windows need cleaning periodically, as illustrated in table 6.1, to secure clarity of vision and maximum daylight penetration, maintain good appearance of building, prevent accumulation of dirt which when washed off by rain may harm wall cladding, and reduce deterioration of glass through attack by pollutants.

Windows may be cleaned in a number of different ways. Ladders may be used for heights up to 9 m and for reasons of safety the feet should be placed at one-quarter the vertical height from the building. Rubber inserts or cups are sometimes provided on the feet as a further safety precaution. Travelling ladders and suspended systems may be used from permanent rails or tracks fixed in front of the parapet or under the eaves, using suspended cradles on wire ropes where the height exceeds 30 m. Other alternatives are demountable rails and walkways.

Some windows can be cleaned from the inside by using projecting hinges or pivoted windows. The maximum safe reach to clean adjoining fixed glazing is 560 mm sideways, 510 mm upwards and 610 mm downwards.

Flooring

Floors suffer from the effects of traffic and soil, and the most damaging soil is usually that carried in by foot traffic. Soil trapping devices should be installed to combat floor damage and reduce cleaning costs, and these include gratings, mechanical matting and walk-off matting. Other floor protection measures include the selection of good furniture glides and the use of plastics strips or corners under filing and stationery cabinets. Protection of floors against chemical damage is important—harsh alkaline cleaners should never be used on resilient floors.

Student surveys collated by the York Institute of Advanced Architectural Studies[33] found lino sheet to be the cheapest floor covering investigated, followed fairly closely by carpet. Other floor coverings which proved reasonably economic in cleaning costs were wood strip, rubber tile and sheet, and wood block. The most costly were clay, composition and terrazzo tiles, with thermoplastic tiles and PVC sheet lying between the two extremes. The colour of floor tiles and sheets is also important since a dark colour shows dusty footprints while a light colour shows black burn marks. A speckled or marble pattern looks well and is easier to clean.

Easton[34] identified the main types of equipment for the removal of dust from floor surfaces as industrial suction cleaners for hard and soft floor coverings and high risk areas, impregnated mop sweepers for large hard floor surfaces; static mops for plastic floors; and damp mopping for removal of surface dirt. Spray cleaning or spray burnishing effectively removes surface dirt. Ingrained dirt removal requires a deep cleansing system (scrubbing hard coverings and shampooing soft coverings). Modern floor-clearing methods rely on a wide range of electrical floor maintenance equipment, including scrubber polishers, polisher/vacuums, scrubber dryers, wet and dry pick-ups, dry shampoo carpet cleaners, water injection/extraction machines, and water pressure machines.

Sanitary Appliances and Kitchens

Health hazards can arise if staff restaurants, kitchens and toilets are not properly cleaned and maintained. The vermin, likely to thrive in dirty conditions, may also endanger building services by gnawing through electric wiring and plugs and plastics pipes. There needs to be a close working relationship between cleaning and maintenance staffs to secure a satisfactory standard of cleanliness. Frequent cleaning will do little to improve the appearance of badly stained, chipped or cracked appliances which have served their useful life and require replacement. Similarly, continuously dripping hot taps should be re-washered.

Kitchens need hard, smooth and non-absorbent wall, floor and ceiling surfaces, with coved angles and rounded corners to facilitate cleaning. From time to time it will be necessary to restore food preparation rooms and appliances to a thoroughly clean and serviceable state compatible with age and general condition. Deep cleaning is specialised work and experienced specialist contractors should be employed.[35]

Graffiti

The action taken will depend on the reason for the graffiti, the surface on which it has been applied and the materials used. The purpose, be it casual, political, offensive or recreational, may determine whether other facilities are needed. Aerosol on brick may be scrubbed off with paint remover and water, oil paint superimposed on oil paint is best overpainted, felt pen and emulsion on brick can be overpainted with anti-graffiti paint, felt pen on concrete is difficult to remove and the application of paint removers, bleaching and overpainting with anti-graffiti paint may be required, and lipstick and chalk on brick scrubbed with detergent. Possible alternative devices include shrub planting, anti-climb paint on downpipes, murals or hoardings, and improved play facilities, according to the particular circumstances.[36]

Pest Infestation

Types of Infestation and Associated Risks

BRE Digest 238[37] describes how pest infestations in buildings can cause risk to health, as well as fire hazard, economic loss and sometimes deterioration of the structure itself. Birds, rodents, insects and other invertebrate pests all carry parasites or pathogenic bacteria.

Rodents may damage the fabric or fittings of buildings, as hard materials are gnawed to wear down teeth and softer substances are shredded to form nests. Rats can gnaw through plastics, lead and copper pipes causing water damage, and gnawed electric cables can result in fires. Furthermore, burrowing by rats can cause localised subsidence. Birds cause indirect damage as their droppings may increase the rate of lichen growth, their nests can block drains and gutters and they can introduce parasitic insects into the building.

In determining what precautions should be taken to prevent pest infestation, the level of risk and the acceptability of infestations should be considered. For example, basic risk would be unacceptable in a hospital, restaurants and food premises would be regarded as high risk and domestic buildings as low risk. In low-risk buildings suitable foundations, good detailing of roofs, tight building-in round pipes and the use of closely fitting doors and windows should prevent pest infestation.

The most important pest species are rats and cockroaches. Birds such as pigeons and starlings may introduce insect pests such as fleas, other biting insects and mites. Fungi can grow if conditions are damp in buildings, and surface mould growths can support book lice, plaster beetles and silverfish. Pests will survive only if the temperature and humidity are suitable and there is adequate food and water together with shelter, nest sites and nesting materials.

Prevention of Infestation

Insects can enter through minute cracks and cannot be excluded. Young mice can pass through holes about 6 mm diameter, and young rats through a 9 mm diameter hole. Common rats can burrow extensively and gain access to buildings in this way, while common sewer rats are good swimmers and can enter buildings through sewers and drains, where the covers are deficient. Birds seek nesting and roosting sites on and in buildings, will nest in roof spaces if they gain access, or use external ledges as roosts or nest sites. Holes to roofs should be kept small enough to prevent their entry; for example, pigeons should be excluded by 40 mm mesh and sparrows by 20 mm.

Foundations will exclude rats if they extend vertically about 900 mm. Good building practice requires cavity closers at the head of the wall for fire stopping, and this will prevent rodents and birds obtaining access to the cavity. The spaces between joists and rafters should be filled at the eaves to prevent birds or rodents from entering the roof space, but providing small ventilation holes or gaps closed with mesh to avoid condensation.

Door should close on to a level threshold, so that there is an insufficient gap to allow access or a gnawing edge. In high-risk areas, external doors not made of metal should be fitted externally with metal kicking plates not less than 300 mm high. These kicking plates should also be fixed to the jambs and linings to prevent gnawing and entry. Internal partitioning and ceiling cavities should be sealed sufficiently to deny access. Hollow spaces behind skirtings, architraves and other mouldings should be filled. Pipes, ducts and trunking should be tightly built in wherever they pass through walls, floors, ceilings or foundations.

Rodents must be prevented from entering lift shafts, as they would have access to all levels of the building and could damage the winding mechanism. Refuse hoppers provide food supplies as well as harbourage for pests, and there is a high risk of infestation in the room where collecting bins are housed. The door should preferably be of metal with a suitably protected frame and a self-closing device. Polythene bags for waste disposal are not sufficiently stout to keep rodents out and may also be damaged by domestic pets or foxes, allowing insects access to food scraps within. Wire mesh stands to these bags should be 5 mm mesh or smaller, surround the bag with ample enclosing space, and be fitted with well-fitting, solid lids.

Other Pests

Much unnecessary alarm is caused by the presence of insects in houses in the mistaken belief that they may be attacking timber, although many of these are associated with materials other than timber. New houses, particularly those of brick and block construction, remain damp for a year or two as the concrete and plaster dry out and some insects are attracted to these conditions.[38] Details of some of the more common pests of this type are now briefly described.

Plaster beetles are very small dark coloured beetles, varying from 1 to 3 mm in length, which feed exclusively on moulds, mildews and other fungi. With heat and adequate ventilation these insects should die within 3 to 4 months as the house dries out.

Silverfish are small silvery insects about 10 mm in length with long antennae and three long bristles at the tail. They cause little damage and are mostly found in dark, damp corners in kitchens, larders and bathrooms, and they dart away when disturbed. Good ventilation and heat should destroy them within a few months of the house drying out.

Wood lice are small grey oval insects 15 to 18 mm long with a hard segmented shell. They feed on decaying wood and other vegetation in the garden and occasionally enter houses looking for damp areas, such as under sinks and baths. The best remedy is to ensure that the house is dried out as quickly as possibly and to remove any old leaves or garden refuse near the house.

Earwigs are dark brown in colour, about 10 to 14 mm in length and have a distinctive pair of pincers at the end of the abdomen. They are usually found in the garden feeding on small insects but may be brought into the house in bunches of cut flowers. They do not attack timber. Creeping plants, such as ivy and Virginia creeper, should be kept away from windows and doors as the insects could find shelter in them.

Sawflies have bright green larvae with a chestnut head about 12 mm long and these bore holes approximately 3 mm in diameter. They prefer low growing weeds but can bore a short depth into timber, but any damage is usually superficial.

Flour beetles are dull reddish-brown in colour and 3 to 4 mm long. They are occasionally found in houses, generally in store cupboards, feeding upon flour, bread or other dried foodstuffs. They do not attack timber.

Bread or *drug store beetles* are reddish-brown in colour, 2 to 3.5 mm long with a dense covering of short yellow hairs. They are occasionally found in houses feeding on hard, dry products such as crispbread, pasta, cereal and nuts. These insects can penetrate paper, cardboard and even aluminium, but do not attack timber. The remedy is to keep cupboards thoroughly cleaned out and to store foodstuffs in containers with tight lids.

Carpet and *fur beetles* produce small larvae up to 5 mm long, which are very hairy and active, while the beetles are dark brown or black and often mottled with yellow or white patches, and about 3 to 5 mm long. Blankets, woollens and furs should be thoroughly cleaned and stored in sealed bags or tightly fitting chests, and the articles shaken regularly. Crevices between floorboards, wardrobes and carpets should also be cleaned regularly.

Larder or *bacon beetles* are oval in shape, black to dark brown in colour, 5.5 to 12 mm long and covered in fine hairs. They feed upon animal matter such as skins, bones and leathers, and can sometimes be found in store cupboards eating bacon, ham, cheese and pet foods. Occasionally they bore into timber but damage is normally only superficial. Cupboards should be inspected to locate the source of infestation which should be thrown away and the cupboards thoroughly cleaned.

Ambrosia or *pinworm beetles* leave circular holes varying in diameter from 0.5 to 3 mm. These insects are forest pests which attack standing trees and wet logs, and occasionally the damage caused by them can be seen in timbers used for joinery or structural work in new houses. The insects die in dry converted timber.

Wood wasps are 10 to 50 mm long and black, yellow or metallic blue in colour. They emerge from flight holes which vary from 6 to 9 mm in diameter. These are forest pests which attack softwood but can emerge from converted timber. They require wet timber with the bark present in which to lay their eggs, but there is no danger of them spreading to adjoining timber in the house.

Bark beetles are red chestnut brown to black in colour, vary from 3 to 6 mm in length and are covered by silky yellow hairs. They attack softwood logs or sawn timber where the bark is present and are occasionally seen in joists, rafters, fence posts and garden sheds where the bark has been retained. They normally disappear after a few years when the bark has been consumed, but it is good practice to remove any bark present on timbers in use.[38]

Repair of Flood Damage

Before starting work on a building that has been flooded, electricity, gas and water services should be tested and isolated if necessary. All water trapped in or around the building should be drained or pumped out as soon as possible. Checks should be made for water trapped in underfloor air ducts, service access pits, cavities in cavity walls, and areas under the building. Mud, silt and debris that have accumulated· against external walls, under boarded floors or in hollow wall cavities should be removed, and underfloor spaces sprayed with disinfectant.

After cleaning, the building should be heated and ventilated as much as possible, taking the following precautions—keep windows and doors open as much as possible, lift floorboards near walls to increase draught under floors, keep furniture and pictures away from affected walls and open cupboard doors. Porous building materials such as brickwork may take months to dry out. Where both sides of a wall have an impervious coating, it may be necessary to remove the covering from one side. All timber must be dried out as soon as possible to minimise the risk of fungal attack. Timbers attached to or embedded in damp walls are very vulnerable and require careful examination.

Structural damage can occur if there has been severe buffeting by flood waters; simple immersion is unlikely to cause such damage. Damage to foundations may occur during drying out if the subsoil is shrinkable clay.

Water trapped in electrical ducts or conduits must be removed by opening inspection elbows and conduit boxes. The electrical installation should be inspected and tested at monthly intervals for the first six months and at least twice in the next six months.

Seawater flooding creates additional problems, as walls contaminated by deliquescent salts in seawater may remain permanently damp. These salts may also cause severe corrosion of metal fastenings and electrical installations.[39]

References

1 I. H. Seeley. *Building Technology*. Macmillan (1986)
2 *BRE Defect Action Sheet 44*. Trussed rafter roofs: tank supports—installation (1984)
3 *BRE Defect Action Sheet 61*. Cold water storage cisterns: overflow pipes (1985)
4 P. J. Davidson and C. J. D. Webster. *BRE Information Paper IP 12/83: Water economy with the Skevington/BRE flush valve for WCs* (August 1983)
5 *BRE Digest 249*. Sanitary pipework: Part 2, Design of pipework (1981)
6 *BRE Digest 81*. Hospital sanitary services: some design and maintenance problems (1967)
7 H. S. Staveley and P. V. Glover. *Surveying Buildings*. Butterworths (1983)
8 *The Building Regulations 1985*: SI 1985 Nr 1065. HMSO (1985)
9 *BRE News 58*. Research on water supply: unvented hot water systems (1982)
10 *BRE Digest 253*. Heat pumps for domestic use (1981)
11 *BRE Digest 205*. Domestic water heating by solar energy (1977)

12 *BRE Digest 254*. Reliability and performance of solar collective systems (1981)
13 C. T. Gosling. Design considerations for easier maintenance of air conditioning equipment. *Building Maintenance* (October 1970)
14 DOE. *The Relationship between Design and Maintenance*. HMSO (1970)
15 *BRE Defect Action Sheet 62*. Electrical services: avoiding cable over-heating (1985)
16 *BRE Digest 272*. Lighting controls and daylight use (1983)
17 D. E. Sexton. *BRE Current Paper 12/73: Studies of refuse compaction and incineration*
18 British Standards Institution. *BS 5906: 1980 Code of practice for the storage and on-site treatment of solid waste from buildings*
19 *BRE Defect Action Sheet 50*. Flexibly jointed clayware drainage pipes: jointing and backfilling (1984)
20 R. Payne. *Drainage Maintenance: Estate Management*. Construction Press (1982)
21 E. D. Mills (Ed.). *Building Maintenance and Preservation*. Butterworths (1980)
22 D. Hughes and P. Bowler. *The Security Survey*. Gower (1982)
23 BRE. *Guidelines for the construction of fire resisting structural elements* (1982)
24 *BRE Digest 230*. Fire performance of walls and linings (1984)
25 *BRE Digests 214 and 215* Parts 1 and 2: Cavity barriers and fire stops (1978)
26 *BRE Digest 260*. Smoke control in buildings: design principles (1982)
27 British Insurance Association. The scale of the fire problem. *The Architect* (October 1973)
28 *BRE Digest 285*. Fires in furniture (1984)
29 *BRE Digest 300*. Toxic effects of fires (1985)
30 E. W. F. Hill. Premises management—European attitudes. *Fourth National Building Maintenance Conference*. HMSO (1973)
31 *DOE Construction 9: Cleaning and design*. HMSO (March 1974)
32 DOE. *Costs in Use: A Study of 24 Crown office buildings*. HMSO (1971)
33 Institute of Advanced Architectural Studies, University of York. *A study of costs in use and performance of floor finishes: analysis of student survey results* (1973)
34 A. V. Easton. *Achieving good standards in the cleaning of buildings*. CIOB (1982)
35 *DOE Construction 7: Maintenance of hygiene*. HMSO (1973)
36 H. Haverstock. Anti-graffiti measures. *Building Design* (29 April 1983)
37 *BRE Digest 238*. Reducing the risk of pest infestations: design recommendations and literature review (1980)
38 Timber Research and Development Association (TRADA). *Pests in Houses* (1986)
39 *BRE Digest 152*. Repair and renovation of flood-damaged buildings (1973)

7 ALTERATIONS AND IMPROVEMENTS

Many older buildings require alterations and improvements to meet the changing needs of occupants. Alterations and extensions often involve temporary supporting works of the forms described in chapter 1. Improvements to dwellings normally attract improvements grants, while districts of older dwellings pose the problem of rehabilitation or demolition and redevelopment, with a third possible alternative involving a mixture of both. This chapter concludes with a study of dilapidations and the preparation of structural and similar reports and proofs of evidence.

Scope of Alterations and Improvements

Dwellings need careful consideration and skilful planning to make the fullest use of their potential for improvement and conversion without at the same time adversely affecting their character. Sound doors and windows should not be replaced by inappropriate modern fittings, nor should pleasant large rooms be divided unnecessarily nor small sunny gardens be drastically reduced by kitchen and bathroom extensions.

Wider frontage dwellings provide scope for inserting a bathroom without losing bedspace, incorporating an entrance hall, or providing an extension without overshadowing rear rooms or garden. Deeper dwellings often permit the economical provision of internal bathrooms. A second entrance to a dwelling may be closed to provide more usable floor space on the ground floor. Stairs may be reconstructed to provide increased safety or a more effective first floor layout.

After improvement, dwellings should normally possess a minimum life of 30 years and careful thought and attention should be given to the following basic criteria.

(1) A main entrance door opening into an entrance porch or hall.

(2) Conveniently arranged bathroom, internal WC and spacious kitchen.

(3) All rooms provided with independent access from circulation areas, except possibly a kitchen from a dining or living room.

(4) Ideally two living spaces should be provided in family dwellings, although in small houses one could be a reasonably spacious dining kitchen.

(5) Use of appropriate materials and components externally to ensure harmony with existing elevations.

A survey of modernisation schemes for pre-1945 houses in 17 former country boroughs in England and Wales showed the following operations to be the most common:

(1) plastering of fair-faced brickwork;
(2) renewal of floors and laying of thermoplastic or PVC tiles;
(3) removal of old fireplaces from living room and bedrooms;
(4) installation of central heating systems;
(5) renewal of electric wiring;
(6) installation of 13 amp power points and 30 amp cooker control unit;
(7) provision of new sink unit;
(8) removal of larder and provision of kitchen floor units and ventilated wall cupboards;
(9) provision of new bath, wash basin and WC suite;
(10) removal of small paned steel windows.[1]

Work completed by one local authority on the renovation of pre-war houses is typical of the scope of improvement work which is frequently undertaken. These houses were gutted downstairs and the interiors rebuilt to provide a bathroom, separate WC, kitchen/dining area and living room, the latter being separated by a half-glazed partition and connecting door. The improvements to each house cost between £6000 and £10 000 in 1987 and included newly installed central heating.

Basic Improvements

There are a number of essential basic improvements that are needed in a large number of dwellings in this country to bring them up to an acceptable minimum standard of provision. These basic improvements include the provision of an internal water closet, bath, wash basin, replanning of kitchens, roof insulation, improvement of services and provision of fittings.[2]

Internal water closet. In 1981 the number of dwellings without internal water closets still approached 0.3 million,[2] and this was particularly serious as many of them were occupied by elderly people. Where there is an existing external WC it should if practicable be retained to provide an additional fitting. If the improvement results in a single internal WC, this should wherever possible be located independently of the bathroom.

Bath. In 1981 approximately 0.3 million dwellings in this country had no bath. There are various ways of making space for a bath, including conversion of a bedroom, extension of the dwelling or conversion of adjoining outbuildings. In a few cases space is so restricted that a bath cannot be provided and a shower may be installed as an alternative.

Wash basin.   In the absence of a wash basin, occupants of dwellings are obliged to wash at the kitchen sink, with consequent inconvenience and lack of privacy. A wash basin should wherever practicable be provided close to the WC. If the bathroom is not large enough to accommodate a bath, wash basin and WC, it may be possible to provide a wash basin and WC together in another part of the dwelling, although this is almost certain to prove more expensive. The substitution of a sliding door or an outward opening door for an inward opening door may sometimes provide the extra space needed for an additional fitting.

Kitchen improvements.   Kitchens to many houses built prior to 1919 and some later dwellings have unplastered brick walls. Any improvements should include plastering or tiling the walls to give improved hygiene and aesthetics. The floors may be of dusty, cold concrete which badly needs covering with a more attractive and functional material. Kitchens are usually replanned to make better use of available space and this may incorporate a former fuel storage area which is no longer needed because of the installation of new means of space heating. The additional space may permit the inclusion of a dining area within the kitchen which will relieve pressure on use of space in other parts of the dwelling. Larders may be replaced by refrigerators and food-storage cupboards.

Roof insulation.   Roofs should be insulated to an acceptable standard to reduce heat loss significantly and provide greater comfort, accompanied by lagging of storage tanks and service pipes.

Improvement of services.   Gas and water services may require partial or complete replacement to serve the new fittings. Most electrical installations to older dwellings are inadequate and need replacement preferably with 13 amp ring mains with ample provision of three-core socket-outlets, carefully sited to avoid breaking up valuable wall space needed by occupants for placing furniture. The installation of a 30 amp cable with consumer unit and cooker control unit are normally considered essential. With gas supplies, the normal minimum provision is a cooker supply and one additional point.
     Hot water supply constitutes a vital part of any modernisation scheme with full consideration given to the relative merits and costs of the different systems. An immersion heater ought desirably to be installed in the hot water cylinder rather than to leave it to the occupant. More recent modernisation schemes usually incorporate full central heating or some form of background heating, often accompanied by the replacement of the living room open fireplace with a gas or electric space heater and renewal of the old fireplace surround.

Fittings.   Many pre-war houses contained shallow earthenware sinks and wood draining boards which have sometimes been replaced with deep white sinks. Modern practice is to replace these fittings with sink units supported by purpose-made cupboards which are more durable, hygienic and attractive. There is a tendency to provide more built-in cupboard space than hitherto and this often makes a significant improvement to living conditions.

Space Heating

There is a wide range of heating systems available for use in improvement schemes and they are generally classified according to the type of fuel used.

Solid fuel systems can be either back boiler and radiators or central boiler and radiators. These systems are particularly popular in coal-producing areas. Their principal problems are provision of adequate fuel storage and conflict with smokeless zone requirements where coal is used.

Gas-fired systems comprise five separate arrangements: boiler and radiators; back boiler and radiators, sometimes combined with a radiant fire; fan-assisted warm air; balanced flue warm air convectors; and flued gas fires. The fan-assisted warm air systems may cause condensation problems and balanced flue warm air convectors may prove difficult to install because of insufficient external wall. Individual radiant gas fires are particularly suitable for connection to existing flues where the building is being extended.

Electric systems are of three main types: off peak radiators (sometimes including warm air distribution mechanism); off peak warm air systems (central heat storage with warm air distribution); and on peak warm air systems which require a high standard of insulation to secure acceptable running costs. Warm air systems can cause condensation problems. Off peak radiators and on peak warm air systems are relatively easy to install.

Oil-fired systems became less popular when oil prices escalated. Furthermore, the tank can be unsightly in small gardens and requires stringent fire precautions.

The *average age* of the housing stock is increasing and most of the dwellings found to be unfit or needing extensive repairs were old and the majority of them privately owned.[3]

Modernisation of Dwellings

Modernisation of older dwellings can take many forms and is obviously influenced by the form, layout, construction, condition and anticipated life of the dwellings themselves. Older dwellings frequently need altering to provide a separate bathroom and a WC entered from inside the dwelling. A thoughtfully arranged house improvement is detailed in figure 7.1.1 showing alterations to houses some seventy years old in New Earswick village, Yorkshire, and contained within a General Improvement Area.[4] The houses were reduced to their basic brick shell before renovation. Ground floors were screeded and all timberwork, including roofs, entirely replaced.

Fireplaces and fuel store were removed and central heating installed. A more convenient living room and separate dining room were provided; also a kitchen replacing the scullery and a food cupboard substituted for the

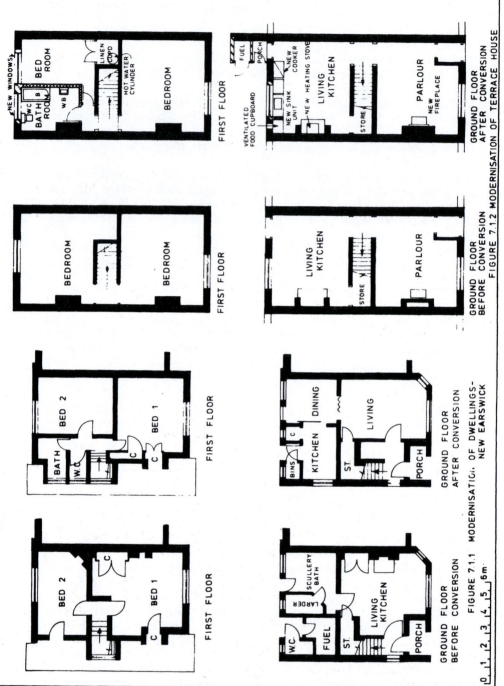

Figure 7.1

FIRST FLOOR

NEW WINDOWS
BED ROOM
WC
BATH B ROOM
WB
LINEN CUPD
HOT WATER CYLINDER
BEDROOM

FIRST FLOOR

BEDROOM
BEDROOM

FIRST FLOOR

BED 2
BATH
W.C.
C
C
BED 1

FIRST FLOOR

BED 2
C
BED 1
C

FUEL
PORCH
NEW COOKER
NEW HEATING STOVE
VENTILATED FOOD CUPBOARD
NEW SINK UNIT
LIVING KITCHEN
STORE
PARLOUR
NEW FIREPLACE

GROUND FLOOR AFTER CONVERSION
FIGURE 7.1.2 MODERNISATION OF TERRACE HOUSE

LIVING KITCHEN
STORE
PARLOUR

GROUND FLOOR BEFORE CONVERSION
FIGURE 7.1.2 MODERNISATION OF TERRACE HOUSE

BINS
C
KITCHEN
DINING
ST
LIVING
PORCH

GROUND FLOOR AFTER CONVERSION
FIGURE 7.1.1 MODERNISATION OF DWELLINGS- NEW EARSWICK

W.C.
FUEL
LARDER
SCULLERY BATH
ST
LIVING KITCHEN
PORCH

GROUND FLOOR BEFORE CONVERSION
FIGURE 7.1.1 MODERNISATION OF DWELLINGS- NEW EARSWICK

0 1 2 3 4 5 6m

212

rather space-consuming larder. Rearrangement of the second bedroom permitted the provision of a bathroom and separate WC on the first floor.

Vehicular access has been re-routed to the backs of houses, thus restoring the pedestrian character devised by Raymond Unwin in the original scheme. The average cost of improvements per dwelling amounted to £12 500 compared with a likely replacement cost of £24 000 (1987 prices), although the replacement houses would have a much longer life.

Another example of a house improvement is illustrated in figure 7.1.2, showing the conversion of a typical two-up and two-down terrace house built before 1914. These by-law houses are often soundly constructed and with appropriate repairs and improvements can be transformed into convenient and useful dwellings.

Before improvement the accommodation consisted of a front parlour, opening directly off the street, a living-kitchen behind, with access to a back yard; and on the first floor, two bedrooms of approximately equal size. Sanitary facilities were limited to an external WC at the far end of the yard, which had been a pail closet prior to 1938. Before conversion the house had no hot water supply and the cold water supply was restricted to the stone sink in the kitchen.

The improvements resulted in the provision of a bath, wash basin and new kitchen sink all supplied with hot and cold water, an inside WC and a ventilated food cupboard in the kitchen (the standard five point improvement). In this way it is possible to modernise small four-room houses without extending the existing structure. It does, however, reduce the size of the rear bedroom to provide space for the bathroom.[5]

Figure 7.2.2 shows a conversion of a small bedroom into a bathroom. The amount of new plumbing and drainage work is restricted to a minimum by positioning the new bathroom over the kitchen. In the example illustrated the landing has been extended to give access to the bathroom and bedroom 1, following the division of the original larger bedroom 1 into two bedrooms. Alternatively, the dwelling could be converted into a 2-bedroom house. An additional WC is required as the existing one is external.[6]

Many of the larger houses erected 60 to 80 years ago are now too large for the needs of present-day families. On the other hand they are often well built and lend themselves to conversion into flats. A number of different layouts are often possible; each of these needs considering and the advantages and disadvantages of each weighed one against the other. Each flat should ideally be as self-contained as possible, both visually and audibly. The former requires good layout and the latter adequate sound proofing. Adequate sound insulation is often difficult and sometimes virtually impossible. A reasonably efficient method of sound insulating upper wood joist floors is to 'pug' the space between the joists with sand or other heavy filling material, which absorbs and deadens sound but will not stop the noise from stereophonic hi-fi or even the average television set. In addition to providing each flat with its own private approach and entrance, the aim should be to plan the accommodation so that the noisier rooms of the various flats are adequately separated. Thus the living room of a first floor flat is better located over a bedroom of a ground floor flat rather than the living room.

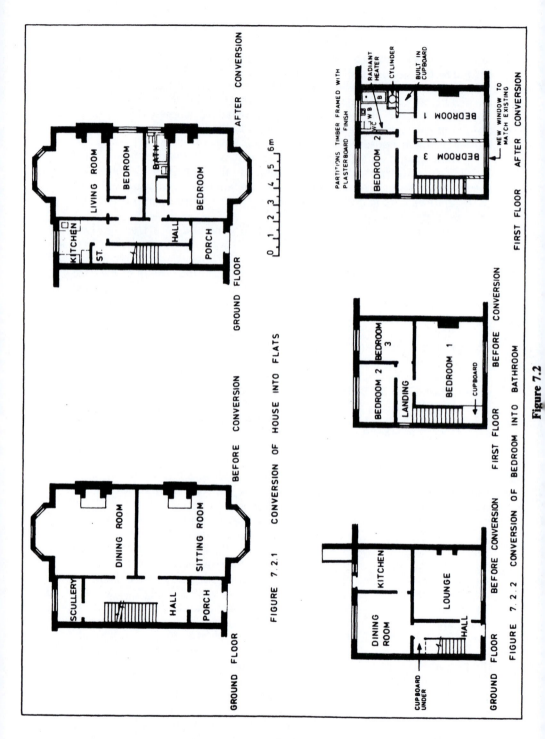

FIGURE 7.2.1 CONVERSION OF HOUSE INTO FLATS

BEFORE CONVERSION

AFTER CONVERSION

GROUND FLOOR

GROUND FLOOR

LIVING ROOM

KITCHEN

ST.

BEDROOM

BATH

HALL

BEDROOM

PORCH

0 1 2 3 4 5 6m

DINING ROOM

SCULLERY

SITTING ROOM

HALL

PORCH

FIGURE 7.2.2 CONVERSION OF BEDROOM INTO BATHROOM

FIRST FLOOR BEFORE CONVERSION

BEDROOM 2

BEDROOM 3

LANDING

BEDROOM 1

CUPBOARD

FIRST FLOOR AFTER CONVERSION

PARTITIONS TIMBER FRAMED WITH
PLASTERBOARD FINISH

RADIANT HEATER

CYLINDER

BUILT IN CUPBOARD

BEDROOM 2

WB

WC

B

BEDROOM 1

BEDROOM 3

NEW WINDOW TO MATCH EXISTING

GROUND FLOOR BEFORE CONVERSION

KITCHEN

DINING ROOM

HALL

LOUNGE

CUPBOARD UNDER

Figure 7.2

214

In the conversion illustrated in figure 7.2.1 it was unnecessary to interfere with the main internal walls. Furthermore, bathrooms (one to each flat) were placed centrally one above the other on each floor with consequent economy of drainage and plumbing services.[7]

Good examples of rehabilitating older housing as sheltered housing and for use by single persons are illustrated in *The Housing Rehabilitation Handbook*.[8]

Loft Conversions

Garages, sunrooms, utility rooms, and extensions to living rooms and kitchens can often be added to a dwelling on the ground floor, but the need for an additional bedroom may cause the family to contemplate a move to a larger dwelling. A loft conversion can often yield the extra bedroom and avoid the need for an expensive move.

Most loft conversions occupy little additional external space and normally come within the permitted extra space provisions of planning legislation, although any dormer at the front of a dwelling will need to blend in with the remainder of the elevation and neighbouring dwellings. Most dwellings with a pitched roof and a ridge height above ceiling joists in excess of 2.40 m can provide additional habitable accommodation through the construction of dormers.

The crucial factors are headroom and access. Wherever possible the staircase should continue over the existing stairwell, thus utilising the dead space over the existing stairs and avoiding taking space from a bedroom below. Once the staircase and dormer are plotted on the drawing, the remainder of the conversion work falls into place. In very old houses the existing roof structure will need to be examined very carefully and if used it may be necessary to form the walls, ceiling and floor as a box supported on the outer walls and largely independent of the roof. The existing ceiling joists are unlikely to be adequate as floor joists for the new room. A suspended floor constructed over the existing ceiling joists and bearing on either external walls, steel joists or trussed purlins is usually necessary. Any internal loadbearing walls will assist in reducing spans and permitting smaller section joists to be used. Rearrangement of purlins, struts, hangers and props will often be required to give unencumbered space.

The new room can be illuminated by a dormer window or a light in the slope of the roof, sometimes supplemented by a window in a gable wall. Dormers are often formed by bolting collars to the existing rafters and extending them outwards, and the dormer is often supported by a trussed purlin or steel joist. With the collars in position the dormer can be decked out and covered, often with built-up felt. The dormer roof can either slope to the front with a new gutter and downpipe or fall back to the main roof and drain into a box gutter. Once the dormer is secured the intermediate rafters can be cut away and the windows fitted.

Internal work consists mainly of erecting studding, inserting insulation and fixing plasterboard. The plasterboard is usually skim coated with any brick or block walls being rendered and set. Joinery work should ideally match the rest of the dwelling. Flooring and doors should both provide half-hour fire resistance. Electric wiring will require repositioning or replacing and the cold water tank may need relocating.

Houses in Multiple Occupation

A government report in 1987[9] indicated that more than 80 per cent of the bedsits, student lets and hostels in England and Wales lacked adequate fire escapes, and nearly half required substantial repairs costing more than £10 000.

Conversion of Other Buildings into Dwellings

Old oasthouses, windmills and watermills can provide most attractive and comfortable dwellings, while many barns, cowsheds, stables, blacksmiths' forges, and disused railway stations and schools have conversion potential. In all cases the structures must be thoroughly examined to determine its stability and suitability for conversion, although the majority of buildings with a good foundation and a sound roof are suitable.

Figure 7.3.1 shows a small village school converted into a two-bedroom bungalow. This conversion required no extensions; indeed a number of dilapidated outbuildings were demolished to provide adequate light and air. The school consisted of two classrooms which conveniently provided a bedroom wing and a living room wing.[10] Figure 7.3.2 illustrates a stable conversion which was carried out without major structural alterations. The architect was able to plan a good cottage within the main external walls of the old building. Drains and roof coverings often need replacement.[11]

Cantacuzino and Brandt[12] have described and illustrated the conversion to housing of warehouses in Amsterdam and Copenhagen, a tannery complex in Massachusetts, telephone laboratories in New York, a mercantile wharf in Boston, riverside granaries in Driffield, a warehouse loft in New York and a piano factory in Boston. These provide a good illustration of the wide range and scope for successful conversions.

Prefabricated Additions

The use of factory-produced bathrooms and other units can result in considerable time savings and suffer fewer problems when carried out in inclement weather. They are often constructed with treated timber stud panel walls, insulated with glass-fibre quilt or expanded polystyrene, and with foil-backed plasterboard internal lining. A number of external finishes are available in a wide range of colours. Floors are often formed of 18 mm plywood with underfloor glass fibre quilt insulation framed up on 50 × 100 mm joists and covered with vinyl asbestos tiles. Flat roofs are most common using 12 mm plywood decking possibly covered with three-layer felt, with suitable under-roof insulation and a foil-backed plasterboard

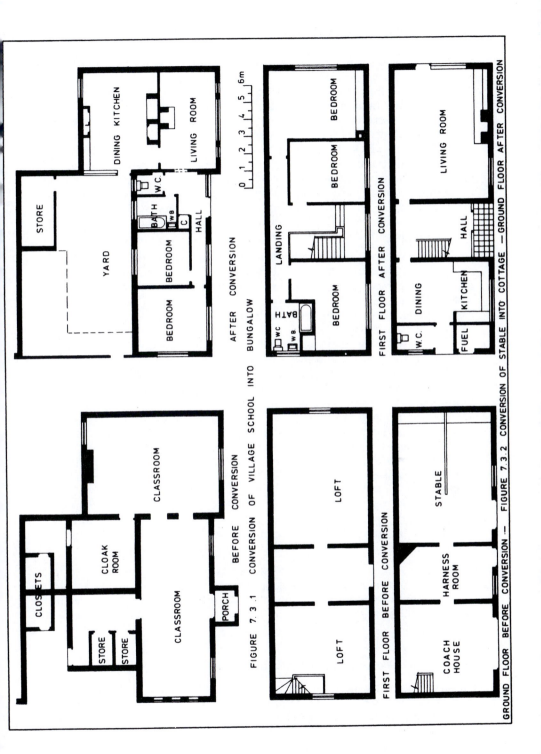

STORE

DINING KITCHEN

YARD

BATH

W.C.

WB

C

BEDROOM

BEDROOM

LIVING ROOM

HALL

AFTER CONVERSION

BEDROOM

LANDING

BEDROOM

BATH

WC

WB

BEDROOM

FIRST FLOOR AFTER CONVERSION

DINING

W.C.

LIVING ROOM

HALL

KITCHEN

FUEL

GROUND FLOOR AFTER CONVERSION

0 1 2 3 4 5 6m

AFTER CONVERSION OF BUNGALOW

CLOSETS

CLOAK ROOM

CLASSROOM

STORE

STORE

CLASSROOM

PORCH

BEFORE CONVERSION

FIGURE 7.3.1 CONVERSION OF VILLAGE SCHOOL INTO

LOFT

LOFT

FIRST FLOOR BEFORE CONVERSION

STABLE

HARNESS ROOM

COACH HOUSE

GROUND FLOOR BEFORE CONVERSION — FIGURE 7.3.2 CONVERSION OF STABLE INTO COTTAGE

217

ceiling. These units can incorporate double glazing units and central heating radiators.

The former Greater London Council designed a prefabricated bathroom unit for use with older dwellings with all works required after delivery of the bathroom unit scheduled to be carried out in seven working days. Preparatory work included removal of party fences and, in some cases, removal of air-raid shelters and/or timber structures situated within 6 m of a rear wall; boring for and forming precast concrete pile foundations; drainage work including building a new manhole and laying branch drains ready to connect up to the bathroom unit; cutting an opening in the rear wall of the existing WC for access to the bathroom unit and other ancillary work.

The work schedule following delivery and positioning of the bathroom unit on prepared foundations follows.

Day 1: Clearing out all redundant fittings in kitchen, preparing for gas and heating engineers and commencing modernisation of kitchen by replacing existing rear access door and wing light with prefabricated window board and splash back. Repositioning rising main in concealed position behind new kitchen units and providing temporary supply to WC.

Day 2: Gas fitters at work.

Day 3: Heating engineers installing gas-fired boiler in kitchen and heating units in passage and living room.

Day 4: Heating engineers complete work and wall linings and plumbing of kitchen commenced. Plumbing between kitchen and bathroom unit connected to bring new WC into use.

Day 5: Kitchen wall lining completed and fitting of kitchen units commenced and taps and water supply connected. A temporary sink is provided to cause tenants as little interference as possible. The new WC is in use and the existing WC suite is removed, branch drain sealed and a new airing cupboard formed, walls to new passage lined and painting commenced.

Day 6: Joinery works in kitchen completed and painting commenced. Decorations completed to new passage.

Day 7: Painting of kitchen completed; floor tiles laid and works cleaned up.

During the 7 days, work will also proceed with connecting drains to waste pipes from bathroom unit, completing rear paving and erecting new screen fences. Many parts required for the internal works will be pre-assembled in the contractor's yard to save time on the site.

Sequence and Management of Alteration Work

The dwelling must first be surveyed and floor plans of the existing accommodation drawn to an appropriate scale, usually 1:100 or 1:50. Alternative alteration schemes can then be planned and costed prior to selection of the approved scheme (chapter 9). Full particulars can then be prepared in the form of working drawings and specification (chapter 8), the necessary statutory approvals obtained (chapter 11) and tenders invited from a number of contractors (chapter 10). An improvement grant application is normally submitted to the local authority and this is usually accompanied by two copies of the drawings and a priced copy of the specification. The local authority before making the grant may request the carrying out of additional work, such as damp-proofing and the making good of defects to the roof and/or external walls. After receipt of all approvals, a tender will be accepted· and the successful contractor will be requested to commence work on the site.

The order of work on the site will be determined to some extent by the nature and scope of the work. In the case of an extension the first operation will be to clear the site and remove the topsoil, after which foundation trenches will be excavated and filled with concrete, which is allowed at least 48 hours to set and harden. External brickwork, usually of cavity construction, and usually incorporating an inner skin of insulating concrete blocks, follows, adequately bonded into the existing brickwork, and with window and door frames built in as and where required. Internal walls are usually built of concrete blocks and from the walls will be built upper floor and ceiling joists. The roof is then constructed followed by rainwater goods, hanging of external doors and glazing of windows. The building is now watertight and electrical and plumbing services and drainage work can proceed, followed by plasterwork, other wall and ceiling finishings and floor screeds.

Internal doors and their linings can now be fixed, followed by architraves and skirtings. The electrician can complete the fixing of switches, pendants and socket-outlets, while the plumber can install the sanitary fittings, connect to the hot and cold water services from the existing building and install waste pipes. Decorations should be carried out as soon as the walls are dry enough, normally comprising two coats of emulsion paint. Woodwork generally receives a priming coat, two undercoats and a gloss finishing coat, and all external joinery should receive suitable preservative treatment. All glazing is cleaned inside and out and floor tiles laid with suitable adhesive. All services are duly tested.

Because of the scale of modernisation work carried out by local authorities, most properties remain occupied while work is being carried out. The main exceptions are where tenants are sick or elderly or where extensive repairs are required, such as the provision of new floors. The primary objective is to execute the work as quickly and efficiently as possible with the minimum of inconvenience to tenants.

The size of a typical local authority modernisation contract varies between 100 and 200 dwellings, and handover rates of ten to fifteen per week are quite common. The time taken for modernisation is normally between eight and twelve weeks per house, so that a contractor can have over 120 houses in his possession at any one time.

Banton[13] has emphasised that speed is of the essence of such projects and that each house is a contract involving many trades. This necessitates very detailed planning and close co-operation between the local authority, sub-contractors, suppliers, statutory authorities and the tenant, to achieve a constant flow of completed properties. Resource planning must implement this aim and minimise delays between trades. Banton used a simple bar chart to establish the average duration of the work content of each dwelling and a second bar chart to show the possession and handover dates in relation to the overall contract period. Tenants must be kept informed of the programme.

The supply of materials on site should be adequate for at least two weeks' work. Materials and components should be stored in a central compound affording ample cover and security, and of sufficient size to permit bulk purchasing and deliveries. The operatives' canteen is often sited in the compound and a nearby house used for offices. Materials are distributed to the houses on a daily basis to reduce losses and minimise inconvenience to tenants.

At the start of the modernisation work it is often necessary to remove the hot water services backboiler and bathroom fittings. A common procedure is to leave the WC and install a 'temporary services unit' in the kitchen. The latter consists of a sink unit with a hot water heater. At the end of the working day this unit and the WC, cooker and living room heating must all be in working order. Houses are without personal washing facilities until the plumbing and central heating have been installed, which must be completed as quickly as possible.

Working on occupied estates requires discipline in terms of cleanliness and safety, having full regard to the needs of occupants. Operatives must be trained to work in a tidy manner and to minimise safety hazards. Rubbish must not be allowed to accumulate and should be removed on a daily basis using skips or lorries.

Crowter[14] has described very effectively both the practical and contractual problems that can arise in connection with modernisation contracts.

Meeting the Needs of Disabled People

The majority of existing buildings have been designed without any thought for disabled people; for example, many public buildings have imposing entrances approached by a large flight of stairs, making access by a chairbound person impossible. Since the early 1970s, designers have become more aware of the problems faced by disabled people. Hence new public buildings are accessible to chairbound people, specially reserved parking places are becoming standard provision and many public lavatories include special facilities.

There are however conflicts between the requirements of the able bodied, the ambulant disabled and the wheelchair user. Some of the more common conflicts are:

(1) *Ramps and steps.* Ramps are essential to the mobility of most wheelchair users, while an ambulant disabled person will normally find a flight of stairs with a handrail easier to manage. Women wearing high-heeled shoes find ramps awkward and ramps can be more hazardous than steps in wet and frosty weather.

(2) *Floor finishes*.   The ambulant disabled favour non-slip, roughened and resilient finishes, while wheelchair users require a smooth, non-resilient and hard finish.

(3) *WC compartments*.   A WC compartment for the wheelchair user needs to be wide and spacious, whereas the ambulant disabled person requires a narrow compartment with support rails on both sides.

A joint government circular[15] proposed that responsibility for identifying, assessing and advising on the housing needs for individual disabled persons, including the need for adaptations to their homes, should remain with the social services departments, in collaboration with health authorities. Housing authorities are responsible for work involving structural alterations to houses owned or managed by them, while social services departments deal with non-structural modifications and the provision of aids and equipment.

It is usually cheaper and more satisfactory to the occupant to adapt an existing house for a chairbound person if it is structurally feasible. Social services authorities have a duty under the Chronically Sick and Disabled Persons Act 1970 to assist those who are handicapped, if it is demonstrated that adaptations will make it easier for the handicapped person to manage.

The following items of work are some of the more common adaptations that may be necessary and are normally grant aided.

(1) Substituting ramped access for steps (normally to a maximum gradient of 1:12).

(2) Widening doorways, normally to 900 mm minimum width, and deepening windows to 800 mm above floor level.

(3) Changing door and window ironmongery (probably to lever safety type about 1 m above floor level).

(4) Extra handrails to staircase (usually 45 to 50 mm diameter on both sides of staircase and extending 300 mm horizontally at top and bottom of staircase).

(5) Support rails to bath and WC (usually 30 to 45 mm diameter).

(6) Raising level of WC seat (often about 50 mm to about 475 mm above floor level).

(7) Extending accommodation to provide ground floor bath, wash basin and WC.

(8) Altering kitchen fittings (worktop surfaces usually 800 mm above floor level and no handles lower than 450 mm; sink units with shallow bowls; and cookers of split level type.[16]

(9) Changing taps from pillar to lever operation.

(10) Provision of ceiling track for hoist.

(11) Installation of stair lift (details of different types in Lockhart[16]).

(12) Moving pre-payment gas and electric meters to about 1 m above floor level.

(13) Moving socket outlets and light switches (preferably 1 m above floor level). Traditional light switches can be replaced by ceiling pull-cord type, or by a rocker plate switch, operated by elbow or shoulder.[16]

(14) Provision, if necessary, of some form of background or central heating in all accessible rooms, with some local fires or heaters to boost

individual room temperatures as desired, complemented by adequate insulation against heat loss.[16]

Grants for Improvements and Conversions

Types of Grant

Grants are payable by local authorities towards the cost of work required for (a) the provision of dwellings by conversion of houses or other buildings, (b) the improvement of dwellings, (c) the repair of dwellings, and (d) the improvement of houses in multiple occupation by the provision of standard amenities, where the provision, improvement or repair is by a person other than a housing authority.

An *improvement grant*, which is mandatory, can be given at the local authority's discretion in respect of works required for the provision of a dwelling by conversion of a building or for improvement of a dwelling beyond the works covered by an intermediate grant or, in the case of a registered disabled person, works required for his welfare, accommodation or employment where the existing dwelling is inadequate or unsuitable.

An *intermediate grant* relates to work required for the improvement of a dwelling by the provision of standard amenities which it lacks or which in the case of a registered disabled person are inaccessible to him. The *standard amenities* include a fixed bath or shower, hot and cold water supply at a fixed bath or shower, a washhand basin, hot and cold water supply at a washhand basin, a sink, hot and cold water supply at a sink, and a water closet. A *special grant* is allowable for works required to improve a house in multiple occupation by the provision of standard amenities or means of escape from fire, together with related repairs.

A *repairs grant* can be given for works of substantial and structural repairs to old dwellings (pre-1919) and is discretionary. The authority must be satisfied that after the completion of the works, the dwelling will be in reasonable repair, having regard to its age, character and locality.

The appropriate percentage to determine the maximum amount of any grant is 75 per cent in a housing action area; 65 per cent in a general improvement area; and 50 per cent in any other case. Where the local authority approving the grant feels that the applicant will not, without undue hardship, be able to finance the cost of works not covered by the grant, they may increase the percentage of 50 per cent to 65 per cent and 75 per cent to 90 per cent.

Improvement Grants

Under the provisions of the Housing Act 1985, to qualify for an improvement grant, the local authority must be satisfied that on completion of the works the dwelling will be provided with all the standard amenities for the exclusive use of its occupants; that it will be in good repair having regard to its age and character and the locality; that it conforms to certain general standards laid down by the Secretary of State; and that it is likely to provide satisfactory housing accommodation for 30 years. In certain circumstances it may be necessary to improve houses which cannot meet these requirements and the local authority is empowered to vary the general conditions relating to grants, and can reduce the 30 years life expectancy to not less than 10 years.

Furthermore the applicant must be a freeholder of the property for which grant aid is required, or hold a lease with not less than five years unexpired at the time of the application. All applicants must provide a certificate relating to the future use of the dwelling. An improvement grant will not be given for a dwelling erected after 2 October 1961; a second or holiday home; an owner-occupied house which has a rateable value in excess of the permissible limit (£400 in Greater London and £225 elsewhere in 1986); or where work has already begun (except where the local authority accepts that there was good reason to start). Where the local authority approves an application for an improvement grant, it will notify the applicant of the amount which, in its opinion, the work should cost and the amount of the grant it has approved.

The amount of improvement grant paid is based on a percentage of eligible expenses, varying with the location of the dwelling, and the 1986 cost limit of £10 200 for single dwellings and £11 800 for conversion of buildings of three or more storeys outside London. The corresponding figures for Greater London were £13 800 and £16 000 respectively. The 1986 limits on grants are shown in Table 7.1

Table 7.1 Improvement grants

Area		Normal	3 or more storeys
Housing action areas	(75%)	£7650	£8850
Hardship cases (HAA)	(90%)	£9180	£10620
General improvement areas and hardship cases (elsewhere)	(65%)	£6630	£7670
Elsewhere	(50%)	£5100	£5900

There is a strong case for periodic reviews of grants so that the prescribed limits can keep pace with inflation.

Intermediate Grants

The intermediate grant is a mandatory grant designed to provide standard amenities, and these are listed in table 7.2 extracted from S.508 of the Housing Act 1985.

The grant towards the cost of these standard amenities will vary according to the area, namely 75 per cent in a housing action area; 65 per cent in a general improvement area; and 50 per cent elsewhere, subject to the provisos with regard to hardship cases. In many cases, before it is practicable

Table 7.2 Intermediate grants

Description of amenity	Maximum eligible amount (£)	
	Greater London	Elsewhere
Fixed bath or shower	450	340
Hot and cold water supply at fixed bath or shower	570	430
Wash hand basin	175	130
Hot and cold water supply at wash hand basin	300	230
Sink	450	340
Hot and cold water supply at sink	380	290
Water closet	680	515

to install an amenity, it may be necessary to carry out other relevant works of repair or replacement. In such cases grant aid may also be approved for works other than the installation of amenities, at rates graded according to the type of area. Claimants of intermediate grants must satisfy the status requirements but the rateable value restrictions do not apply.

Applicants must specify the standard amenity or amenities which it is intended to provide, and where some only are listed must state whether the dwelling is already provided with the remainder. In addition they must state whether the dwelling has been without that amenity for a period of not less than 12 months. Before approving an intermediate grant the local authority must be satisfied:

(a) that on completion of the relevant works the dwelling(s) will be fit for human habitation, or

(b) it seems reasonable in all the circumstances to approve the application even though the dwelling(s) will not reach that standard on completion of the relevant works: as observed by Arden,[17] this is a curious provision.

Where an application for improvement grant is approved the local authority will separately determine the amount of expenses to be incurred in repair and replacement, and for the installation of standard amenities. Except where the Secretary of State approves higher expenses, the grant for repair and replacement will be limited to the appropriate proportion of a sum not exceeding £800 in addition to that for the installation of standard amenities, increased to £2000 where it appears to the authority that the applicant could not without undue hardship finance the cost of the works necessary to put the dwelling into reasonable repair.

Special Grants

Special grants are designed to enable a house in multiple occupation to be improved by the provision of standard amenities or means of escape from fire. Both standard amenities and repairs expenses are as detailed for intermediate grants and a limit of £6750 is prescribed for provision of means of escape from fire.

Repairs Grants

A repairs grant, which is discretionary, may only be given in connection with a dwelling built before 1919 and S.491 of the Housing Act 1985 expressly excludes works to improve an existing dwelling or to provide additional dwellings. The application must be accompanied by a certificate of owner-occupation and if not situated in a housing action area, the rateable value must not exceed the prescribed limit. The relevant works must be of a substantial and structural character, resulting in a standard of reasonable repair. The limit for eligible expenses is £800.

House Insulation Grants

An occupier of a dwelling completed before 1 January 1976 can apply for a home insulation grant under the home insulation scheme 1984, for insulation installed in the loft of a pitched roof which was not previously insulated to a depth of 30 mm or more. Grants are payable upon completion of the works and include work done by the occupier. The grant paid to applicants in special need is the lesser of 90 per cent of the cost or £95, and to other applicants, the lesser of 66 per cent of the cost or £69.[18]

Environmental Works

Owners in housing action areas only may apply for grant aid to bring the external appearance of their houses up to a suitable standard. The grant may be paid after completion of the work, or part of it may be paid in instalments as the work is executed and the balance after completion, except that the aggregate of instalments paid shall not at any time before completion exceed one-half of the aggregate cost of the works executed up to that time. A local authority is empowered to provide materials, such as paint, for carrying out environmental works.

Assistance for Owners of Defective Housing

Part XVI of the Housing Act 1985 makes provision for reinstatement grants to be paid to those who purchased 'system-built' housing, which has been designated by the Secretary of State, from local or public authorities and which is defective, but which can be reinstated to give a life of 30 years. The percentage of the cost payable is 90 per cent or 100 per cent where financial hardship would occur. Another alternative is repurchase by the authority at 95 per cent of the defect-free value of the property and incidental expenses in full.

Payment of Grants

A local authority may prescribe a time (not less than 12 months) within which the relevant works must be completed. The payment of an instalment automatically sets a time for the completion of the works—12 months from the date of payment of the instalment—even where no time limit was originally prescribed. Such time limits can be varied by the local authority but failure to complete within the prescribed time can result in the clients being required to refund the instalment plus accrued interest.

A source of irritation to builders has been the fact that local authorities pay the grant to the applicant and not to the builder. On occasions the client not only fails to pay the builder on receipt of the grant when the works are complete, but then proceeds to spend the grant money. It is in the builder's interest to obtain agreement from the client before work commences that the local authority will pay the grant directly to the builder. The most he would then have to recover would be 25 per cent in housing action areas and 50 per cent for the average intermediate grant. Where the increased grant is operative in cases of 'undue hardship', a prior agreement with the client is vital as his restricted personal finances hold out little hope of recovery by recourse to law.

Distinguishing between Improvements and Repairs

It may not always be practicable to separate improvement work from repairs associated with the improvements and in these circumstances the repairs would rank for grant aid. Nevertheless many other works of repair are usually eligible for grant such as reslating or retiling roofs and renewing roof timbers and rainwater goods; replacing defective plaster; renewing or repairing defective walls; replacing defective floors and infected timbers; removal of disused chimney stacks and making good roof; renewal or repair of defective joinery; replacement of obsolete sanitary fittings and plumbing systems; renewal of perished or obsolete electric wiring and essential repairs to drainage systems.

Rehabilitation or Redevelopment

Advantages of Rehabilitation

A government white paper[19] asserted that in the majority of cases, comprehensive development is no longer the answer to problems associated with bad housing. Some of the social costs of redevelopment can be quantified —for example, the extra houses needed to rehouse people elsewhere. Others are less tangible but no less real—the personal misery and distress suffered by people who have to abandon surroundings with which they are familiar and friends for whom no substitute can be found. Many of the residents are elderly with low incomes and their lack of skills makes them least able to cope with the upheavals which follow major redevelopment.

Research has shown that people in this situation can be very satisfied with their houses even though they may have WCs in the backyard or be substandard in other ways. In this situation, well-planned programmes of rehabilitation would be a better solution than redevelopment. It takes many years for an area of comprehensive redevelopment to mature into a community.

The majority of older houses which are in sound basic condition are a valuable asset to the community and can be made into good homes with many years of useful life ahead of them at a smaller cost in money and other resources than equivalent new houses. Where rehabilitation can be carried out over large areas of housing, linked with environmental improvements, considerable social advantages can be gained through the preservation of local communities, and by preventing the distressing process of decay and neglect by which older houses become slums. Existing houses can make a valuable contribution to the character and variety of our cities, towns and villages.

Rehabilitation has been aptly defined as "a carrying out of building work to any property, or series of properties beyond normal routine maintenance, thus extending its life to provide a building or buildings which are socially desirable and economically viable."[20] It has, in many cases supplanted the terms 'conversion' and 'modernisation' and implies a broader approach to embrace the environment as well as interiors of dwellings. Hence proposals may include car parking, open spaces, children's play spaces and better facilities for storage and collection of refuse.

Financing of Housing Rehabilitation

People will invest in an enterprise if they can see a reasonable return. Rent restrictions and controls have for many years prevented the operation of normal market forces with the consequent neglect of many private rented dwellings as landlords were unable effectively to maintain their properties, because of the high level of cost involved and the low return. This constitutes the source of most of our present rehabilitation problems.

There are three principal methods of financing rehabilitation schemes in the private sector.

(1) Schemes of a non-profit producing character normally carried out by housing associations funded by public money through provisions in the Housing Acts. The financial viability of a scheme is controlled by an arbitrary upper limit to the total cost per unit of accommodation produced.

(2) Profit producing schemes executed by housing trusts, public and private companies and some individuals. Generally the rehabilitation of outdated houses for rack renting on completion is unattractive to the private sector because of the low return, sometimes as low as 5 per cent.

(3) Schemes of conversion by owner-occupiers taking advantage of the provisions of the Housing Acts and the declaration by local authorities of general improvement areas is likely to give a considerable boost to this activity.[20,21]

The financial viability of rehabilitation is influenced by a range of factors.

(1) Structural condition of property and level of accrued dilapidations.

(2) Anticipated residual life.

(3) Adaptability of existing layout.

(4) Acquisition costs—affected by pressures exerted by various agencies attracted by inherent rehabilitation potential.

(5) Cost of building work, particularly as ratio of labour to material cost is higher than in new work.

(6) Planning requirements—density limitations can result in a loss of 30 to 50 per cent on redevelopment of three and four-storey Victorian and Edwardian houses.

(7) Effect of legislation and subsidies.

(8) Rental levels.

(9) Interest rates and funding periods—high interest rates and relatively short funding periods have had an inhibiting effect on the rehabilitation plans of some local authorities.

(10) Annual maintenance costs—with simple brick buildings where rehabilitation work has been well down, annual maintenance should not be significantly higher than for new buildings, whereas stucco-faced properties with moulded cornices and similar features are likely to be considerably more expensive to maintain.[20]

In a cost appraisal it will be necessary to assess the value of rehabilitation compared with other potential uses of the site. The economics of the decision whether or not to proceed is usually determined in the private sector on the basis of percentage return on capital invested, but with local authorities and some larger property companies, a further calculation based on the resultant capital value is often considered. With capital value, rehabilitation would appear to be preferable if the cost of rehabilitation for a life of x years plus the cost of redevelopment deferred x years is less than the cost of present redevelopment. This approach however takes no account of quality and in many cases the modernised dwelling is worth only a percentage of the new equivalent because of poor layout, lack of modern amenities and open space, and similar inadequacies. The assessment of quality can often be resolved by open market prices and rents. Having determined the viability of the scheme in capital terms, the next step is to make a comparison of the annual income and expenditure of rehabilitation and redevelopment.[20]

Redevelopment permits amortisation over a longer period and probably lower maintenance costs with less constraints on design. On the other hand, financial restrictions tend to produce minimum standards of accommodation and appearance resulting in loss of character and construction generally takes longer.[20]

On rehabilitation it is often necessary to improve the environment by a certain amount of selective demolition to create open space and provide garages, redirect roads to give improved traffic circulation and to eliminate the worst property. Improvement works may include the removal of rear additions, the conversion of three houses into two in some cases to provide three or four-bedroom houses and the removal of chimney stacks.[22]

Cost–benefit Studies

There is increasing support for the application of cost–benefit analysis to comprehensive redevelopment proposals with, in particular, consideration of social costs. Cost–benefit analysis has been defined as "a technique of use in either investment appraisal or the review of the operation of a service for analysing and measuring costs and benefits to the community of adopting specific courses of action and for examining the incidence of those costs and benefits between different sections of the community." The technique involves attributing monetary values to present and future costs and benefits, including social costs and benefits which are not normally expressed in monetary terms. Future costs and benefits are discounted to present-day figures by using an appropriate long-term rate of discount. The result is either to compare the present day values of costs and benefits or to express the net benefit as a rate of return on the capital investment involved.

This technique has not been used very extensively in comparing the economics of redevelopment and rehabilitation. The primary reason has been that when an area is considered for clearance, legal and technical considerations (notably public health aspects) are uppermost and it is frequently stated by public health or environmental control officers that no reasonable alternative to clearance exists. A government white paper[23] supported the use of cost–benefit analysis in these situations.

An interesting study was undertaken at Leeds to find the most economical way of providing acceptable living standards (both housing and environmental) in existing areas of sub-standard housing. The study concluded that assuming that the improved property had a life of 20 years and considering the cost to public funds as a whole, improvement was more favourable at all discount rates over 4 per cent. This highlights one of the main criticisms of cost–benefit analysis, that the results can be varied as desired by using different discount rates, and that the values attached to social factors are often quite arbitrary. In cost–benefit studies in the public sector, a discount rate of 10 per cent is usual.

Needleman[24] suggested that the test of replacement or repair is whether the cost of replacement exceeds the cost of modernisation plus the present worth of rebuilding at the end of the renewed life of the old asset plus the present worth of the difference in annual maintenance expenditure. Thus in terms of an equation, it is worth modernising where:

$$b > m + b(1 + i)^{-n} + \frac{r}{i}[1 - (1 + i)^{-n}]$$

where b = cost of demolition and rebuilding
m = cost of adequate modernisation
i = rate of interest
n = useful life of modernised property in years
r = difference in annual repair costs

Kilroy[25] refined this approach to incorporate:

(1) quality of rehabilitated dwelling expressed as a percentage of one redeveloped;
(2) respective lives of development;
(3) discount rate of interest;
(4) difference in maintenance costs;
(5) effect of changes in price levels;
(6) for publicly subsidised housing; the subsidies receivable.

The main considerations are:

(1) how much is it worth spending on improvement compared with redevelopment?
(2) how can areas for improvement be selected to secure best value for the resources used?

Improvement normally costs considerably less than redevelopment but, after improvement, the standard of the accommodation is likely to be lower than that provided by a newly-built flat or house and its useful life will be shorter. The maximum amount worth spending on improvement, given the cost of redevelopment, will depend *interalia* on the standard of accommodation after improvement, the useful life and the rate of interest. Present high rates of interest may well sway the balance in favour of improvement which will lead to smaller calls on capital.

MHLG (DOE) Circular 65/69 contained the data shown in table 7.3.

Table 7.3 Quality of improved dwellings as compared with new dwellings

Useful life (years)	Quality of improved dwellings as a percentage of that of a new dwelling at 8 per cent					
	100	90	80	70	60	50
40	0.96	0.87	0.77	0.68	0.58	0.48
30	0.91	0.82	0.73	0.64	0.55	0.45
20	0.79	0.71	0.63	0.56	0.48	0.40
15	0.69	0.62	0.55	0.48	0.42	0.35

These calculations are based on an interest rate of 8 per cent which could be considered low. The Needleman formula provided a rough and ready method of comparison but contained several fundamental weaknesses. It ignored the effects of changes in price levels and excluded social costs, such as the social effects of displacements, although these are admittedly extremely difficult to evaluate. Two other important social factors are the degree of satisfaction which residents express with their existing conditions and the extent of housing choice available to them. In practice a combination of approaches will probably be desirable; as adopted in Camden, by the gradual process of redevelopment and rehabilitation by the local authority, linked with a certain amount of controlled improvement by private owners in general improvement areas, the standard of housing in an area of 3 km² was steadily raised and the traditional population was largely kept together at rents it could afford.

Dilapidations

Liability for Repairs

The term 'dilapidations' denotes a condition of disrepair which has been caused or allowed to develop in the property, and which will involve the person responsible in legal liabilities. The person whose acts of omission or commission has caused the dilapidations, is normally one with a limited interest in the property, such as a tenant for life or a lessee under a lease, whose neglect to keep the property in a good state of repair will have detrimental consequences for those who are to take over possession of the property when his interest terminates.[26]

A lease usually contains a number of terms and conditions agreed upon by the parties. Certain covenants may be implied and these are often referred to as 'the usual covenants'. Among the usual covenants are those by the tenant that he will keep and deliver up the premises in repair, and allow the landlord to enter and view the state of repair. The tenant may however expressly covenant 'to repair the premises and to yield them up in good and substantial repair and condition'. To repair implies that the structure, fixture or installation is rendered fit to perform its proper function. Repair often involves the replacement of a part but it cannot be extended to encompass complete rebuilding.[26]

'Good tenantable repair' has been defined as "such repair as having regard to the age, character and locality of the house, would make it fit for the occupation of a reasonably minded tenant of the class who would be likely to take it." This definition must however be qualified. It has been contended that where a neighbourhood has seriously declined during the period of the lease, the tenant need only repair to an extent necessary to bring the house up to the new debased standard.

Often the tenant is made liable for repairs 'fair wear and tear excepted' This means that the tenant will not be liable for disrepair resulting from the normal actions of the elements (wind and rain), or to normal use by the tenant. He would not therefore have to repair worn stair treads, replace broken sash cords or renew slates or tiles which have slipped from the roof. He will, however, be liable for exceptional damage caused by the elements, such as hurricanes or floods, and for damage arising from the improper use of the building, such as over-loading the upper floor of a warehouse.[26]

Under the lease the responsibility for repairs may be shared between landlord and tenant; for example, the landlord may be made expressly liable for external repairs and the tenant for internal repairs. Whether the tenant will be required to repair buildings erected subsequent to the grant of the lease will depend largely upon the wording of the covenant. Responsibility for repair of fire damage will normally be covered by insurance provisions in the lease. These will prescribe who is to insure and for what amount, production of receipts for premiums, and an undertaking to expend any sum received from the insurance in rebuilding.

Painting presents a problem in classification since it can serve two purposes. It may be undertaken to preserve woodwork and metalwork from decay and thus be classed as repair, or it may be used solely for purposes of decoration to improve appearance and comfort. Some painting will serve both purposes and is termed 'decorative repair'. Leases commonly contain a covenant requiring the tenant 'in every third year to paint all outside woodwork and metalwork with two coats of suitable oil colours in a workmanlike manner . . . and in every sixth year paint other outside works, now or usually painted, all internal woodwork and metalwork . . . and also paint with two coats of emulsion paint, such parts of the said premises as are now plastered'. In practice it is better to state the particular years in which the work is to be done and to provide for it to be undertaken in the last year of the term. Where the covenant is not specific and the tenant is liable for repairs, he must paint as necessary to preserve woodwork and metalwork from decay.

Preparing Schedules of Dilapidations

A surveyor who is instructed to prepare a schedule of dilapidations will first examine all the relevant documents and particularly the repairing covenants in the lease to establish the extent of the lessee's obligations. He will then inspect the premises and compile a list of defects and also take measurements to build up an estimate of the probable cost of repairs where the lease is about to expire. The schedule of dilapidations is then prepared and served on the lessee by a solicitor. The schedule must contain the location of the property, the date of inspection and the name and address of the person preparing it. A RICS guidance note[27] describes how the schedule should clearly and concisely list the defects and the necessary repairs.

Schedules of dilapidations are of two kinds.

(1) Interim schedules served during the currency of a lease with a notice to repair, the object being to enforce a right of entry or forfeiture if the tenant fails to carry out repairs.

(2) Terminal schedules prepared towards the end of a lease which form the basis for a monetary claim against the tenant for his breach of covenant to keep and leave in repair.

In the first case the surveyor prepares a list of defects for which the tenant is liable under the repairing covenant and which he is required to repair and make good. In the second case the surveyor prepares a schedule of claim, containing the quantities and cost of each item, in addition to a list of defects. The schedule of claim is not produced unless required by an order of court or for some special reason. A schedule of dilapidations and a schedule of claim are included later in the chapter under Proofs of Evidence. A Scott schedule may be prepared for negotiation purposes as illustrated in *Building Surveys, Reports and Dilapidations*.[26]

The inspection of the premises should be carried out in a methodical manner to avoid the omission of items and to make it easier for other persons to work through and check the schedule. Most surveyors start with the interior of the building on the uppermost floor, commencing with the rooms at the front of the building. One means of identification of the rooms is to describe them as front room right, front room middle and front room left, the handing taken from a position looking out of the windows. The same procedure is adopted for the rear; thus the front room right is on the same side of the house as the back room left.

In each room a logical sequence of items should be followed such as ceiling, cornice, frieze, walls, wood trim, doors, windows, fireplace, fittings, floor and electrical installation, with all defects carefully noted, even down to the extent of cracks. When examining doors and windows, the ironmongery should receive special attention, doors checked to see whether they close properly and whether windows can be opened and closed. The investigation of the interior is completed with the inspection of the staircase, corridors, lobbies, porches and cupboards. The exterior then follows, taking each elevation in turn and bearing in mind that some roof areas may be out of sight. Finally outbuildings, fences, paved areas and drainage work are inspected. The underlying causes of defects must be identified in all cases.

Schedules of Condition

A schedule of condition is a report on the condition of a property at a specified date, set out in sufficient detail so that any part of the structure, finishings or fittings which subsequently becomes defective or missing can be readily identified. It is good practice for both parties to a lease to have a professionally prepared and jointly agreed schedule of condition of the property at the commencement of the lease to prevent subsequent disputes. They are often prepared in the form of abridged specifications as illustrated in *Building Surveys, Reports and Dilapidations.*[26]

Technical Reports

Nature of Reports

Surveyors are often instructed by property owners or prospective purchasers to inspect and report on the condition of property. The report should contain all the relevant technical information set out in an orderly manner in terms that can be understood by a layman.

The basic requirements of a good report are as follows.

(1) Accuracy—a report must be accurate in all respects as errors or vague statements will detract considerably from the value and credibility of the report.

(2) Simplicity—freedom from technical terms as far as practicable, and where used they need explaining.

(3) Clarity—presentation and arrangement of information should be in a logical order with sufficient headings and sub-headings to act as signposts. Each paragraph should be complete in itself and yet so related as to lead to an ultimate conclusion through a series of steps.

(4) Systematic approach—reports normally comprise three component

parts: (a) introduction often containing client's brief or object of report; (b) main body of report or recital of facts, including a full description of conditions as they exist; (c) conclusions and recommendations, often including an estimate of cost of remedial works.

(5) Conciseness and completeness—the report should cover all matters coming within its scope, yet should be kept as concise as possible.

(6) Good grammar and correct spelling.

(7) Neatness—reports should be clear and legible, particularly in examinations, and be free from abbreviations and contradictions.

(8) Certainty—advice and opinions should be definite or the reasons for any uncertainty stated (a saving clause is advisable to cover hidden parts).

(9) Recommendations—advice as to action to be taken.

Reports on the Structural and Sanitary Condition of Property

This type of report may be required by a prospective purchaser of a property or by an owner to determine the extent of maintenance and repair work. Such a report might include the following matters.

(1) Situation—situation and general details of property.

(2) Accommodation—particulars and sizes of rooms.

(3) Construction—details of construction and materials used.

(4) Repairs—defects and repairs needed.

(5) Services—particulars of gas, electricity and water services.

(6) Heating and hot water—form and adequacy of heating and hot water supply.

(7) Drainage—adequacy and condition of drainage system.

(8) Estimate—probable cost of essential repairs and redecorations.

A good example is contained in *Building Surveys, Reports and Dilapidations*.[26]

Report on a Building Defect

Background information. A substantially brick-built house has a parapet supporting a lead parapet gutter to a tiled roof. The owner of the property, who is responsible for repairs, has requested a report on the tenant's complaint that rain has caused a damp ceiling in the front bedroom. The report is incorporated in a letter to the client.

Report
J. P. Isaacs, Esq.   15 September 1987
Homeland
Little Rising
Lincolnshire

Dampness at 28 Willow Way, Marchby

Dear Sir

Cause of complaint
Following the complaint of dampness in the front bedroom of this house, I examined the parapet wall, gutter and adjoining roof on 14 September 1987 and found them to be soundly constructed. Several defects have however developed which have resulted in rainwater penetrating the front bedroom ceiling. These defects are as follows.

(1) A length of lead flashing forming the junction between the parapet wall and the lead gutter has come away from the wall, thus permitting rainwater to pass down the inner face of the wall at this point.

(2) A length of lead gutter covering has been damaged by some sharp object being forced through the lead, possibly the feet of a ladder.

(3) The gutter has become badly choked with decaying leaves and other debris, causing water to build up in the gutter and ultimately flow over the top edge of the lead covering to the gutter, where it passes up the tiled roof slope. The position has been aggravated because the lead gutter covering does not extend as far up the roof slope as is really desirable, nor is it dressed over a tilting fillet (triangular piece of timber) below the bottom course of tiling, which would have produced a much sounder job.

Remedial works
I recommend that the following remedial works be carried out by a selected builder

(1) Hack out the brick joints and rewedge and repoint the loose length of lead flashing.

(2) Replace the defective length of lead gutter covering.

(3) Clear all debris from the gutter.

It is not considered necessary to replace the lead gutter covering with wider sheets or to fit a timber tilting fillet, provided the gutter is cleaned out regularly, preferably annually.

The estimated cost of carrying out the repair work to the parapet gutter is £110.

Additionally, the decorations to the ceiling of the front bedroom are badly discoloured and need redecorating with two coats of emulsion paint, after the external repairs have been carried out and the plasterboard ceiling has adequately dried out.

The estimated cost of the redecorations is £45.

C. T. Arrowsmith
Chartered Surveyor

Proofs of Evidence

Expert Witnesses

When a building dispute arises it may be settled by an action in the courts or by arbitration. The parties in dispute may employ legal advisers (counsel) who will prepare evidence and argue their case before the tribunal. On matters of a technical nature, the assistance of an expert witness may be required, and with building disputes this could be a surveyor. An expert witness is not restricted to giving statements of fact and can explain technical matters and also express an opinion, if requested, based upon his special knowledge and experience.

Counsel will need to know the technical arguments and how they are to be presented and developed. Hence the expert witness prepares a document, termed a proof of evidence, for the benefit and use of counsel when arguing the case. The contents of the proof of evidence are copied into the counsel's brief and the expert witness will subsequently be questioned and cross-examined upon it.

Preparation of Proofs of Evidence

A proof of evidence must contain a concise statement of facts and a logical statement of the line of argument. It is usually typed in double spacing on A3 size paper, folded down the centre, with a margin on the left to take notes, additions or corrections inserted by counsel. Paragraphs are usually referenced by numbering and valuations, schedules and other exhibits are generally inserted at the end of the proof. Words with special significance may be underlined to catch the eye of counsel.

A proof of evidence contains a suitable title, name and qualifications of the expert witness, subject of reference, particulars of survey or inspection, evidence—giving findings of expert witness and presenting and developing arguments, and finally the conclusions.

Specimen Proof of Evidence, incorporating a Schedule of Dilapidations

Background information. Mr P. J. Beddington was a tenant occupying premises at 12 Norfolk Street, Haverton, Hampshire, for 21 years under a lease dated 21 March 1966. The tenancy agreement included a covenant to keep and leave the house and grounds in good and tenantable repair and condition, fair wear and tear excepted. The landlord, Homeville Enterprises Ltd, submitted to Mr Beddington a schedule of dilapidations six weeks prior to the expiration of the lease. The tenant disputed the schedule, refused to comply with it and instructed a solicitor to contest it. The landlord has commenced legal proceedings and a surveyor, Mr Ronald Sharpe, has been instructed to prepare a proof of evidence incorporating the schedule of dilapidations and a schedule of claim for damages for breach of the repairing covenant.

Proof of Evidence

Homeville Enterprises v. *Peter John Beddington*
Claim for damages for breach of repairing covenant in respect of lease of 12 Norfolk Street, Haverton, Hampshire.

RONALD SHARPE

will say

Qualifications

(1) I am a Fellow of the Royal Institution of Chartered Surveyors in practice on my own account at 53 High Street, Haverton. I have had thirty-two years' experience in the profession of a surveyor and have undertaken a large amount of work involving dilapidations throughout this period.

Subject of appeal

EXH 1   (2) I produce the lease of 12 Norfolk Street, Haverton, granted by the plaintiff to the defendant for a term of twenty-one years from 21 March 1966. (*Note*: EXH refers to exhibit)

(3) EXH 1 repairing covenant
The lessee agrees at all times during the said term to keep the premises, including all fixtures and additions, in good and tenantable repair and condition and to deliver up the same in such good and tenantable repair and condition to the lessor at the expiration or sooner determination of the said term.

EXH 2   (4) I produce a schedule of dilapidations EXH 2 and a
EXH 3 schedule of claim EXH 3 signed by the plaintiff's agent, showing the sum of £6888 as the cost of the items of repair therein set out and constituting the plaintiff's claim for damages.

(5) The defendant was presented with the schedule of dilapidations EXH 2 six weeks before the expiration of the lease. None of the repairs listed in the schedule has been carried out.

(6) The plaintiff claims damages for breach of covenant to keep the premises in repair in accordance with the terms of the lease EXH 1. The basis of the claim is exhibit EXH 3 (schedule of claim).

Inspections

(7) I inspected the premises on 4 February 1987, 12 March 1987 and 22 April 1987.

(8) I have examined the lease of the property directing special attention to the repairing covenant and I consider that the schedule of dilapidations EXH 2 has been prepared in strict accordance with the terms of the lease.

Evidence of dilapidations

(9) The repair of the property has been neglected for a

considerable time and I could find no evidence of any recent repairs to the premises. I submit that the want of repairs listed in the schedule of dilapidations EXH2 is evidence that the defendant has not carried out his obligations under the repairing covenant EXH 1.

EXH 2   (10) *Schedule of dilapidations*

A. No decorations have been carried out for a number of years and certainly not within the last year of the lease as required by the repairing covenant. The wallpaper in the hall and lounge is badly discoloured and disfigured. The paintwork to plastered walls and ceilings elsewhere is also badly discoloured. Paintwork to woodwork both inside and out has deteriorated badly, with bare wood showing in a number of places.

B. Rainwater gutters and downpipes are rusting badly with some cracked sections and leaking joints, causing damp brickwork and the growth of algae.

C. Putties to windows on the south and west sides of the house have perished and there are several cracked panes of glass.

D. Twenty roofing tiles are missing from the south and west roof slopes, permitting some rainwater penetration into the roof space.

E. Pointing to extensive areas of brickwork on the south wall of the house has perished.

F. The WC pan in the bathroom is cracked and the flushing mechanism is defective.

G. Two inspection chamber covers are badly cracked and corroded.

H. The close boarded fence on the southern boundary is badly decayed and in a state of collapse.

EXH 3   (11) *EXH Schedule of claim*

	£
A. Prepare plastered walls and ceilings and apply two coats of emulsion paint	
walls 550 m^2 @ £3.90	2145.00
ceilings 160 m^2 @ £4.80	768.00
Strip off existing wallpaper and replace with new	
80 m^2 @ £4.70	376.00
Rub down woodwork, touch up with primer and apply three coats of oil paint	
90 m^2 @ £7.60	684.00
B. Take down defective rainwater gutters and pipes and replace with new	
gutters 30 m @ £11.80	354.00
downpipes 15 m @ £17.00	255.00
Prepare, prime and paint gutters and pipes with three coats of oil paint 60 m @ £3.80	228.00
Remove algae growth from brickwork 6 m^2 @ £2.50	15.00
C. Rake out defective putties to windows and replace with new	
40 m @ £3.80	152.00
Hack out cracked glass and reglaze	
8 m^2 @ £19.50	156.00

D. Replace missing roofing tiles 　　20 @ £5.60	112.00
E. Rake out joints of brickwork and repoint 　　70 m² @ £10.00	700.00
F. Replace WC suite	145.00
G. Replace two inspection chamber covers @ £80	160.00
H. Replace decayed close-boarded fence 　　22 m @ £29.00	638.00
	£6888.00

(12) The estimate of £6888 is based on current prices.

Date:　7 May 1987　　　　Signed　Ronald Sharpe

References

1　D. A. Kirby. The maintenance of pre-war council dwellings. *Housing and Planning Review* (January/February 1972)

2　*English House Condition Survey.* HMSO (1981)

3　Royal Institution of Chartered Surveyors. *Housing: The Next Decade* (1986)

4　New Earswick General Improvement Area No. 1. *The Architects' Journal* (30 January 1974)

5　J. H. Cheetham. New homes from old—some typical schemes. *Building Trades Journal* (9 September 1966)

6　J. A. Foreman. Conversions—bedroom into a bathroom. *Building Trades Journal* (3 May 1968)

7　J. H. Cheetham. New homes from old—houses into flats. *Building Trades Journal* (23 September 1966)

8　J. Benson, B. Evans, P. Colomb and G. Jones. *The Housing Rehabilitation Handbook.* Architectural Press (1980)

9　*1985 Physical and Social Survey of Houses in Multiple Occupation in England and Wales.* HMSO (1987)

10　J. H. Cheetham. New homes from old—specialised schemes. *Building Trades Journal* (30 September 1966)

11　A. Edgar. Rehabilitation standards. *Housing Review* (July/August 1975)

12　S. Cantacuzino and S. Brandt. *Saving Old Buildings.* Architectural Press (1980)

13　J. H. Banton. *Management of Modernisation.* CIOB Site Management Information Service (1980)

14　H. Crowter. Housing modernisation: a trap for councils. *Chartered Quantity Surveyor* (May 1984)

15　DOE, DHSS and Welsh Office. Circular 59/78. *Adaptations of Housing for People who are Physically-handicapped.* HMSO (1978)

16　T. Lockhart. *Housing Adaptations for Disabled People.* Architectural Press (1981)

17   A. Arden. *Housing Act 1985*. Sweet and Maxwell (1986)
18   C. J. Wright. *Housing Improvement and Repair*. Sweet and Maxwell (1986)
19   DOE. Cmnd. 5280. *Widening the Choice: The Next Steps in Housing*. HMSO (1973)
20   Royal Institution of Chartered Surveyors. *Rehabilitation of Houses and Other Buildings* (1973)
21   DOE. *Good Practice in Area Improvement*. HMSO (1984)
22   I. H. Seeley. *Building Economics*. Macmillan (1983)
23   DOE. Cmnd. 5339. *Better Homes: The Next Priorities*. HMSO (1973)
24   L. Needleman. Rebuilding or renovation? A reply. *Urban Studies 5.1* (1968)
25   B. Kilroy. Housing—the rehabilitation v. redevelopment seesaw—the effect of high land prices. *Local Government Finance* (April 1973)
26   I. H. Seeley. *Building Surveys, Reports and Dilapidations*. Macmillan (1985)
27   Royal Institution of Chartered Surveyors. *Building Surveyors Guidance Note: Dilapidations* (1983)

8 SPECIFICATION OF MAINTENANCE WORK

The specification forms an extremely important document on building maintenance, conversion and improvement contracts, as it constitutes a schedule of instructions to the contractor and prescribes the materials and workmanship requirements. When read in conjunction with the contract drawings, and bill of quantities where one is prepared, it provides all the information that a contractor will need to price the work and carry it out.

In addition, if often contains general clauses which set out the more important rights and obligations of the parties to the contract. For larger contracts, conditions of contract are normally used which prescribe in considerable detail the general rights, duties and liabilities of parties to building contracts, and these are described in more detail in chapter 10.

Sources of Information

Information for use in building specifications can be obtained from a variety of sources, and the principal sources are now described.

(1) *Previous Specifications*

In the majority of cases specifications for past contracts are used as a basis in the preparation of a new specification for a contract of similar type. This procedure speeds up the task of specification writing considerably, but care must be taken to bring the specification clauses up to date by the incorporation of latest developments, techniques, materials and components. It is also necessary to be constantly on the alert for any changes of specification needed to cope with differences of design, construction or site conditions, as few contracts are really identical. Care must be taken to omit details which are not applicable and to insert information on additional features.

(2) *Drawings*

The contract drawings must form the basis of any specification as they show the nature and scope of the work and frequently contain a great deal of descriptive information. A close examination of the drawings will indicate the matters that are to be covered in the specification. The drawings will also distinguish between new and existing work.

(3) *Employer's Requirements*

The employer may prescribe certain requirements in connection with the work and these will probably need to be included in the specification. Typical requirements of this kind are programming of the works to provide for completion of certain sections at specified dates and the taking of various precautions to minimise interference with productive processes in the employer's existing premises. It is essential that requirements of this kind are brought to the notice of the contractor, as they may quite easily result in increased costs.

(4) *Site Conditions*

With extensions to existing buildings, it is necessary to obtain information on soil conditions, groundwater level and the extent of site clearance work. With alteration work, full details of the existing construction are needed to specify the full nature and extent of the new work. The contractor should be supplied with the fullest information available, to reduce to a minimum the risks that he must take and the number of uncertain factors for which he must make allowance in his tender.

(5) *British Standards*

British Standards are issued by the British Standards Institution, an organisation recognised by the government and industry as the sole body responsible for the preparation of national standards. The institution has a general council which controls a number of divisional councils, one of which is concerned with building, and in addition there are many industry standards committees. These committees are largely responsible for developing industrial standardisation as they decide the subjects of new standards and their scope, and approve the draft standards which are prepared by the various technical committees. The technical committees are made up of experts on the subject of the particular standard and consist of representatives of the user, producer, research and other interests.

Thousands of British Standards have now been prepared covering a wide range of subjects. Furthermore, they are kept constantly under review in order that they shall be kept up to date and abreast of progress. They have proved to be an efficient means whereby the results of research can be made available to industry in practical form.

These standards lay down the recognised minimum standards of quality for materials and components and also define the dimensions and tests to which they must conform. British Standards are of great value in the drafting of specifications as they reduce considerably the amount of descriptive work that is required. At the same time they ensure the use of a good-quality product and generally meet the 'deemed to satisfy' requirements of the Building Regulations. The standards incorporate the most searching requirements that the latest stage of technical development and knowledge can produce.

Manufacturers and contractors are intimately involved with British Standards and thus can reasonably be expected to have a fair knowledge of the contents of appropriate standards. They will be generally freed from the necessity to examine carefully lengthy specification clauses relating to materials and components. It is however often necessary to specify the class or grade required where a British Standard incorporates classes or grades. For instance BS 3921 recognises two classes of engineering bricks (classes A and B).

A selection of some of the most common British Standards covering building materials and components used in maintenance work follows.

BS 4	Structural steel sections
BS 12	Portland cement
BS 65	Vitrified clay pipes, fittings and joints
BS 402	Clay plain roofing tiles and fittings
BS 416	Cast iron spigot and socket soil, waste and ventilating pipes and fittings
BS 417	Galvanised mild steel cisterns and covers, tanks and cylinders
BS 437	Cast iron spigot and socket drainpipes and fittings
BS 459	Doors
BS 460	Cast iron rainwater goods
BS 473 and 550	Concrete roofing tiles and fittings
BS 497	Manhole covers, road gully gratings and frames for drainage purposes
BS 544	Linseed oil putty for use in wooden frames
BS 585	Wood stairs
BS 644	Wood windows
BS 680	Roofing slates
BS 699	Copper direct cylinders for domestic purposes
BS 743	Materials for damp-proof courses
BS 747	Roofing felts
BS 882	Aggregates from natural sources for concrete
BS 890	Building limes
BS 913	Wood preservation by means of pressure creosoting
BS 952	Glass for glazing
BS 988	Mastic asphalt for building (limestone aggregate)
BS 1010	Draw-off taps and stopvalves for water services (screwdown pattern)
BS 1125	WC flushing cisterns
BS 1178	Milled lead sheet for building
BS 1181	Clay flue linings and flue terminals
BS 1186	Quality of timber and workmanship in joinery
BS 1188	Ceramic wash basins and pedestals

BS 1189   Cast iron baths for domestic purposes
BS 1191   Gypsum building plasters
BS 1195   Kitchen fitments
BS 1196   Clayware field drain pipes
BS 1197   Concrete flooring tiles and fittings
BS 1198–1200   Building sands
BS 1206   Fireclay sinks
BS 1230   Gypsum plasterboard
BS 1243   Metal ties for cavity wall construction
BS 1244   Metal sinks for domestic purposes
BS 1245   Metal door frames
BS 1286   Clay tiles for flooring
BS 1289   Precast concrete flue blocks for domestic gas appliances
BS 1336   Knotting
BS 1566   Copper indirect cylinders for domestic purposes
BS 1567   Wood door frames and linings
BS 1722   Fences
BS 1881   Methods of testing concrete
BS 2592   Thermoplastic flooring tiles
BS 2760   Pitch-impregnated fibre pipes and fittings
BS 2870   Rolled copper and copper alloys, sheet, strip and foil
BS 2871   Copper and copper alloys: tubes
BS 3260   PVC (vinyl) asbestos floor tiles
BS 3505   Unplasticised PVC pipe for cold water services
BS 3794   Decorative laminated plastics sheet
BS 3921   Clay bricks
BS 3958   Thermal insulating materials
BS 4092   Domestic front entrance gates
BS 4305   Baths for domestic purposes made from cast acrylic sheet
BS 4449   Hot rolled steel bars for the reinforcement of concrete
BS 4483   Steel fabric for the reinforcement of concrete
BS 4514   Upvc soil and ventilating pipes, fittings and accessories
BS 4576   Upvc rainwater goods
BS 4660   Upvc underground drainpipes and fittings
BS 4756   Ready mixed aluminium priming paints for woodwork
BS 4787   Internal and external wood doorsets, door leaves and frames
BS 4873   Aluminium alloy windows
BS 5224   Masonry cement
BS 5328   Specifying concrete, including ready-mixed concrete
BS 5503 and 5504   WC pans
BS 5803   Thermal insulation for use in pitched roof spaces in dwellings
BS 5872   Locks and latches for doors in buildings
BS 5911   Precast concrete pipes and fittings for drainage and sewerage
BS 6073   Precast concrete masonry units
BS 6340   Shower units
BS 6398   Bitumen damp-proof courses
BS 6431   Ceramic floor and wall tiles
BS 6510   Steel windows, sills, window boards and doors

BS 6515   Polythylene damp-proof courses
BS 6577   Mastic asphalt for building (natural rock asphalt aggregate)

There are nearly 2000 British Standards covering building materials and components and a number of these are constantly being revised and amended, whilst at the same time new standards are formulated. The British Standards Handbook 3, published annually, contains useful summaries of British Standards for building.

(6) *Codes of Practice*

Codes of Practice are also issued by the British Standards Institution and these cover workmanship requirements and methods of carrying out various classes of work. The following Codes of Practice may be of particular value when drafting specifications for building maintenance and alteration work. The more recent codes of practice are given British Standard references, as shown at the end of the following list.

CP 99   Frost precautions for water services
CP 101   Foundations and substructures
CP 102   Protection of buildings against water from the ground
CP 111   Structural recommendations for loadbearing walls
CP 112   The structural use of timber
CP 114   Structural use of reinforced concrete in buildings
CP 143   Sheet roof and wall coverings
CP 144   Roof coverings
CP 151   Doors and windows, including frames and linings
CP 153   Windows and rooflights
CP 201   Flooring of wood and wood products
CP 202   Tile flooring and slab flooring
CP 203   Sheet and tile flooring
CP 204   In-situ floor finishes
CP 209   Care and maintenance of floor surfaces
CP 310   Water supply
CP 312   Plastics pipework
BS 5234   Internal non-loadbearing partitions
BS 5262   External rendered finishes
BS 5268   Structural use of timber
BS 5385   Wall tiling
BS 5395   Stairs, ladders and walkways
BS 5449   Central heating for domestic purposes
BS 5492   Internal plastering
BS 5516   Patent glazing
BS 5534   Slating and tiling
BS 5572   Sanitary pipework
BS 5589   Preservation of timber
BS 5618   Thermal insulation of cavity walls
BS 5628   Structural use of masonry

BS 6150   Painting of buildings
BS 6229   Flat roofs
BS 6262   Glazing for buildings
BS 6297   Design and installation of small sewage treatment works and cesspools
BS 6367   Drainage of roofs and paved areas
BS 6465   Sanitary installations
BS 8110   Structural use of concrete
BS 8301   Building drainage

(7) *Agrément Certificates*

Since the early 1960s the building industry has become increasingly involved in the use and development of a perplexing array of new materials, products and associated techniques. The proliferation of unfamiliar and novel products makes evaluation and selection an increasingly difficult process. The situation is further aggravated by the establishment of firms with little or no experience of building, producing materials or products that have been developed without a detailed knowledge of required performance standards or guidance on actual conditions in use. Nevertheless, the pace of development dictates that some products and processes will need to be used before they have been proved over long periods.

To overcome this problem, the Agrément Board was established by the government in 1966 to investigate new materials, products, components and processes and new uses of established products, and to issue certificates of worthiness where appropriate. The Board has established a close working relationship with the European Union and is contributing to the development of common methods of assessment and standards of performance and testing. At the same time, membership of the European Union offers the valuable facility of speedy acceptance among other member countries of products granted certificates in the United Kingdom.

Since its inception, the Board has issued many hundreds of certificates. Each of these involves an assessment of the product, its performance related to suitable usage, and devising and carrying out appropriate tests. The samples tested are deliberately chosen as representing the lowest level. Strict quality control has to be maintained to ensure that overall production is at least as good as the tested samples. The certificate contains details about the proprietary product, the design data appropriate to its use, handling on site and subsequent maintenance. The certificates issued cover a wide range of products.

The Department of the Environment notifies local authorities whenever a certificate is issued and is available for consultation on the standard of the product under the Building Regulations. Furthermore, the Approved Documents accompanying the Building Regulations 1985 contain references to materials and products that are likely to be suitable for the purposes of the regulations. They mainly refer to materials or products covered by British Standards or by Agrément certificates. Abstracts of certificates are available on a subscription basis, together with a regular distribution of information sheets from the Agrément Board.

The independent safety check in the Agrément system, based on satisfactory performance in use requirements, minimises the risk of damage in use and subsequent repair, and it ensures compliance with the currently accepted comfort standards where appropriate. The assessment of durability introduces one of the more difficult problems with new materials, especially those for external use. It is now generally accepted that attempts to accelerate the breakdown of products can be very misleading unless there is an adequate background of research; the alternative of trying to improve the accuracy of observation on short-term natural exposures is generally more satisfying. It seems likely that the realistic prediction limit is in the range of 10 to 20 years without maintenance other than perhaps periodic washing. The maintenance aspects of the design data on an Agrément Certificate provide a sound guide to the kind of maintenance which may be needed and the procedures to be adopted for this work.

(8) *Trade Catalogues*

Where proprietary articles are specified, reference should be made to the manufacturers' catalogues for the extraction of the necessary particulars for inclusion in the specification. It is often advisable to quote the catalogue reference when an article is produced to a number of different patterns.

This procedure will reduce the length of specification clauses and will ensure the use of a specific article with which the specification writer is familiar and in which he has confidence. Some public bodies object to this practice on the grounds that it restricts the contractor's freedom of choice and in some cases prevents the use of local products. Furthermore, it may prevent the contractor from using his regular source of supply and may thus result in higher prices.[1]

(9) *Publications*

Other publications can be used for reference purposes when compiling a specification such as booklets issued by trade associations, as for example the very useful publications of the Cement and Concrete Association. The annual publication *Specification*[2] contains a wealth of useful information.

Form of the Specification

When the specification is accompanied by drawings it should amplify, not repeat and certainly never contradict, the information given on the drawings. The specification should explain the purpose and intent of the drawings more fully and clearly, so that the two documents when taken together leave no doubt as to the work to be executed. Most building operations require both drawings and specification but the specification alone will be adequate for some works, such as redecorations and repairs.

The specification is a highly technical document and should be written in technical language using appropriate building terms. In this respect it differs greatly from a structural or sanitary report prepared for a client, which is kept as free from technical terms as possible. The architect or surveyor preparing a specification must have a thorough knowledge of the materials and forms of construction that he is specifying and must know exactly how they will be used, in order to draft an entirely satisfactory specification.

A specification should be concise and comprehensive, and avoid duplication of particulars and the inclusion of vague or ambiguous details. Excessively long and involved specifications are apt to produce highly priced tenders. In these circumstances the contractor experiences difficulty in assimilating the document and gains the impression that the requirements of the contract may be more far-reaching than is customary, and so tends to increase his price accordingly. Similarly where the specification places unreasonable risks upon the contractor, he is almost certain to increase the price to safeguard his position.

Building specifications normally start with general clauses or preliminaries which relate to the contract as a whole and define the contractor's general liabilities. Typical general clauses are given later in the chapter. The remainder of a building specification is normally subdivided into trades or works sections, as detailed in the Standard Method of Measurement of Building Works,[3] and each subsection is frequently subdivided into materials and workmanship. The customary trades and works sections are listed in table 8.1 and are not entirely coincident, although significant changes to work section classifications are anticipated in SMM7.

Table 8.1   Comparison of traditional trades and work sections

Traditional trades	Work sections
	Demolition
Excavator	Excavation and Earthwork
	Piling
Concreter	Concrete work
Bricklayer	Brickwork and Blockwork
	Underpinning
	Rubble Walling
Mason	Masonry
Asphalter	Asphalt work
Roofer	Roofing
Carpenter and Joiner	Woodwork
Steel and Iron Worker	Structural Steelwork
	Metalwork
Plumber	Plumbing and Mechanical Engineering Installations
Heating and Ventilating Engineer	
Electrician	Electrical Installations
Plasterer	Floor, Wall and Ceiling Finishings
Pavior	
Glazier	Glazing
Painter	Painting and Decorating
Drainlayer	Drainage
	Fencing

When preparing specifications for works of alteration, conversion and repair, it is often advantageous to depart from the accepted format previously described. Each contract needs to be considered separately and the most logical format adopted. In some cases the natural sequence of work as executed on the site offers the best approach, in others a room by room approach or subdivision of the building(s) into specific parts on a locational basis, or the work may be partly or entirely grouped into trades or work sections as for new work. Some flexibility in approach is desirable to accommodate wide variations in the character, scope and extent of the works.

National Building Specification

The basic concept of the National Building Specification (NBS) first published by a RIBA company in 1973 is to provide a master list of concise specification clauses from which a specification writer can make his own choice. It aimed to provide improved documentation, with good technical content, clarity, flexibility and better documentation in association with other documents. It is produced in loose leaf form and is continually updated.

In essence, NBS is a collection of coded standard specification clauses that describe building materials and building operations. A typical NBS work section is subdivided into two subsections—commodities and workmanship. There are 94 work sections, including preliminaries, and within each section there are a number of clauses which describe the alternative materials used and others that describe the related alternative operations. Together they cover a remarkably large number of materials and operations. A column on the left-hand side of all clauses contains guidance notes of all kinds, including technical explanations, cross-references to other documents and commentary on the content of the clauses. The sections have been coded in accordance with the CI/Sfb system but users have the choice of CI/Sfb or SMM sequence. An abridged edition, known as the Small Jobs Version, is suitable for small contracts using specifications, drawings and/or schedules of works.[4]

In the introduction it is suggested that NBS should be used to produce project specifications and also a standard specification for each office. It is also suggested that most specifiers will begin by checking their existing specifications against NBS with a view to adopting a NBS standard clause whenever possible, although there does not seem much evidence of this procedure being followed in practice and NBS has certainly not been universally adopted. Doubtless, architects and surveyors would benefit by comparing their current specification clauses with the corresponding ones in NBS to identify any deficiencies in the office documents.

The procedure for using NBS is described in the document:

(1) Select the relevant work section, such as brick/block walling.

(2) Select the materials clauses required and insert additional information where necessary.

(3) Delete materials clauses not required.

(4) Insert new materials clauses required to cover any special or particular items.

(5) Select the relevant workmanship clauses to complement the selected materials.

(6) Delete workmanship clauses not required; together with any alternative or conflicting clauses.

Performance Specifications

A performance specification is a method of defining a component, product or system not by composition or form, but by the needs of the user which it must satisfy. Expressed in another way, it states what is needed rather than how the need shall be met. This concept has been used on a limited scale for many years; for example, it is customary to specify ventilation plant to be operated by a fan with a capacity of x m³/min, steel may be specified as having a prescribed yield point and concrete a prescribed strength. Yet no performance specification has been written for a complete building, except of the simplest requirements, largely because of the complexity of the process. The difficulties can be illustrated by reference to the performance requirements of a window, which would include methods and extent of opening, type of glazing, cleaning, compatibility of materials and jointing with adjoining materials, strength related to size and use, long-term maintenance and appearance. These would need backing up with suitable means of testing, quality control and subsequent certification.

A BRE paper[5] describes how a performance specification makes purchasers' needs explicit by listing essential properties required in a systematic form—with their values or limit values—and methods of testing or evaluation. It leaves the supplier and his designer maximum freedom to innovate in terms of materials, form, method of manufacture and assembly. In fact, many British Standards for building materials and components contain a range of specified values coupled with prescribed tests designed to measure them. Also as described earlier, in order to assess the probable behaviour of new products in use, the Agrément Board prepares check lists of performance requirements and methods of assessment and test (MOATS).

In its simplest form, a performance specification constitutes a shopping list for choosing already designed products; it is used to establish an equitable and systematic basis for the choice between available products. Alternatively, the more common current approach is as a tool for commissioning the design of products for a specific market.

Not all properties of a component, element or building are physically quantifiable. Furthermore, a product meeting the basic performance requirements may still be unacceptable for other reasons, and these further design constraints must be listed.

The Building Research Establishment[6] has shown just how complicated the preparation of a performance specification for a building can be by reference to a suggested procedural approach.

(1) Decide on the overall strategy for design of the building and its parts.

(2) Decide the scope of the performance statement—the range of contexts in which components will be used and the basic geometry of the spaces that will be occupied. Any resulting design constraints will then be identified.

(3) At the appropriate levels, decisions must be made on functional requirements of the building, its spaces or the packages.

(4) A list of relevant properties should be determined, for instance with cladding panels it will be necessary to decide whether thermal insulation is an appropriate property to consider.

(5) The relative importance of desired properties must be assessed and, where appropriate, weighted.

(6) A decision must be taken as to how the requisite properties are to be measured—by inspection, calculation or test, and the units of measurement. Possibly a BS test or Agrément MOAT may be appropriate.

(7) The limiting values for each quantifiable property must be considered. For example, in the case of a window, is the upper limit of air infiltration to be nil or is some air flow needed for health reasons? At the other extreme, what should be the greatest permissible leakage under extreme conditions, representing the lowest quality of performance to be endured in practice?

(8) Within a single performance specification it may be beneficial to list different limits (step levels) in a range of values for different use situations.

(9) A statement must be made about other criteria governing the acceptance of a design, such as labour content for site erection, contractual obligations, supply, delivery, storage and other administrative aspects. Cost is a consideration running through all stages, particularly the later ones.

Drafting of Specifications

Material Descriptions

Considerable care must be exercised in drafting a specification to prepare clauses which are concise, complete and free from ambiguity. For instance, when drafting materials clauses it is desirable to adopt some pre-arranged order of grouping the particulars, to avoid missing an important detail. The formulation of a specification description for engineering bricks in table 8.2 will serve to illustrate the approach.

In practice it would be much move convenient to make reference to BS 3921 for class B engineering bricks. The following alternative methods of describing materials, or possibly a combination of them, can be used in a specification.

(1) A full description of the material or component is given with details of desirable and undesirable properties and appropriate test requirements.

(2) The relevant British Standard reference, together with details of

Table 8.2   Brick criteria

Criteria	Requirements
Material	Bricks
Type	Southwater red Nr 2 engineering bricks
Name of manufacturer or source of supply	Messrs X of Y
Prime cost	£225 per thousand
Desirable characteristics	Well burnt, of uniform shape, size and colour, and sound and hard
Undesirable characteristics	Freedom from cracks, stones, lime and other deleterious substances
Tests	Average compressive strength of not less than 48.5 MN/m^2
	Average water absorption by weight not greater than 7 per cent

class or type where appropriate, is given. The contractor can then refer to the British Standard for fuller information.

(3) The name of the manufacturer, proprietary brand or source of supply is stated and the contractor can obtain further particulars from the manufacturer or supplier.

(4) A brief description of the material is given together with the prime cost for supply and delivery of a certain quantity of the material to the site.

Typical quantities are a thousand bricks, a cubic metre of sand, a tonne of cement and 5 litres of paint, and these normally represent the units in which the materials are sold. This latter method ensures that all contractors are tendering on the same basis, without the need to obtain quotations from manufacturers or suppliers. It also permits the client to defer the choice of the material if he so wishes.

Avoidance of Unsatisfactory Descriptions

In practice, use is often made of a number of wide and embracing terms which are not sufficiently precise in their meaning and can be interpreted in different ways. This leads to inconsistencies in pricing with consequent undesirable effects. Some examples of undesirable terms follow.

(1) The word 'best' is widely used in specifications, where best-quality materials or workmanship are obviously not required. If this term is frequently and loosely used throughout the specification, without any real regard to its true intent and meaning, then the contractor will be tempted to disregard it. It is important to prevent this happening by using the term only when materials and workmanship of the highest quality are required. Materials are frequently produced in a number of grades and it is essential that a clear indication is given of the particular grade required. For instance, it would be pointless and costly to specify British Standard normal quality vitrified clay pipes for surface water drains when an appropriate British Standard Surface Water classification exists in BS 65.

(2) The word 'proper' is also frequently misapplied, particularly in descriptions of constructional methods. As a general rule it is far more satisfactory to include full instructions in the specification, and so leave the contractor in no doubt as to the actual requirements of the contract. With minor items of work a comprehensive description of the method of construction may not be essential and in these circumstances the use of the word 'proper' may be acceptable.

(3) The term 'or other approved' usually represents an undesirable feature in any specification, as it introduces an element of uncertainty. The contractor cannot be sure whether the materials or components which he has in mind will subsequently prove acceptable to the architect or surveyor. All specification requirements should be clear and certain in their meaning and be entirely free from doubt or ambiguity.

(4) The term 'as specified' is often used without specifying anything.

Workmanship Clauses

Specification clauses covering constructional work and workmanship requirements are generally drafted in the imperative, for instance 'lay manhole bases in concrete, class B, 225 mm thick', or alternatively 'The contractor shall lay . . .'. All workmanship clauses should give a clear and concise description of the character and extent of the work involved.

The sequence of clauses within a section will normally follow the order of constructional operations on the site. This procedure reduces the possibility of omission of items from the specification and assists the contractor in working to its requirements on the site. The specification writer must avoid specifying standards of workmanship which are completely out of keeping with the class of work involved. He may need to choose between repair, replacement and cleaning.

Typical Specification Clauses (Alterations and Repairs)

Title

The title could be produced in the following format.

SPECIFICATION

of

works to be executed and materials to be used in alterations and repairs to Nr 24 Baxter's Close, Pendlebury for Mr P. T. Johnson under the direction and to the reasonable satisfaction of:

J. T. Hanbury, FRICS
52, High Street
Pendlebury

Date . . . .

Preliminaries

The materials and workmanship requirements of the specification will be preceded by preliminaries or general clauses and some typical clauses follow. In specifications for alteration and repair work, the materials and workmanship requirements are often scheduled on a locational basis, for instance room by room, and provision may be made for pricing the preliminaries items and the works items that follow on the right-hand side of the specification.

Tenders	1	Tenders should be submitted not later than . . . on the form of tender.
Drawing	2	The work consists of alterations and repairs to the detached house, 24 Baxter's Close, Pendlebury and the alterations are shown in Drawing R1 62.
Visit Site	3	The contractor is advised to visit the site and familiarise himself with working conditions, access to and general extent of the works.
Conditions of Contract	4	The form of contract shall be the JCT Agreement for Minor Building Works, 1985. The contractor shall make allowance in his tender for complying with these conditions.

The defects liability period shall be six months from the date of certified practical completion. Insurance against injury to persons and property shall provide cover for up to £30 000 for any one incident, the number of incidents for which cover is provided being unlimited.

Interim certificates will be issued monthly with a retention of 5 per cent.

The period of final measurement and valuation shall be three months from the date of practical completion.

The tender is to be firm price and no fluctuations will be permitted for increased costs of labour, materials, plant, and other components and services.

Materials and Workmanship	5	Materials, components and workmanship shall be of good quality and in accordance with the British Standards and Codes of Practice prescribed.
Extent of Works	6	The contractor shall do everything necessary for the proper execution of the works, whether or not shown on the drawing or described in the specification, provided it may be reasonably inferred.
Figured Dimensions	7	Figured dimensions shall be followed in preference to scaled dimensions and particulars shall be taken from the actual work where possible.
Setting Out, Notices, Fees and Compliance with Regulations	8	The contractor shall be responsible for the correct setting out of the works. He shall give all necessary notices to local and service authorities, pay all

appropriate fees and charges, and carry out all work in compliance with the Building Regulations.

Nominated Suppliers 9 The contractor shall allow for all expenses in connection with the unloading, storing and return of packings of materials and components, including those listed under prime cost items.

Care of the Works 10 The contractor shall take reasonable care of the existing premises and of all new work and shall take steps to reduce interference with the occupants of the building to a minimum. The contractor shall be responsible for any damage arising from the weather, carelessness of workmen, loss, theft or other cause, and shall make good such damage or loss at his own expense.

Attendance 11 The contractor shall allow for the general attendance of one trade upon another.

Screens and Hoardings 12 The contractor shall provide all necessary screens, hoardings and similar protective devices for the benefit of the occupants and adjoining properties.

Clearing Away 13 The contractor shall remove all temporary works, rubbish, debris and surplus materials from the site as they accumulate and at completion; and shall clean all surfaces internally and externally, remove stains and touch up paintwork, dry out the new works as necessary, and leave the works clean and to the reasonable satisfaction of the surveyor.

Contingency Sum 14 Allow the provisional sum of £500 to cover any unforeseen contingencies. This sum is to be expended in part or in whole at the discretion of the surveyor.

Acceptance of Tenders 15 The employer does not bind himself to accept the lowest or any other tender.

Materials

Cement 16 The cement shall comply with BS 12, be delivered in the original sealed bags of the manufacturer, be stored in a proper manner to avoid deterioration and used in correct sequence.

Aggregate 17 The fine aggregate shall comply with BS 882 and shall consist of well-graded coarse sand mainly passing a 5 mm test sieve. The coarse aggregate shall also comply with BS 882 and shall consist of natural gravel, crushed gravel or crushed stone, well graded with a maximum size of 20 mm.

Concrete 18 Concrete for foundations shall consist of 1 part Portland cement, 3 parts sand and 6 parts coarse aggregate, all measured by volume.

Concrete for new floors, paved areas and lintels shall be 1 part Portland cement, 2 parts sand and 4 parts coarse aggregate, all measured by volume.

Mixing shall be carried out in an approved batch type mixer and shall continue until there is a uniform distribution of materials and the mix is uniform in colour and consistency.

Bricks and Blocks 19 Facing bricks shall be 65 mm Himley mixed russet wirecut facing bricks, extending to at least 75 mm below ground level. Other brickwork below ground and inner leaves of cavity walls below damp-proof course shall be common bricks to BS 3921. Precast concrete blocks for the inner leaf of cavity walls and internal partitions shall be Thermalite, Celcon or other approved complying with BS 6073, with a density of less than 1500 kg/m^2.

Wall Ties 20 Metal wall ties for use in cavity walls shall be of mild steel coated with zinc of the butterfly type complying with BS 1243.

Lime 21 Lime shall be non-hydraulic or semi-hydraulic lime complying with BS 890.

Mortar 22 Mortar used below damp-proof course shall consist of one part of Portland cement, one half part of lime and four and a half parts of sand $(1:\frac{1}{2}:4\frac{1}{2})$ by volume. In all brickwork and blockwork above damp course the mortar shall be 1:1:6.

Damp-proof Course 23 The damp-proof course shall be of bitumen sheet with hessian base and lead complying with BS 743, type D, lapped at least 75 mm at angles and joints and laid on a level bed of cement-lime mortar $(1:\frac{1}{2}:4\frac{1}{2})$ and neatly pointed where exposed.

Timber 24 All hardwood and softwood shall comply with BS 1186. Where timber is to be primed it shall be coated with a thick mixture of red or white lead and linseed oil. The moisture content of the timber used for internal joinery shall not exceed 10 per cent and that used for external doors and frames shall not exceed 17 per cent when the joinery is delivered to the site, and these moisture contents are to be maintained until the work is complete. Carpentry timber shall be supplied and fixed to the lengths and sizes shown on the drawing, be clean, straight and reasonably free from knots and shakes.

Plasterboard for Ceilings 25 Ceilings shall be formed with 12.5 mm gypsum lath plasterboard, of gyproc lath or other approved type, faced with aluminium foil on its upper face in accordance with BS 1230, and fixed with 40 mm galvanised clout headed nails to underside of softwood joists, to give a true plane surface. Nail each lath to every support using not less than 4 nails equally spaced across the width and driven no closer than 13 mm from its edges. The laths shall be finished with one coat of neat gypsum plaster to a thickness of 5 mm.

Plaster for Walls   26 Plastered walls and reveals to openings shall be finished with two coats of plaster. The floating coat shall consist of Carlite high suction browning plaster, complying with BS 1191, Part 2, 11 mm thick, ruled to an even surface and lightly scratched to form a key for the finishing coat. The finishing coat shall be Carlite finish plaster, 2 mm thick, and trowelled to a smooth surface. All plaster shall be supplied by British Gypsum Limited, and used in accordance with the manufacturer's printed instructions.

Glazed Wall Tiling   27 The wall tiles shall be 152 × 152 × 5 mm glazed ceramic tiles of low porosity and of thick glaze of selected colours and from an approved manufacturer. The tiles shall be close jointed with an approved resilient adhesive with 6 mm expansion joints at the corners. Provide plastic trim at top edges, external angles, sills and reveals.

Paints   28 All paints shall be supplied by Messrs Berger, I.C.I. or other approved manufacturer. The materials shall be used strictly in accordance with the manufacturer's instructions. Paint shall be applied to woodwork and metalwork in three coats; two undercoats and a gloss finishing coat of approved colours. Wall and ceiling surfaces shall be painted with two coats of emulsion paint.

Glass and Putty   29 All glass is to be to BS 952 of British manufacture and be free from defects. The putty for glazing to timber frames to be to BS 544 and to metal frames to be of approved manufacture.

Paintwork

Painting   30 Knot, prime, stop and rub down all new woodwork before painting. Burn off and rub down existing painted wood surfaces preparatory to repainting as for new work. Wash down existing walls, clean off old paint finishes, cut out and fill cracks and rub down existing plastered surfaces before redecorating.

No painting or other decorations shall be commenced before all other work has been inspected and approved by the surveyor. Pull out all old nails, screws, hooks and other obstructions from walls, ceilings and woodwork, and make good to surfaces.

Living Room

Floor

31 Lift existing tile floor and hack up brick sub-floor and lime concrete base. Take out soil to average depth of 75 mm and lay and consolidate 150 mm bed of hardcore consisting of stone rejects blinded with gravel. Lay on hardcore, 100 mm bed of concrete, as specified, with a trowel finish. Remove existing tiled skirting, hack away existing wall plaster to a height of 300 mm above floor level, and make good with a cement and sand (1:3) render coat finished flush with the plaster. Lay 1000 gauge polythene membrane and 38 mm cement and sand (1:3) screed on the concrete bed, and on the screed lay PVC tiles of approved make, colour and pattern (p.c. £10.00/m² laid).

Supply and fix matching PVC skirting trim, 75 mm high all round room. The PVC tiles and skirting shall be laid by an approved specialist flooring sub-contractor.

Cast Iron Range/ Fireplace

32 Take out existing cast iron range and remove. Supply and install in opening continuous burning open fire and welded steel back-boiler of approved design (p.c. £180 complete), build in and form new connections to existing flue and copper tubing. Build up front of range opening in 102.5 mm brickwork. Supply and fix new tiled surround (p.c. £90) Render and set as necessary to chimney breast and returns, and make good to old plaster.

Decorations

33 Prepare plastered walls and ceiling as previously described and decorate with two coats of emulsion paint. Prepare existing woodwork as previously described and apply two undercoats and one finishing coat of hard gloss paint to both existing and new woodwork.

Kitchen

Alterations

34 Take out existing sink, copper and shelving. Cut out flue stack from copper and seal off with triangular concrete pad as shown on drawing, and make good brickwork where disturbed.

Take out existing sash window and stone sill. Hack off all loose and cracked plaster and rake out joints of brickwork for replastering.

Take down existing half-brick wall between fuel store and external WC. Brick-up opening to fuel store. Take out existing WC pan and flushing cistern and seal off existing drain.

New Doors and Windows

35 Form new door openings in rear wall of existing scullery and in side wall of existing WC to give access to bathroom extension, and build in 150 × 215 mm precast reinforced concrete lintels and 100 × 75 mm softwood rebated frames to receive 726 × 2040 × 40 mm thick semi-solid flush doors to BS 4787.

Fit 14 × 45 mm chamfered architraves in the positions shown on the drawing. Hang doors from a pair of 100 mm pressed steel butts and fit mortice locks and door furniture (p.c. £11 per door). Make good to skirtings.

Enlarge side window opening, build in precast reinforced concrete boot lintel and new steel casement windows to the dimensions and type shown on the drawing. Form two course tile sill externally. Glaze windows with 4 mm clear sheet glass.

Render and set to window and door opening reveals and make good to old plaster.

New Cupboards

36 Build and bond into existing walls 75 mm concrete block partitions to form new cupboards. Build in cupboard fronts in accordance with the details shown on the drawing, and fix timber shelving as shown. Build in 2 nr 215 × 215 mm terra cotta airbricks to food cupboard in positions shown and finish internally with 215 × 215 mm fibrous plaster vents. Render and set block partitions and make good ceilings. Fit skirtings to new partitions to match existing.

Gas Water Heater

37 Fit and connect Ascot multi-point water heater (p.c. £200) and build in asbestos flue outlet.

Sink Unit

38 Install and connect up 1050 × 525 mm stainless steel combined sink and drainer with cupboard unit below (p.c. £160).

Decorations

39 Decorate plastered walls and ceilings and paint woodwork as for living room.

Bathroom

Extension

40 Build new bathroom extension to the dimensions shown on the drawing. Excavate over site and spread soil in back garden. Excavate for and lay concrete foundations, 550 × 150 mm in section. Build 255 mm cavity walls of 100 mm insulating

concrete blocks internally and half-brick faced skin externally tied together with 4 wall ties/m². The face of brickwork, with flush joints, shall be kept clean and the ties clear of mortar droppings. Seal the cavity with brickwork and vertical felt d.p.c. at window opening, and build the walls solid at eaves. Fill the cavity with 50 mm glass fibre slab insulation.

Lay 100 mm bed of concrete on hardcore, polythene waterproof membrane and 38 mm cement and sand (1:3) screed on the concrete bed, and on the screed lay PVC tiles as for the living room. Lay a boarded and joisted roof as shown on the drawing to falls, and finish with three layers of built-up felt covered with white spar chippings. The roof shall be insulated with 100 mm fibre glass insulation on the plasterboard ceiling. Fit fascia board and soffit boarding to eaves as shown.

Fix cast iron rainwater goods as shown. Build in metal casement of the dimensions and type shown, together with boot lintel, tile sill and window board, and one air brick as for kitchen. Fit 14 × 70 mm chamfered softwood skirting.

Render and set walls and fit plasterboard ceiling with skim coat as previously specified. Lay two courses of 150 × 150 mm wall tiles over bath and wash basin.

Decorate walls and ceilings and paint woodwork as for living room. Paint external woodwork and metalwork to match existing. Glaze window with 4 mm patterned glass.

Sanitary Appliances 41 Install and connect up bath, wash basin and WC (p.c. £300) in the positions shown on the drawing, including provision of hot and cold services and waste pipes.

Electrical Work

Electrical Supply and Fittings 42 Allow the sum of £280 to cover the provision of three additional socket outlets in kitchen, two additional socket outlets in living room and lighting and power circuits and fittings in bathroom, as shown on drawing. Allow for attendance on electrician and necessary builder's work.

Drainage

Drains 43 Excavate for, backfill as necessary and lay 100 mm flexible jointed vitrified clayware pipe drains and fittings to BS 65, to the lines and gradients shown on the drawing. Foul drains shall be of British Standard quality and surface water of British Standard Surface Water quality. Provide vitrified clay-ware back inlet trapped gullies with 150 × 150 mm galvanised wrought iron hinged flat grating and frames, set on and surrounded with 150 mm concrete. All drains shall be tested with water to the satisfaction of the local authority.

Inspection Chamber 44 Excavate for, backfill as necessary and construct inspection chamber in one brick walls in class B engineering bricks in cement mortar (1:3), 900 × 675 mm internally on a 150 mm concrete base. Corbel the sides of the chamber to take a coated cast iron cover and frame, size 600 × 450 mm complying with BS 497 (grade C—light duty—reference C6—24/18).

Note: The installation of a gas fired central heating system including boiler, piping, radiators, pump and controls could be in the range of £2000 to £2500 (1987 prices).

Specifications for New Work

When specifying new building work, including that contained in extensions to existing buildings like the bathroom extension described earlier, it may be deemed necessary to prepare much fuller workmanship clauses to ensure a good standard of work, even although the quantity of work may be quite small. The following specification clauses for the trade of bricklayer relating to the work in the bathroom extension are given to show a possible approach.

Bricklayer

Brickwork and Blockwork Generally
Build the whole of the brickwork and blockwork to the dimensions and heights shown upon the drawings. Cavity walls are to have a 50 mm cavity and wall ties are to be provided spaced 900 mm horizontally and 450 mm vertically and staggered and with extra ties at reveals and openings; the ties to be carefully laid so that they do not fall towards the inner leaf. Keep the cavity clear by lifting battens or other means, leave openings at the base, clean out cavity at completion and subsequently brick up the openings uniformly with the surrounding brickwork.

In dry weather the suction rate of the bricks shall be adjusted by wetting before use and the tops of walls left exposed shall be wetted before work is recommenced.

All bricks shall be well-buttered with mortar before being laid and all joints shall be thoroughly flushed up as the work proceeds.

Brickwork shall be carried up in a uniform manner, no one portion being raised more than 900 mm above another at one time. All perpends and quoins shall be kept strictly true and square and the whole properly bonded together and kept level.

No brickwork shall be carried out during frosty weather without the consent of the surveyor and subject to his requirements. All brickwork laid during the day shall, in seasons liable to frost, be adequately covered up at night with suitable protective material. Should any brickwork be damaged by frost the brickwork shall, at the discretion of the surveyor, be pulled down and made good at the contractor's expense.

Half-brick walls shall be in stretcher bond. Bats shall not be used except where required for bond. No four courses shall rise more than 300 mm. Build cavity walls solid at head of wall as shown on drawings. Rake out joints of all brickwork which is to receive plaster.

Concrete blocks shall be jointed in cement-lime mortar (1:1:6). Construction joints must be formed at intervals not exceeding 6 m with a 38 mm × 200 mm long galvanised mild steel strip across joints in alternate courses and with the render and plaster coat severed.

External Facework

Face the brickwork externally with facing bricks as described, taking care to obtain an even mixed distribution of bricks of the various colours (red, brown and purple) and to avoid patches of bricks predominantly of one colour. Joint the bricks with cement-lime mortar, as described, and point with a curved recessed joint (bucket handle) as the work proceeds. Bricks are to be carefully handled to prevent damage.

All faced work shall be kept perfectly clean and no rubbing down of brickwork will be allowed.

Scaffold boards shall be turned back during heavy rain and at night to avoid splashing.

Damp-proofing

Lay over the whole of the walls, for the full thickness at a minimum height of 150 mm above finished ground level and under copings, a bitumen hessian base and lead damp-proof course as previously described.

Cover the cavities of hollow walls under wood sills with bitumen damp-proof course, as described, 200 mm wide.

Build solid the reveals of hollow walls at window and door openings with brickwork, 100 mm thick and insert bitumen hessian base and lead vertical damp-course.

Sundries   Cut and fit brickwork and blockwork around waste pipes; provide all necessary chases for pipes and conduits; build any oversailing courses that are required; build in or cut and pin ends of sills and thresholds; bed plates; bed door and window frames in cement mortar (1:3) and point in mastic; secure door and window frames in brickwork with stout galvanised mild steel ties (six to each opening) and build in fixing bricks, wherever else required, for fixing joinery. Cut away as required, and make good after all trades.

Further Sources of Information

Gardiner[7] has formulated a systematic approach to the scheduling of works of building repair, improvement and conversion, while Bernstein and Richardson[8] illustrate a logical method of formulating specification clauses for rehabilitation and conversion work, and useful examples are provided by Scott.[9]

References

1 I. H. Seeley, *Civil Engineering Specification*. Macmillan (1976)
2 *Specification*. Architectural Press, published annually
3 Royal Institution of Chartered Surveyors and Building Employers Confederation. *SMM6: Standard Method of Measurement of Building Works: sixth edition* (1979)
4 A. J. Willis and C. J. Willis. *Specification Writing for Architects and Surveyors*. Granada (1983)
5 *BRE Current Paper 37/69: Performance Specifications for Building Components* (1969)
6 *BRE Report 32. Performance Specifications for Whole Buildings* (1983)
7 L. Gardiner. *Standard Method of Specifying for Minor Works*. Lewis Brooks (1986)
8 L. Bernstein and A. Richardson. *Specification Clauses for Rehabilitation and Conversion Work*. Architectural Press (1982)
9 J. J. Scott. *Specification Writing: An Introduction*. Butterworths (1984)

9 MEASUREMENT AND PRICING OF MAINTENANCE WORK

Measurement of Building Work

Historical Background to Measurement

Up to the middle of the nineteenth century it was normal practice to measure and value the work after completion. This practice gave rise to various problems as, for instance, when some of the craftsmen's surveyors made extravagant claims for waste of material in executing the work on the site, and the architects felt obliged to engage surveyors to contest these claims.

During the period of the Industrial Revolution there was a large increase in the volume of building work and general contractors became established who submitted inclusive estimates covering the work of all trades. They also engaged surveyors to prepare bills of quantities on which their estimates were based. As competitive tendering became more common the general contractors began to combine to appoint a single surveyor to prepare a bill of quantities, which all the contractors priced. In addition, the architect on behalf of the building owner usually appointed a second surveyor, who collaborated with the surveyor for the contractors in preparing the bill of quantities, which was used for tendering purposes.

In later years it became the practice to employ one surveyor only who prepared an accurate bill of quantities, based on the architect's drawings, which was used for tendering. This surveyor also measured any variations that arose during the execution of the work. This was the origin of the present-day independent professional quantity surveyor, who is an expert on all matters relating to building costs.[1]

Purpose of a Bill of Quantities

The bill of quantities is a schedule of all the items of labour and materials needed to carry out a building contract. Each item is suitably referenced with a full description of the work and the quantity involved. Rate and pricing columns are also contained in the bill so that each contractor tendering can insert prices against each billed item, and the total of all these prices provides the tender sum.

If any variations arise in the work, resulting for instance from changes of design or substitution of alternative materials, the rates in the bill will normally form the basis for valuing the varied items of work. One of the primary functions of the bill of quantities is to enable all tenders to be computed on an identical basis and it invariably forms a contract document.

The main object of preparing a bill of quantities is to determine the cost of the particular contract, and so everything that is likely to affect cost must be included. The principles on which such a bill of quantities is to be prepared are detailed in the Standard Method of Measurement of Building Works.[2] For domestic alterations and small extensions, the Code for the Measurement of Building Works in Small Dwellings[3] is more appropriate.

In the absence of a bill of quantities, each contractor has to prepare his own bill of quantities, or possibly an abridged version, in the limited amount of time allowed for tendering. With larger contracts this places a heavy burden on each contractor and also involves him in additional cost which must be spread over the contracts in which he is successful.

Bill Preparation

The traditional process of bill preparation can conveniently be broken down into two main processes—taking-off and working-up.

In *taking-off*, dimensions are scaled or read from drawings and entered in a recognised form on specially ruled paper, called 'dimensions paper', as illustrated in table 9.1.

Table 9.1   Dimensions paper

1	2	3	4	1	2	3	4

Each page of dimensions paper is split vertically into two identically ruled parts, each consisting of four columns, which are used for the following purposes.

Column 1—timesing column, in which multiplying figures are entered where there is more than one of the particular item being measured.

Column 2—dimension column, where the actual dimensions are entered. There may be one, two or three lines of dimensions, depending on whether it is a length, area or volume.

Column 3—squaring column, where the length, area or volume, obtained by multiplying together the figures in columns 1 and 2, is recorded, ready for transfer to the abstract or bill.

Column 4—description column, in which the written description of each item is entered. The right-hand side of this column is known as waste, in which preliminary calculations, build-up of lengths, locational references and other explanatory information can be entered. Abbreviations are used extensively in writing descriptions at the taking-off stage; a comprehensive list is given in *Building Quantities Explained*.[1]

The dimensions must always be recorded in the order of (1) length, (2) width or breadth and (3) depth or height; all taken to two places of decimals (to nearest 10 mm), although dimensions in waste are taken to three places of decimals.

Typical entries on dimensions paper are shown in table 9.2.

Table 9.2   Entries on dimensions paper

				Explanatory notes
	5.78		Exc. topsoil av. 150 dp.	Superficial item for stripping topsoil from site of small building (5.78 m × 5.48 m)
	5.48			
	19.50		Exc. fdn. tr. ex 0.30 m wide, n.e. 1.00 m max. depth, startg. at stripd. lev.	Cubic item of foundation trench excavation, taken in the depth stages listed in the Standard Method, 19.50 m long × 750 mm wide × 750 mm deep.
	0.75			
	0.75		&	
			Fillg. previously excvtd. mat. to excavns.	All taken as filling in the first instance and subsequently adjusted when concrete and brickwork below ground are measured.

The order of taking-off must be logical and normally follows fairly closely the sequence of work on site as follows.

Carcass—foundations, brickwork and facework; blockwork; fireplaces, chimney breasts and stacks; floors; and roofs.

Finishings—wall, ceiling and floor finishings; windows, including adjustment of openings; doors, including adjustment of openings; fittings; stairs; plumbing installation; drainage work; other services; and other external works, such as roads, paths, fences and landscaping.

One of the first principles of measurement to be mastered is the girthing of buildings, measured on the centre lines of the main enclosing walls. Taking for instance a rectangular building of 255 mm cavity walls, measuring 15 m × 7 m externally , the girth of the perimeter wall can be built up in the following manner.

	15.000	
	7.000	
	2/22.000	
	44.000	
less corners 4/255	1.020	(4 times thickness of wall)
girth of building	42.980	(measured on centre line of enclosing walls)

Working-up generally consists of squaring the dimensions, transferring the resultant lengths, areas and volumes to the abstract, where they are arranged in a convenient order for billing and reduced to the recognised units of measurement; and finally the billing process, where the various items of work making up the complete building are listed in full with their quantities in a suitable order under appropriate work section headings. The traditional working-up process is both lengthy and tedious and various alternative approaches have been introduced with a view to accelerating this stage of the work. One of the older methods was 'billing direct', involving the transfer of items direct from the dimension sheet to the bill. This system operates quite well where the number of similar items is limited and the work is not too complicated, as with drainage. Other developments include 'cut and shuffle' and computerised systems, both of which are particularly well suited for larger projects.

In *abstracting* the dimensions are transferred from the dimension sheets on to abstract sheets which are double width A3 sheets ruled vertically into columns about 25 mm wide. Each sheet is headed with the contract reference or title and the particular works section and subdivided into appropriate subsections as necessary. The order in each section of the abstract is cubes, supers, linear items (runs) and lastly enumerated items, so that they are in the correct sequence for billing, normally using the abbreviations C, S, L and Nr.

The description of each item is normally spread over two columns as shown in table 9.3. It is now customary to omit metric symbols from the descriptions as far as possible.

Table 9.3   Abstract entries

BRICKWORK			PROJECT REFERENCE	
S/ H.b. skin of comms. in g.m. (1:1:6)	holl. wall in stretcher bond in		L/ Hor. d.p.c. 102 wide of single layer of hessian base bit. to BS 743, ref. A, lapd. 100 at jts. (mesd. net) & bedded in c.m. (1:4)	
81.36(11) 132.14(12) 61.04(12) 9.21(14) 25.82(15) 12.48(21)	*Ddt.* 8.32(28) 13.54(54) 17.73(58) 28.10(60) ——— 67.69		41.56(8) 52.48(9) 37.52(10) 12.81(10) 23.59(11)	
322.02 67.69 ——— 254.33 = 254 m²			167/96 ——— = /168 m	

These items will be crossed through on the dimension paper as they are transferred to the abstract. Any deductions, such as window and door openings from brickwork, which are measured overall in the first instance, are entered in the second column, and the numbers appearing in brackets after the dimensions are the page or column numbers of the dimension sheets. Each total in the abstract will be reduced to the recognised unit of measurement. Finally each item will be crossed through on its transfer to the bill.

The final stage of *billing* takes place on bill paper usually in the form of single right-hand billing, in accordance with BS 3327 (Stationery for Quantity Surveying). This form of billing is shown in table 9.4.

Table 9.4   Billing entries

Item Nr	Description	Qty	Unit	Rate	£	p
A	*Excavation* Excavate topsoil average 150 mm deep.	50	m^2			
B	Excavate foundation trench, exceeding 0.30 m wide, not exceeding 1.00 m maximum depth starting at stripped level.	30	m^3			

The most common referencing arrangement is alphabetical commencing on each page, thus an item reference could be 24D (page 24, item D). All descriptions are entered in full with no abbreviations permitted. The monetary totals at the bottoms of pages are normally transferred to a collection at the end of the section, and the total for each section is transferred to a summary at the end of the bill. The total of the summary will represent the tender sum. The units are normally expressed as m, m^2, m^3, t and nr.

Preambles are introductory clauses entered at the head of each work section in a bill of quantities, relating to matters which affect the contractor in pricing the bill and which ought to be drawn to his notice. Their main purpose is to help contractors when tendering for projects by making the task of pricing as straightforward as possible. They often assist in reducing the length of billed descriptions by avoiding repetitive entries and they often contain descriptions of materials and workmanship, as found in specifications.[1]

The *Preliminaries Bill* is the first sectional bill in a bill of quantities and covers many important financial matters which relate to the contract as a whole and are not confined to any particular work section, and the contractor is thus given the opportunity to price them. Section B of the Standard Method of Measurement of Building Works[2] describes most of the items which would appear in such a bill. Matters to be inserted in a Preliminaries Bill include the names of the parties to the contract; description of the site; list of drawings; contract conditions clause headings; water for the works; lighting and power for the works; contractor's and employer's liabilities; temporary works such as temporary buildings, roads, screens, hoardings, scaffolding and telephones; nominated sub-contractors' work; goods from nominated suppliers; protection and drying out of works; clearing the site on completion and contingencies.

Principles and Units of Measurement

The general principles of measurement are expounded in the Standard Method of Measurement of Building Works,[2] as slightly modified by the Code for the Measurement of Building Works in Small Dwellings[3] which covers small dwellings and alteration and repair work. An outline of the operative units of measurement applicable to building work follow. The modifications to the Standard Method of Measurement[2] contained in the third edition of the Small Dwellings Code[3] are minimal. SMM7 was expected to be published in late 1987.

Excavation and Earthwork

Excavating *topsoil* which is to be preserved shall be given in m^2 stating the average depth. Disposal of the soil is measured separately in m^3.

Reduce level excavation is given in m^3 stating the appropriate maximum depth range.

Excavation of *foundation trenches* shall be given in m^3 stating the starting depth, such as stripped or reduced level, and the maximum depth of excavation in appropriate stages, for instance not exceeding 0.25 m, 1.00 m, 2.00 m, 4.00 m and thereafter in 2.00 m stages

Earthwork support is measured in m^2 to vertical faces and sloping faces exceeding 45° and exceeding 0.25 m in height.

Levelling or *grading* the bottom of excavation shall be given in m^2.

Filling to excavations and *disposal* of surplus soil shall be given in m^3.

Concrete Work

Formwork is measured in m^2 in the various categories listed in the Standard Method.

Concrete in *foundation trenches* shall be given in m^3, stating the thickness in stages of not exceeding 100 mm, 100 to 150 mm, 150 to 300 mm and exceeding 300 mm thick.

Concrete in *beds*, *roads*, *pavings*, *walls* and *suspended slabs* shall be given in m^3, stating the thickness in the ranges previously listed.

Concrete in *beams* and *attached columns* are deemed to be included with the suspended slabs and walls respectively.

Bar reinforcement shall be given in tonnes stating the diameter, with each given separately, and classified according to location.

Fabric reinforcement shall be measured the net area covered in m^2, stating the mesh, the weight per m^2 and the minimum extent of side and end laps.

Precast concrete steps, sills and lintels shall be enumerated stating the size, and giving the appropriate particulars.

Brickwork and Blockwork

Descriptions of *brickwork* shall include the kind, quality and size of bricks, type of bond, and composition and mix of mortar. Brickwork shall be measured in m^2, stating the thickness under various classifications, such as walls, filling existing openings, and isolated piers and chimney stacks. Skins of *hollow walls* are measured separately in m^2 as is also forming of the cavity, giving the width of cavity and type and spacing of wall ties.

Various *labours* such as eaves filling, chases, and rough and fair cutting are deemed to be included with the brickwork.

Facework shall be measured as extra over brickwork in m^2 (no deduction is made from common brickwork measurements), stating the kind and quality of bricks, type of bond, composition and mix of mortar for pointing and method of pointing. Half-brick walls and one-brick walls built fair both sides or entirely of facing bricks shall each be given separately in m^2 stating the thickness, as for example the outer skin of a hollow wall.

Blockwork shall be given in m^2 stating the thickness under various classifications, as for brickwork.

Damp-proof courses generally over 225 mm or one-brick wide shall be given in m^2, while those of narrower width shall be given in metres stating the width.

Underpinning

This forms a separate section in the Standard Method of Measurement of Building Works.

Rubble Walling

Rubble *walling* shall be measured separately in several categories in m^2 stating the thickness, including faced work.

Rough cutting is not measured but other labours shall be given as linear items.

Masonry

Stonework shall be measured separately in several categories in m^2 stating the thickness.

Various *labours* on superficial items of stonework, such as ends, reveals, external angles, fair raking and curved cutting, grooves, rebates and sinkings, shall each be given separately in metres.

Columns, lintels, sills, mullions, transoms, copings, cornices and band courses shall each be given separately in metres stating the size and profile.

Asphalt Work

Asphalt *coverings* over 300 mm wide shall be given in m² while those not exceeding 300 mm wide shall be given in metres stating the width in stages of 150 mm, to various classifications.

Various *labours*, such as fair edges, rounded edges, drips, arrises, and turning asphalt nibs into grooves shall each be given separately in metres, while skirtings, aprons, fascias and gutter linings shall each be given separately in metres stating the width on face.

Roofing

Coverings shall each be given separately with a full description in m².

Work to *edges of roofs*, such as eaves, verges, valleys and hips, shall each be given separately in metres as extra over roof coverings.

Underfelting shall be given in m² stating the extent of laps and method of fixing.

The *sheet metal* roofing section lists allowances to be made for drips, rolls, seams, welts and upstands.

Sheet metal flashings, aprons and cappings shall each be given separately in metres stating the profile. The supply of soakers is enumerated stating the size.

Woodwork

Carcassing timbers shall be given in metres stating the cross-section dimensions under various classifications, such as floors, partitions, flat roofs, pitched roofs and bearers.

Roof boarding shall be given in m², stating the thickness.

Bolts and *straps* shall be enumerated and described stating the method of fixing.

Flooring shall be given in m² with raking and curved cutting each given separately in metres.

Boarding to *eaves*, *verges* and the like shall be given in metres stating the cross-section size.

Doors shall be enumerated and described.

Casements and *sash windows*, including their frames, shall be enumerated and described.

Door frames and *linings* shall be enumerated stating the overall size and sizes or cross-section dimensions of the various parts.

Skirtings, *architraves*, *picture rails*, *cover fillets*, *shelves* and the like shall each be given in metres stating the cross-section dimensions.

Fittings, *fixtures* and *staircases* shall be enumerated and supported by component details.

Structural Steelwork

Structural steelwork shall be given in tonnes under the headings of fabricated steelwork, unfabricated steelwork and erection, and appropriately described.

Metalwork

Handrails shall be given in metres stating the size and *balusters* and *railings*, other than fencing, given in metres, stating the height and describing the posts or other supports.

Metal windows and *doors* shall each be enumerated separately stating the overall size, nature of construction, finish and number of opening portions.

Plumbing Installations

Gutters and *pipes* shall be given in metres stating the type and nominal size.

Joints to pipes shall be given in the descriptions of the relevant pipes. *Pipe fittings*, such as bends and junctions, shall be enumerated as extra over the pipes in which they occur.

Sanitary appliances shall be enumerated giving adequate particulars.

Water storage tanks and the like shall be enumerated giving adequate particulars.

Valves shall be enumerated.

Electrical Installations

Cables for lighting and heating, earthing conductors and conduit shall be measured in metres.

Lighting fittings, switches, socket-outlets and the like shall be enumerated and described.

Plasterwork and Other Finishings

Wall, floor and *ceiling finishings* shall be given in m² and work not exceeding 300 mm wide shall be so described. Work in compartments not exceeding 4 m² on plan shall be given separately.

Glazing

Glass shall be given in m², classified as to sizes of panes (not exceeding 0.10 m², 0.10–0.50 m², 0.50–1.00 m² and over 1.00 m²), giving a full description of the glass and method fixing.

Curved cutting shall be given in metres.

Painting and Decorating

Painting work on *general surfaces* over 300 mm girth, shall be given in m^2 with isolated surfaces not exceeding 300 mm girth given in metres. The girth shall be stated in stages of 150 mm, giving a full description of the work and keeping internal and external work, and new work and redecorations separate.

Similar rules apply to work on wood frames, wood and metal windows and glazed wood and metal doors, with work on edges of opening casements in metres. Descriptions of windows and glazed doors shall include the sizes of panes, classified as small (not exceeding 0.10m^2), medium (0.10–0.50 m^2), large (0.50–1.00 m^2) and extra large (over 1.00 m^2).

Work on railings, fences, gates, pipes and gutters shall be measured as for general surfaces.

Polishing shall be measured as for *painting* and *paperhanging* in m^2, stating the number of pieces, classified as on walls and columns or ceilings and beams (both grouped together). Raking and curved cutting (grouped together) and border strips shall be given in metres.

Drainage

Excavating pipe trenches shall be given in metres, stating the starting level and the depth range in stages of 2 m and the average depth to the nearest 0.25 m.

Beds, *benchings* and *coverings* shall each be given separately in metres, stating the necessary dimensions.

Pipes shall be given in metres stating the kind and quality of pipe, nominal size and method of jointing. Pipe fittings, such as bends and junctions, shall be enumerated as extra over the pipes in which they occur. Pipes in runs not exceeding 3 m long shall be so described giving the number.

Gullies shall be enumerated and fully described.

Manholes and *soakaways* shall be given in detail under an appropriate heading, stating the number.

Fencing

Each type of *fencing* shall be given separately in metres with a full description of the materials and method of fixing and the height.

Gates shall be enumerated stating the size and method of construction, and gateposts shall also be enumerated stating the size.

Special fencing posts shall be enumerated as extra over the fencing and excavation for post holes enumerated, stating the size and depth.

Refurbishment and Alteration Work Documentation

A RICS publication[4] which contains some useful examples, describes how the type of quantity surveying documentation will depend largely upon the project, as illustrated in the following guidelines:

(1) A quantified specification is ideal where there are extensive minor alterations.

(2) A priceable schedule is useful where there is extensive repetition of similar items.

(3) Elemental bills of quantities could be the best approach for a one-off project involving fairly extensive alterations.

Building Estimates

Building estimates are prepared by different categories of people for various purposes and using a variety of approaches. Surveyors frequently prepare approximate estimates of building work at the design stage, to indicate to the client his probable financial commitments. Builders preparing estimates will be influenced in their approach by the nature and extent of the information supplied by the designer. It may consist of a bill of quantities accompanied by working drawings, reasonably comprehensive drawings and specification, or possibly just annotated drawings for a small alteration contract or a schedule of repairs for repair work.

Approximate Estimates

Surveyors are frequently required to prepare approximate estimates of the cost of building projects before the detailed schemes have been prepared. A variety of approaches are available each with their own particular advantages and disadvantages. The most commonly used is the floor area method, but other methods occasionally employed are the unit, cube, storey-enclosure and approximate quantities.

Floor area method.   The total floor area of the building is measured within the internal faces of the enclosing external walls, with no deduction for partitions, lift shafts, internal walls, stairs, landings and passages. A unit rate is then calculated per square metre of floor area and the probable total cost of the building is obtained by multiplying the total floor area by the calculated unit rate.

This is a popular method of approximate estimating as it is relatively easy to compute, and most published cost data is expressed in this way in terms which can be understood by a building client. It has a number of inherent weaknesses and, in particular, it cannot directly take account of changes in plan shape or total height of the building, or of variations in finishings, number and quality of fittings and related factors. A few typical rates are shown in table 9.5 to illustrate ranges but it must be emphasised that wide variations occur in the unit rates for any given class of building.

Table 9.5 Floor area price rates

Type of building	Typical costs/m² (1987)
Factory workshop (for owner occupation, including all services)	£250 to £350
Shop (shell)	£260 to £400
Private detached house (built singly and including central heating, garage and external works)	£350 to £500
Local authority 4/5 person two-storey house	£270 to £360

Cube method. The cubic content in m³ of the building is obtained by multiplying the length, width and height (external dimensions) of each part of the building. Heights are taken from the top of foundations to a point half-way up the roof slope in the case of an unoccupied pitched roof and 600 mm above a flat roof. If the roof space of a pitched roof is to be occupied then the height measurement is taken three-quarters of the way up the roof slope, and with mansard roofs it is customary to take the whole of the cubic contents. If a flat roof is surrounded by a parapet wall, then the height measurement is taken to the top of the parapet wall, but where the wall height is less than 600 mm, the minimum height of 600 mm applies. All projections, such as porches, steps, bays, dormers, projecting roof lights, chimney stacks and tank compartments on flat roofs, are measured and added to the cubic content of the main building. A small part of the foundations may be deeper than the remainder, and the unit rate is then best adjusted to account for this variation, rather than attempting to alter the cubic content of the building.

The assessment of the price per cubic metre of a building calls for the exercise of careful judgement coupled with extensive knowledge of current prices and trends. Unit prices show large variations between different classes of building and will even vary considerably between buildings of the same type, where the proportion of walling to floor area, quality of finishings and fittings and amount of partitioning and other components vary. The greater the proportion of wall in relation to the cubic contents of the building, the greater will be its cost per m³. With single-storey industrial buildings, wide variations in storey height can occur and costs will not vary directly in proportion to height. Features such as piling, lifts, external pavings, approach roads, external services and similar works, which bear no relation to the cubic unit of measurement, should be covered separately by lump sum figures or approximate quantities. Some typical cubic metre rates for various types of buildings are shown in table 9.6 to give a rough guide.

Table 9.6 Cube price rates

Type of building	Typical costs/m³ (1987)
Church halls	£80 to £130
Hotels	£200 to £260
Offices	£180 to £210
Shops with one floor of offices over (excluding shop fronts)	£80 to £140
Houses (of various types)	£75 to £230

Table 9.7 Composite price rates

Composite items	Unit	£
Strip foundations. Excavating trench 1 m deep in heavy soil; levelling; compacting; earthwork support; backfilling; disposal of surplus material from site; concrete 11.50 N/mm^2 foundations 300 mm thick; hollow brickwork in cement mortar (1:3) to 150 mm above ground level; bitumen hessian-based horizontal d.p.c.; and facing bricks externally. (£18.00 for each additional 300 mm in depth).	m	80.00
Hollow ground floor construction. Excavation; disposal of surplus; hardcore 100 mm thick; concrete bed 150 mm thick; half-brick sleeper walls, honeycombed, at 2 m centres; horizontal dpc; 100 × 50 mm plates; 50 to 100 mm joists at 400 mm centres; and 25 mm tongued and grooved softwood boarded flooring.	m^2	52.00
Upper floor. 50 × 175 mm joists at 400 mm centres with ends creosoted and built into brickwork; 50 × 25 mm herringbone strutting; trimming to openings; 25 mm tongued and grooved softwood boarded flooring; plasterboard; one coat of 5 mm gypsum plaster; and two coats of emulsion paint.	m^2	50.00
Pitched roof (measured on flat plan area). 75 × 40 mm plates; TRADA trussed rafters at 2 m centres; 32 × 100 mm rafters and ceiling joists at 500 mm centres; 40 × 150 mm purlins; 125 × 50 mm binders; 25 × 150 mm ridge; 40 × 19 mm battens; felt; and concrete interlocking tiles nailed every second course.	m^2	46.00
Stairs. 900 mm wide; treads and risers, winders, balustrade one side; plasterboard and one coat of gypsum plaster to soffit and painting; rising 2600 mm. (£78.00 for each additional 300 mm in height).	Nr	575.00
Wash basin. White glazed vitreous china, waste fitting, cantilever brackets and pair of taps (p.c. £55 complete); trap; and copper waste pipe. (Comparable costs of other sanitary appliances would be £215 for a sink; £406 for a bath; £170 for a ground floor WC and £520 for a WC on an upper floor, including a cast iron soil pipe and connection to drain.)	Nr	135.00
Electrical installations. These are most conveniently priced on a cost per point basis, for example lighting points at £45.00 each and double 13 amp switched socket outlets wired in a ringmain circuit at £57.00 each.		
Drainage. 100 mm g.v.c. pipes and fittings to BS 65; excavating trenches average 1 m deep in heavy soil; grading bottoms; earth work support; backfilling; removal of surplus spoil; and 150 mm concrete 11.50 N/mm^2 beds and benchings. (£4.00 for each additional 250 mm depth of trench not exceeding 2 m deep and £4.30 for each 250 mm between 2 m and 4 m deep).	m	24.00
Manhole 686 × 457 × 900 mm deep internally in one-brick walls in engineering bricks on 150 mm thick concrete base; with 100 mm half-section channel and branches; concrete benchings; 610 × 457 mm cover and frame; and all necessary excavation, backfill, disposal of surplus and earthwork support. (£90.00 for each additional 300 mm of depth up to 2 m deep internally.)	Nr	360.00

Unit method. This allocates a cost to each accommodation unit of a building, be it persons, seats, beds, car spaces or whatever. The total estimated cost is then found by multiplying the total number of units accommodated in the building by the unit rate. The process of calculation is very simple but the computation of the unit is exceedingly difficult, and is based on the analysis of the unit costs of a number of fairly recently completed buildings of similar type, adjusted as necessary. Hence at best it can only be a rather blunt tool for establishing general guidelines, as required for budget estimating on a rolling programme and has little relevance to maintenance and improvement work.

Storey-enclosure method. Takes the area of the external walls, floor and ceiling enclosing each storey of the building, and then proceeds to weight some of them to allow for foundations, upper floors and the extra cost of work below ground level. The method represented a great advance on previous single rate estimating methods but as it has been little used since it was introduced in 1954, there is not much data for comparison purposes, and the method has little relevance to maintenance and improvement work. It is well described in *Building Economics*.[5]

Approximate quantities method. Abridged quantities of the work are measured and priced to give an estimate for the work. It involves more work and is superior to any of the other methods previously described, but there are occasions when lack of information precludes its use. Composite price rate items are obtained by combining or grouping bill items; for example, a brickwork item measured in m^2 will normally include all incidental labours and finishings to both wall faces. Doors and windows are usually enumerated as extra over the walling and associated finishings, thus avoiding the need to make adjustments to the walling. Furthermore, a door item will be a comprehensive one, including the frame or lining, architraves, glazing, ironmongery and decoration. Details of a wide range of composite items with current costs are given in *Spon's Architects' and Builders' Price Book*.[6] Some composite items and costs (1987) are shown in table 9.7 to amplify the approximate quantities approach.

Preparation of Estimates by Contractors

Matters Affecting Building Prices

It is customary to approach several contractors for a price when a client wishes to undertake some building work. The actual number of firms approached will depend largely on the size of the contract and for maintenance and conversion work under normal conditions could vary between three and six. The prices submitted may vary considerably and some of the more important reasons for these price variations follow.

(1) Some contractors may be already heavily committed with work and may submit 'cover prices'.

(2) Extent of availability of labour, plant and materials in the locality.

(3) Time of the year when the work is to be undertaken.

(4) Amount and quality of information supplied to contractors.

(5) Quality of work likely to be expected.

(6) Time allowed for execution of work.

(7) Time allowed for tendering.

Action by Contractors on Receipt of Tender Documents

With larger contracts, the contractor is advised to follow the procedure outlined in the CIOB Code of Estimating Practice.[7] On receipt of a bill of quantities, it is advisable to list all PC items, work to be sublet and major materials and components, and to send out enquiries to suppliers and sub-contractors. The site must be visited to see whether there are any unusual or special features or obstacles which could affect the price; examples being work to be carried out in extremely confined space; site close to a hospital or other establishment where restrictions on noise emission may be imposed during certain hours; site within a river flood plain which might cause problems with excavation and drainage work.

When pricing a bill of quantities, usually on a short time scale, it is best to start with the largest quantity items by detailed cost analysis. These items are probably only about 25 per cent in number of all items in the bill but may well account for about 80 per cent of total cost.

Most contractors approach the pricing of preliminaries in different ways. Some use a percentage of the total sum for measured work, others price almost every item in considerable detail, while some include the cost of preliminaries in the measured rates and insert a lump sum in the summary of preliminaries for the sake of appearances. There are, however, four major items that must be considered as they are not necessarily in direct cost ratio to the measured items—wages of site staff; mechanical plant and scaffolding; travelling time, fares and expenses paid to operatives; and site accommodation.

Site staff consist of agents, foremen, site clerks and similar personnel. Their cost will depend on the length of the contract and can be assessed on a monthly basis.

The contractor's plant will be costed over its effective life (often three or four years). The costs are made up of the purchase price, interest on capital over the life of the plant, repairs and servicing costs, and insurance and road fund licence where appropriate. This total cost can be divided by the estimated number of working hours to give an hourly rate, to which must be added haulage costs to and from sites and fuel costs to give a comprehensive rate. Similarly hire costs of plant need increasing to include haulage and fuel. The costs of the plant operators must not be overlooked.

The contractor is required to pay his own employees' travel time and expenses in travelling to building sites and these will be dependent on the distance involved.

Site hutting can be a fairly expensive item. Even where the contractor has the huts in his possession, he is faced with the cost of depreciation, rates, transport, erection, some repair work to keep them watertight and resistant to pilfering, and dismantling. These costs can amount to as much as £350 to £600 per hut. In addition there is probably the cost of telephone, temporary lighting, heating and labourer attendance to huts used as offices.

Many repair and conversion contracts are too small to warrant a bill of quantities, and they are usually based on drawings and specification. The cost of preliminaries on this type of contract is disproportionately high compared with large contracts or new work. Every contract should be considered on its merits and the kinds of factor needing investigation are excessive amounts of scaffolding, additional site accommodation, unavoidable unproductive time of operatives to meet occupiers' needs, protective screens, and lack of storage facilities.

When estimating for new work, contractors normally take off and price their own abridged form of quantities. Alteration work in a cramped existing building can present many additional problems, particularly when the premises will be occupied throughout the contract period. One approach is to take a notebook ruled in question and answer form to the site, when typical questions might be as follows.

(1) Name and address of project.
(2) Nature of access and any traffic restrictions.
(3) Ground and site conditions, including location of services.
(4) Whether hoist required.
(5) Amount of scaffolding required.
(6) Availability of storage space both inside and outside the premises.
(7) Number of rooms available at any one time.
(8) Any special difficulties.

If the building is empty and likely to remain so until final completion, the foreman will be able to distribute different trades in the various rooms, there will be ample storage and office space and no interference by occupants. Occupied buildings can, on the other hand, create various difficulties as the whole of the work may have to be completed during a holiday closing period or outside normal working hours, or maybe only one or two rooms will be available at a time. Where the specification is not clear on these aspects, the estimator must make allowance for any matters which are likely to hinder the progress of the work. Narrow fronted shop premises in particular frequently create difficult working and access conditions, which are best assessed as lump sums based wherever possible on the estimated number of man-hours needed.

The contractor or his estimator should take sufficient dimensions on the site to permit the pricing of every item of work involved. An additional check is to assess the number of man-hours required for each operation on the site. For example, in forming a new window opening a number of trades are involved—bricklayer and labourer forming the opening and building in the window, joiner, glazier and painter on the window itself and plasterer making good—in all a total time commitment approaching 40 hours. The removal of an old sink and its replacement with a new sink unit may involve a plumber, bricklayer, plasterer and painter for a total time of around 16 hours. Furthermore, allowance must be made for labourers' time unloading materials and components on the site and clearing away on completion.

Many contractors, particularly the smaller ones, use estimating and price books[6] as a basis for their estimates. It is important to check the materials' prices, which are changing quite frequently, and the operative labour rates, which are generally London based and make no allowance for payments over the recognised wage rates. Labour constants (time taken to execute a specific task) in price books generally relate to average conditions, whereas few projects are ever average. Work may be in confined spaces or at excessive heights involving higher labour costs because of lower productivity. For example, ceilings with a total area of 200 m^2 are to be painted with two coats of emulsion paint. The ceilings may be 2.50 m or 5 m above floor and although the materials requirements remain constant, the labour involvement will vary appreciably.

This problem can be approached by assessing the additional time allowances to be included for difficult working conditions. They occur mainly with painter's work but could also be applied to other trades.

(1) Room by room in a private house. This involves moving furniture to the centre of rooms and covering and protecting it—15 per cent addition.

(2) Single room flatlets or bedsitters. This form of accommodation is often in large properties of three or more storeys divided into units of one or two roomed furnished apartments. These are very congested—30 per cent addition.

(3) External work to buildings more than two storeys high, involving more time negotiating ladders and hoisting materials. Additional time might be roughly assessed as follows.

full scaffold—2 to 4 storeys	5 per cent
full scaffold—over 4 storeys	10 per cent
ladder work—2 to 4 storeys	20 per cent
ladder work—over 4 storeys	50 per cent

(4) Shop or factory premises outside normal working hours. If this involves work in the evenings or at weekends, the labour costs could be doubled.

(5) School work during holidays. The premises are unoccupied and the work is quite straightforward but there is invariably a target date for completion. Labour shortage during the crucial period at the end of the school summer holiday may involve additional labour costs of the order of 10 per cent.

Price Build-up

A contractor often builds up the price for each item of work in detail, working from first principles, in order to compute a realistic price for the complete contract. The price of any item of work is made up of certain components from amongst the following.

(1) *Labour*—craft operatives' and labourers' time at recognised rates or at actual rates where these are higher. It is customary to use all-in hourly rates when associated labour costs are added to the basic wage rate to include such items as public and annual holidays, NWR allowances, productivity bonus payments, non-productive overtime, sick pay, employer's national insurance contributions, employer's liability and third party insurance, severance pay, trade supervision, training and CITB levy. These labour oncosts can amount to as much as 100 per cent or more of the basic hourly wage rate.

(2) *Materials*—including waste, transportation, unloading, stacking, storing and distributing around the site, and return of crates or packages where appropriate.

(3) *Plant*—either owned or hired calculated on an hourly rate in either case, and including fuel and operator.

(4) *General overheads and profit*—often expressed as a percentage of the sum of the appropriate previous items; it could be in the range of 10 to 20 per cent according to the organisation of the building firm and the method of pricing. As an alternative a higher general overheads and profit percentage may be added only to the all-in labour rates, which is rather more realistic and makes analysis and recovering by the contractor much simpler. General overheads include such items as office salaries, supervision by office staff, rent and rates, insurances, running and maintenance expenses of premises and plant, printing and stationery, postages and telephone charges, legal and accountancy charges, bad debts, depreciation of office equipment and car expenses.

Site or project overheads such as site supervisory staff, clearing site, site transport services, scaffolding and gantries, site accommodation, small plant and hand tools, temporary services, welfare, first aid and safety provision, defects liability costs, transport of men to site and abnormal overtime are likely to be priced separately in preliminaries items as they will vary considerably from one contract to another.

Some typical examples of the build-up of prices for some of the more common items of building work are given in this chapter. For more information on this extremely complex subject, readers are referred to Smith.[8] The amount of labour required to perform a given unit of work is termed a labour constant and most estimating and pricing books contain many hundreds of such constants. In practice the labour constants or outputs should be computed from analyses of actual performance on past contracts, varied as appropriate for differences in quantity, working conditions and other relevant matters.

Excavation

In building up prices for excavated soil disposal allowance must be made for increases in bulk on excavation. This varies with different types of soil but an average figure is 25 per cent. Typical labour constants applicable to normal ground conditions follow.

With earthwork support to sides of excavation it is necessary to determine just how much support is required having regard to the depth of excavation and type of soil. It is customary to allow seven uses of timber, so that one-seventh of the cost of timber appears in the calculations.

When pricing hardcore, allowance must be made for loss of material on consolidation, as shown in the example on pricing hardcore.

Excavation work	*Unit*	*Labourer/hours*
Excavate oversite, average 150 mm deep	m^2	0.50
Excavate foundation trench, not exceeding 1 m deep	m^3	3.00
Backfill excavated material	m^3	1.50
Level and compact bottom of excavation	m^2	0.15

The following examples show the price build-up of typical hand excavation items incorporating 1987 all-in hourly rates.

Excavate foundation trench not exceeding 1 m deep/m^3

	£
Labourer—3 hours @ £4.10	12.30
Add general overheads and profit (15%)	1.85
Cost/m^3	£14.15

Filling to excavations/m^3

	£
Labourer—1.50 hours @ £4.10	6.15
Add general overheads and profit (15%)	0.92
Cost/m^3	£7.07

Bed of hardcore, 150 mm thick/m^2

	£
0.16 m^2 of hardcore @ £7.50/m^3	1.20
Allow for 20% consolidation	0.24
Labourer—0.20 hour @ £4.10	0.82
	2.26
Add general overheads and profit (15%)	0.34
Cost/m^2	£2.60

Concrete work

Concrete can be hand or machine mixed on site or ready mixed. Approximately 90 per cent of all *in situ* concrete work consists of ready mixed concrete, as it offers many advantages, and hand mixing is rarely used.

The following example shows the price build-up for *in situ* machine mixed concrete in foundations.

In situ concrete grade 20 in foundation trenches 100–150 thick/m³

	£
(1) *Material costs/m³*	
320 kg cement @ £76.50 per tonne	24.48
720 kg fine aggregate @ £5.90 per tonne	4.25
1080 kg coarse aggregate @ £6.00 per tonne	6.48
	35.21
Add 5% waste	1.76
	£36.97(1)

(2) *Mixing and placing of concrete*
Assume 14/10 mixer output of 0.28 m³ and 10 discharges per hour (2.80 m³/h)

	£
1 mixer operator @ £4.30	4.30
1 labourer filling @ £4.10	4.10
Hire of dumper per hour plus fuel, etc.	12.30
1 dumper driver @ £4.25	4.25
Hire of 14/10 mixer per hour plus fuel, etc.	9.60
1 spreader @ £4.10	4.10
	£38.65

$$\text{Cost/m}^3 = \frac{£38.65}{2.80} = £13.80 \ (2)$$

	£
Total cost/m³ = (1) + (2) = £36.97 + £13.80 =	£50.77
Add general overheads and profit (15%)	7.62
Cost/m³	£58.39

The comparable cost of ready mixed concrete would be similar.

Brickwork

With brickwork it is necessary to be able to calculate the number of bricks per square metre in different bonds and wall thicknesses, allowing 8 per cent waste.

a one-brick wall contains 114 + 9 (waste) = 123 bricks/m²
a half-brick wall contains 57 + 5 (waste) = 62 bricks/m²
(using 65 mm thick bricks with 10 mm joints).

In calculating the number of facing bricks

Flemish bond = 76 + 7 (waste) = 83 bricks/m²
English bond = 86 + 9 (waste) = 95 bricks/m²

Labour constants	Bricklayer	Labourer
	hours/m²	
One-brick walls	2.30	2.50
Half-brick walls	1.40	1.50
Fair face to brickwork	0.30	0.30
Extra for facework and pointing as work proceeds	0.50	0.50

One brick wall in commons in cement mortar (1:3)/m²

		£
123 bricks (including waste) @ £97 per 1000		11.93
Unloading: Labourer—0.12 hour @ £4.10		0.49
0.08 m³ of cement mortar @ £56		4.48
Bricklayer—2.30 hours @ £4.70		10.81
Labourer—2.50 hours @ £4.10		10.25
		37.96
Add general overheads and profit (15%)		5.69
	Cost/m²	**£43.65**

An alternative approach to the pricing of the last item follows.

	Qty	Per m² Rate £	Per m² Labour £	Per m² Materials £
Bricks (including waste)	12.3	97.00/ thsd.		11.93
Unloading—Labourer	0.12 h	4.10	0.49	
Mortar (including waste)	0.08 m³	56.00		4.48
Labour Bricklayer	2.30 h	4.70	10.81	
Labourer	2.50 h	4.10	10.25	
			£21.55	£16.41
General overheads on labour	15%		3.23	24.78
				£41.19
Profit	6%			£ 2.47
	Cost/m²			**£43.66**

The variation in price between the two approaches for specific items of work will depend upon the respective labour contents. Both methods are used extensively in practice, and the percentages will vary according to the circumstances and policy of the firm.

Extra over common brickwork (£97 per 1000) for facework (£252 per 1000) in English bond and pointing with a neat struck joint as the work proceeds/m²

	£
Cost of facing bricks	252
Cost of common bricks	97
Extra cost/1000	£155

	£
95 bricks (including waste) @ £15.50 per 100	14.73
Bricklayer—0.50 hours @ £4.70	2.35
Labourer—0.50 hours @ £4.10	2.05
	19.13
Add general overheads and profit (15%)	2.87
Cost/m²	£22.00

Woodwork

Labour constants	Carpenter	Labourer
	\<center\>*hours/m*\</center\>	
Bearers (100 × 75 mm)	0.32	0.04
Floor joists (50 × 100 mm)	0.20	0.04
Rafters and ceiling joists (50 × 100 mm)	0.24	0.04
Ridge boards, hip and valley rafters (38 × 175 mm)	0.30	0.04

Softwood in bearers (100 × 75 mm)/m

	£
1 m of softwood bearer @ £1.65	1.65
Allow 10% waste	0.17
Carpenter—0.32 hours @ £4.70	1.30
Labourer—0.04 hours @ £4.10	0.16
	3.28
Add general overheads and profit (15%)	0.49
Cost/m	£3.77

With floorboarding, allowance has to be made for waste in tongued edges and this can vary from 15 to 25 per cent according to the width of the boards. Joiner's time varies from 0.70 to 0.95 hours per m², depending on the form of jointing and the width of boards. With doors and windows it is necessary to calculate the quantity of timber required and joiner's time works out at about 1.2 to 2 hours per m², depending on the site and complexity of the component. Labourer's time is calculated at about half the craft operative's time with windows and around one-eight with doors.

Plastering

Plasterer's time varies from $\frac{1}{2}$ to $1\frac{1}{2}$ hours per square metre depending on the type, thickness and location of plaster, with labourer's time calculated at one-half the craft operative's time. Typical prices (1987) are £7.40/m^2 for 9 mm baseboard ceilings, fixed, scrimmed and skimmed with gypsum plaster, and £7.60/m^2 for two coat gypsum plaster to walls.

Glazing

A typical example of pricing glazing work follows.

4 mm float glass and glazing to wood with putty, exceeding 0.10 and not exceeding 0.50 m²/m²	
	£
1 m^2 of 4 mm float glass	12.60
Allow 5% waste	0.63
2 kg of putty @ 40p	0.80
Glazier—0.70 hours @ £4.70	3.29
Labourer—0.05 hours @ £4.10	0.21
	17.53
Add general overheads and profit (15%)	2.63
Cost/m^2	£20.16

Painting

Examples of price build-up for painting follow.

Twice emulsion paint plastered walls/m^2	
Cost/100 m^2	£
1 coat sealer: 3.75 litres	
2 full coats: 2 × 7.50 = 15 + 3.75 = 18.75	
18.75 litres of emulsion paint @ £8.50/5 litres	31.88
Painter—22 hours @ £4.70	103.40
Add use and waste of brushes/rollers (5% of labour costs)	5.17
	140.45
Add general overheads and profit (15%)	21.07
Cost/100 m^2	100)161.52
Cost/m^2	£ 1.62

Prepare, knot, prime, stop and paint two undercoats and one finishing coat of oil paint on general wood surfaces/m²

	£
Cost/100 m²	
0.75 litres of knotting @ £4.00/litre	3.00
2½ kg of putty @ 40p	1.00
8 sheets of glasspaper @ 10p	0.80
10 litres of wood primer @ £11.50/5 litres	23.00
2 × 9 = 18 litres of undercoat @ £11.00/5 litres	39.60
9 litres of gloss paint @ £11.00/5 litres	19.80
Painter—18 + 2 × 13 + 14 = 58 hours @ £4.70	272.60
Wear of brushes, ladders etc. (5% of labour costs)	13.63
	373.43
Add general overheads and profit (15%)	56.01
Cost/100m²	100)429.44
Cost/m²	£ 4.29

It must be stressed that the labour constants used must appertain to the particular job and those applied by an inexperienced person merely consulting a price book are unlikely to produce realistic prices. For example, repetitive work takes less time to perform—to lay floor boarding in a small room will cost more proportionately than in a large room, and more still if the small room is of irregular shape.

Pricing Alteration Works

Table 9.8 shows one approach to the build-up of an estimate for a rear extension to a house with an extended floor area of 10 m² and accommodating a bathroom and ventilated lobby. The bathroom contains a white bath, wash basin and WC. The prices given are typical 1987 provincial rates.[9]

Table 9.8   Estimate of alteration work

Item	Description	Quantity	Rate £	Cost £
1	Excavate over site	11 m²	1.50	16.50
2	Excavate foundation trenches and dispose of surplus soil	4 m³	14.00	56.00
3	Concrete foundations	2 m³	72.00	144.00
4	100 mm hardcore, 150 mm concrete floor and damp-proof membrane	10 m²	16.00	160.00
5	255 m cavity wall including outer skin of facings and bonding new work to old	28 m²	64.00	1792.00
6	100 mm concrete block partition	7 m²	20.00	140.00
7	Damp-proof course (felt—102 mm wide)	30 m	0.80	24.00
8	Steel lintels	4 Nr	24.00	96.00
9	Softwood wall plate (100 × 75)	6 m	3.60	21.60

10	Softwood roof joists (50 × 100)	20 m	2.80	56.00
11	19 mm roof boarding	12 m^2	16.00	192.00
12	Fascia and soffit to eaves	3 m	11.50	34.50
13	Roofing felt in 3 layers	12 m^2	11.00	132.00
14	Upstands to roof	10 m	2.80	28.00
15	Casement windows	2 Nr	48.00	96.00
16	Window reveals	10 m	4.00	40.00
17	Flush doors	2 Nr	47.00	94.00
18	Door frame and architraves	10 m	9.00	90.00
19	Ironmongery	Item	21.00	21.00
20	Cutting opening through existing wall for new door	Item	26.00	26.00
21	Plasterboard ceiling and plaster skim coat	10 m^2	8.00	80.00
22	Plaster walls (including reveals)	47 m^2	8.30	390.10
23	Thermoplastic floor tiles	10 m^2	12.00	120.00
24	Softwood skirting	18 m	2.00	36.00
25	Bath complete	1 Nr	350.00	350.00
26	Wash basin complete	1 Nr	180.00	180.00
27	WC suite complete	1 Nr	215.00	215.00
28	Copper water services	10 m	8.00	80.00
29	PVC gutter	3 m	6.80	20.40
30	PVC downpipe	3 m	8.00	24.00
31	Obscure glass and glazing	2 m^2	20.00	40.00
32	Twice emulsion paint walls and ceiling	60 m^2	2.00	120.00
33	Painting doors, frames, etc.	10 m^2	5.00	50.00
34	Painting windows	4 m^2	6.00	24.00
35	Manhole complete	1 Nr	275.00	275.00
36	Excavate drain trench	4 m	24.00	96.00
37	Drain pipe on concrete bed	4 m	20.00	80.00
38	Connection to existing manhole	1 Nr	15.00	15.00
39	Connection to WC	1 Nr	11.00	11.00
40	Gully trap and surround	1 Nr	24.00	24.00
41	Electrical work	Item	250.00	250.00
42	Preliminaries	Item	350.00	350.00
43	Contingencies	Item	250.00	250.00

Total estimated cost   £6340.10

say £6340

Table 9.9 shows a detailed price build-up for forming a new doorway opening in a half-brick partition, plastered both sides, to show the large number of operations involved.

Table 9.9 Price build-up for new doorway

Item	Description	Quantity	Rate £	Cost £
1	Remove and set aside skirting	4 m	1.20	4.80
2	Ditto picture rail	4 m	1.20	4.80
3	Demolish brickwork and plaster and remove	3 m^2	12.00	36.00
4	Cut, tooth and quoin up reveal	4 m	12.00	48.00
5	102 ×150 precast RC lintel	1^1/$_3$ m	18.00	24.00
6	Cut and pin ends of lintel	2 Nr	1.50	3.00
7	Wedge and pin up over lintel	^{1}/$_5$ m^2	15.00	3.00
8	Level brickwork across threshold and lay screed	1 m	12.00	12.00
9	Hack brickwork and render and set 150 mm wide, including making good new to old at one edge	8 m	6.00	48.00
10	Ditto 225 mm wide	2 m	7.50	15.00
11	Refix existing skirting to wall	2 m	2.40	4.80
12	Ditto picture rail	4 m	1.80	7.20
13	Wash off, stop and twice emulsion paint frieze	2 m^2	3.60	7.20
14	Ditto wall filling	4 m^2	4.80	19.20
15	Wash off, stop, touch up and apply two coats of oil paint to skirting	4 m	1.50	6.00
16	Ditto picture rail	4 m	1.20	4.80
17	Contingencies and preliminaries			15.00
		Total estimated cost		£262.80

Small items of repair and alteration work are often disproportionately expensive as they are very labour-intensive, require small quantities of a variety of materials involving considerable waste, and the general overheads are also high. Normal price book unit rates have little relevance to this class of work, although sections covering minor works and alterations and repairs are helpful.

The essence of estimating for rehabilitation is to ensure that the estimate covers realistically the work content specified. One risk in rehabilitation work is the content and/or extent may not be clearly defined. The general condition of the building; existing defects and the contractual responsibility for their correction must be determined. Ambiguous or insufficient descriptions must be clarified or qualified before entering into a contract and all risks identified and quantified.[10]

Checking Builder's Accounts

Surveyors are often called upon to check builders' accounts for works of repair and alteration. These accounts normally list the materials used and amount of labour employed with the costs involved plus allowances for overheads and profit. This type of work is often carried out on a daywork basis under some form of cost reimbursement contract. The reader is referred to chapter 10 for more detailed information on contractual arrangements.

On receipt of an account, the surveyor should examine it against the background of his own notes of site inspections made while supervising the work, where these are available. A further site visit to clarify any doubtful points prior to arranging a meeting at the builder's office is often helpful. It may be that alternative materials were substituted for those previously specified resulting in changes in price.

The quantities of materials listed in the account can be checked by measuring the work done and calculating the materials required, incorporating suitable waste allowances. The labour component is more difficult to assess, although the constants listed in price books and included earlier in this chapter form a useful guide. The surveyor must be prepared to adjust the constants to suit the particular site conditions. Materials prices can be checked against manufacturers' price lists with due allowance made for extra costs attributable to small quantities.

At the meeting at the builder's office, queries will be discussed and the builder will produce invoices and time sheets to substantiate details in the account. It is possible that materials have been incorrectly charged to the contract, that some are overpriced or that labour from another site has been inadvertently allocated to the contract. On occasions labour may be rendered unproductive while one trade waits for another, and some adjustment to cost may be necessary if this delay is considered excessive. Frequent transfer of labour from one job to another can result in heavy travelling costs and further unproductive time. Labour costs usually include a variety of labour-related overheads as described earlier in the chapter. Finally, the percentage addition for general overheads and profit requires scrutiny to ensure that no ancillary costs are charged twice and that the allowance is not excessive. For instance, foreman's time in supervision may be included both as a labour-related charge and as a general overhead. Some builders apply a single percentage to both labour and materials while others use separate figures for each. The surveyor's task is not to reduce the account to an irreducible minimum but to ensure that the sum paid by the client is fair and reasonable.

References

1 I. H. Seeley, *Building Quantities Explained*, Macmillan (1979)
2 Royal Institution of Chartered Surveyors/Building Employers Confederation. *Standard Method of Measurement of Building Works, sixth edition*: SMM6 (1979)
3 Royal Institution of Chartered Surveyors/Building Employers Confederation. *Code for the Measurement of Building Works in Small Dwellings, third edition* (1979)
4 Royal Institution of Chartered Surveyors. *Refurbishment and Alteration Work: Quantity Surveying Documentation* (1982)
5 I. H. Seeley, *Building Economics*. Macmillan (1983)

6   Davis, Belfield and Everest (Eds). *Spon's Architects' and Builders' Price Book*. Spon (1987)
7   Chartered Institute of Building, *Code of Estimating Practice* (1983)
8   R. C. Smith. *Estimating and Tendering for Building Work*. Longman (1986)
9   Building Maintenance Information Ltd. *Building Maintenance Price Book* (1987)
10   CIOB Estimating Information Service Nr 42. *Estimating for Rehabilitation* (1981)

10 TENDERING PROCEDURES AND CONTRACT ADMINISTRATION

This chapter is concerned with the contractual arrangements for the execution of building work, with particular reference to works of alteration and repair.

Nature and Form of Contracts

The law relating to building contracts is one aspect of the law of contract and tort (civil wrongs). It is accordingly desirable to have an appreciation of the law relating to contracts generally before considering the main characteristics and requirements of building contracts.

A simple contract consists of an agreement entered into by two or more parties, whereby one of the parties undertakes to do something in return for something to be undertaken by the other. A contract has been defined as an agreement which directly creates and contemplates an obligation. The word is derived from the Latin *contractum*, meaning drawn together.[1]

We all enter into contracts almost every day for the supply of goods, transportation and similar services, and in all these instances we are quite willing to pay for the services we receive. As our needs in these cases are comparatively simple and we do not need to enter into lengthy or complicated negotiations, no written contract is normally executed. Nevertheless, each party to the contract has agreed to do something and is liable for breach of contract if he fails to perform his part of the agreement.

In general, English law requires no special formalities in making contracts but, for various reasons, some contracts must be in a particular form to be enforceable and, if they are not made in that special way, then they will be ineffective. Notable among these contracts are contracts for the sale and disposal of land, and 'land' for this purpose, includes anything built on the land such as houses.

It is sufficient to create a legally binding contract if the parties express their agreement and intention to enter into such a contract. If, however, there is no written agreement and a dispute arises in respect of the contract, then the Court which decides the dispute will need to ascertain the terms of the contract from the evidence given by the parties, before it can make a decision on the matters in dispute.

On the other hand if the contract terms are set out in writing in a document which the parties subsequently sign, then both parties are bound by these terms even if they do not read them. Once a person has signed a contract he is assumed to have read and approved its contents, and will not be able to argue that the document fails to set out correctly the obligations which he actually agreed to perform. Thus by setting down the terms of a contract in writing one secures the double advantage of affording evidence and avoiding disputes.

The law relating to contracts imposes upon each party to a contract a legal obligation to perform or observe the terms of the contract, and gives to the other party the right to enforce the fulfilment of these terms or to claim 'damages' in respect of the loss sustained in consequence of the breach of contract.[1]

Enforcement of Contracts

An agreement can only be enforced as a contract if

(1) The agreement relates to the future conduct of one or more of the parties to the agreement.

(2) The parties to the agreement intend that their agreement shall be enforceable at law as a contract.

(3) It is possible to perform the contract without transgressing the law.

Validity of Contract

The legal obligation to perform a contractual obligation only exists where the contract is valid. In order that the contract shall be valid the following conditions must operate.

(1) There must be an offer made by one person (the offeror) and the acceptance of that offer by another person (the offeree), to whom the offer was made. Furthermore, the offer must be definite, and made with the intention of entering into a binding contract. The acceptance of the offer must be absolute, be expressed by words or conduct, and be accepted in the manner prescribed or indicated by the person making the offer.

An offer is not binding until it is accepted and, prior to acceptance, the offer may come to an end by lapse of time, by revocation by the offeror or by rejection by the offeree, and in these cases there can be no acceptance unless the offer is first renewed.

(2) The contract must have 'form' or be supported by 'consideration'. The 'form' consists of a 'deed' which is a written document, which is signed, sealed and delivered, and this type of contract is known as a formal contract or contract made by deed.

If a contract is not made by deed, then it needs to be supported by 'consideration', in order to be valid, and this type of contract is known as a simple contract. Consideration has been defined as some return, pecuniary or otherwise, made by the promisee in respect of the promise made to him.

(3) Every party to a contract must be legally capable of undertaking th obligations imposed by the contract. For instance, persons under 18 years c age may, in certain cases, avoid liability under contracts into which the have entered. Similarly a corporation can only be a party to a contract if it i empowered by a statute or charter to enter into it.

(4) The consent of a party to a contract must be genuine. It must not b obtained by fraud, misrepresentation, duress, undue influence or mistake

(5) The subject matter of the contract must be legal.

Remedies for Breach of Contract

Whenever a breach of contract occurs a right of action exists in the Courts t remedy the matter. The remedies generally available are as follows:

(1) Damages
(2) Order for payment of a debt
(3) Specific performance
(4) Injunction
(5) Rescission

Each of these remedies will now be considered further.

(1) *Damages*.   In most cases a breach of contract gives rise to a right o action for damages. The damages consist of a sum of money which will, a far as it is practicable, place the aggrieved party in the same position as if th contract had been performed.

The parties to a contract, when entering into the agreement, may agree that a certain sum shall be payable if a breach occurs. This sum is usually known as liquidated damages where it represents a genuine estimate of the loss which is likely to result from the breach of contract. Where, however the agreed sum is in the nature of a punishment for the breach of contract then the term penalty is applied to it, and penalties are not normally recoverable in full.

For instance, in building contracts it is often stipulated that a fixed sum shall be paid per day or per week, if the contract extends beyond the agreed contract period. If this sum is reasonable it constitutes liquidated damages and, unlike a penalty, is recoverable in full.

(2) *Order for payment of a debt*.   A debt is a liquidated or ascertained sum of money due from the debtor to the creditor and is recovered by an 'action of debt'.

(3) *Specific performance*.   The term 'specific performance' refers to an order of the Court directing a party to a contract to perform his part of the agreement. It is now only applied by the Courts on rare occasions when damages would be an inadequate remedy but specific performance con- stitutes a fair and reasonable remedy and is capable of effective supervision by the Court. This remedy will not be given if it requires the constant supervision of the Court.

(4) *Injunction.*   An injunction is an order of Court directing a person not to perform a specified act. For instance, if A had agreed not to carry out any further building operations on his land, for the benefit of B, who owns the adjoining land, and B subsequently observes A commencing building operations, then B can apply to the Court for an injunction restraining A from building. Damages, in these circumstances, would not be an adequate remedy.

(5) *Rescission.*   Rescission consists of an order of Court cancelling or setting aside a contract and results in setting the parties back in the position that they were before the contract was made.

Main Characteristics of Building Contracts

Most contracts entered into between builders or contractors and their employers are of the type known as 'entire contracts'. These are contracts in which the agreement is for specific work to be undertaken by the contractor and no payment is due until the work is complete.

In an entire contract, where the employer agrees to pay a certain sum in return for building work which is to be executed by the contractor, the contractor is not entitled to any payment if he abandons the work prior to completion, and will be liable in damages for breach of contract. Where the work is abandoned at the request of the employer, or results from circumstances which were clearly foreseen when the contract was entered into and provided for in its terms, then the contractor will be entitled to payment on a *quantum meruit* basis, that is, he will be paid as much as he has earned.

It is, accordingly, in the employer's interest that all contracts for building work should be entire contracts, to avoid the possibility of work being abandoned prior to completion. Contractors are usually unwilling to enter into any contracts, other than the very smallest, unless provision is made for interim payments to them as the work proceeds. For this reason the standard form of building contract[2] provides for the issue of interim certificates at various stages of the works, with the proviso that payment, or the issue of a certificate as a preliminary to payment, shall not be taken as approval of the work performed up to the time of payment.

It is usual for the contract to further provide that only a proportion of the sum due on the issue of a certificate shall be paid to the contractor. In this way the employer retains a sum, known as retention money, which will operate as an insurance against any defects that may arise in the work. The contract does, however, remain an entire contract, and the contractor is not entitled to demand payment in full until the work is satisfactorily completed, the defects liability period expired and the final certificate of completion issued.

That works must be completed to the satisfaction of the employer, or his representative, does not give to the employer the right to demand an unusually high standard of quality of work, in the absence of a prior express agreement. Otherwise the employer might be able to postpone indefinitely his liability to pay for the works. The employer is normally only entitled to expect a standard of work that would be regarded as reasonable by competent men with considerable experience in the class of work covered by the particular contract. The detailed requirements of the specification will have a considerable bearing on these matters.

The employer normally determines the conditions of contract, which define the obligations of the contractor. He often selects the contractor for the job by some form of competitive tendering and any contractor who submits a successful tender and subsequently enters into a contract, is deemed in law to have voluntarily accepted the conditions of contract adopted by the employer, and any requirements embodied within the invitation to tender. For example, the employer will not be bound to accept the lowest or indeed any tender and this is often stated in the advertisement. A tender is, however, normally required to be a definite offer, and acceptance of it gives rise legally to a binding contract.[1]

Types of Building Contract

There are a variety of employer/contractor relationships and the choice will be influenced considerably by the particular circumstances. They range from 'cost reimbursement' or 'cost plus' contracts at one end of the scale to truly lump sum contracts at the other.

The essential difference between the two extremes devolves upon which party is to carry the risk of making a loss (or profit) and the incentives which are built into the contract to encourage the contractor to provide an efficient and economic service to the employer or client. An examination and comparison of the most commonly used contractual arrangements follows.

(1) *Cost Plus Contracts*

These contracts are sometimes described as 'cost reimbursement' or 'prime cost' contracts. In practice they can take any one of three quite different forms.

(a) Cost plus percentage contracts are those in which the contractor is paid the actual cost of the work plus an agreed percentage of the actual or allowable cost to cover overheads and profit. They are useful in an emergency, when there is insufficient time available to prepare detailed schemes prior to commencement of the work, but it will be apparent that an unscrupulous contractor could increase his profit by delaying the completion of the works. No incentive exists for the contractor to complete the works as quickly as possible or to endeavour to reduce costs. They are quite commonly used in maintenance work.

(b) Cost plus fixed fee contracts are those in which the sum paid to the contractor will be the actual cost incurred in carrying out the work plus a fixed lump sum, which has been previously agreed upon and does not fluctuate with the final cost of the work. No real incentive exists for the contractor to secure efficient working, although it is to his advantage to earn the fixed fee as quickly as possible and so release his resources for other work. This type of contract is superior to the cost plus percentage type of contract.

(c) Cost plus fluctuating fee contracts are those in which the contractor is paid the actual cost of the work plus a fee, the amount of the fee being determined by reference to the allowable cost by some form of sliding scale. Thus the lower the actual cost of the works, the greater will be the value of the fee that the contractor receives. An incentive then exists for the contractor to carry out the work as quickly and as cheaply as possible, and it does constitute the best form of cost plus contract from the employer's viewpoint.

(2) *Target Cost Contracts*

Target cost contracts are used on occasions to encourage contractors to execute the work as cheaply and efficiently as possible. A basic fee is generally quoted as a percentage of an agreed target estimate often obtained from a priced bill of quantities. The target estimate may be adjusted for variations in quantity and design and fluctuations in the cost of labour and materials. The actual fee paid to the contractor is determined by increasing or reducing the basic fee by an agreed percentage of the saving or excess between the actual cost and the adjusted target estimate. In some cases a bonus or penalty based on the time of completion may also be applied. Target cost contracts can be useful when dealing with unusual or particularly difficult situations, but the real difficulty lies in the agreement of a realistic target and they are expensive to manage.[3]

(3) *Fixed Price Contracts*

These contracts include those based on schedules of rates, approximate quantities and bills of quantities. Their great merit lies in the predetermined nature of the mechanism for financial control provided by the pre-contract agreed rates. The risk of making a profit or loss rests with the contractor.

Schedule of Rates Contracts may take one of a number of different forms. The employer may supply a schedule of unit rates covering each item of work and ask the contractors, when tendering, to state a percentage above or below the given rates for which they would be prepared to execute the work. Alternatively, and as is more usual, the contractors may be requested to insert prices against each item of work, and a comparison of the rates so entered will enable the most favourable offer, and a comparison of the rates so entered will enable the most favourable offer to be ascertained. Approximate quantities are sometimes included to assist the contractors in pricing the schedules and the subsequent comparison of the tendered figures.

This type of contract is very suitable for maintenance and repair contracts, where it is impossible to give realistic and accurate quantities of the work to be undertaken. In this form of contract it is extremely difficult to make a fair comparison between the figures submitted by the various contractors, unless approximate quantities are inserted in the schedules, as there is no total figure available for comparison purposes and the unit rates may fluctuate extensively between the various tenderers. Occasionally schedules of rates are used as a basis for negotiated contracts.

Approximate Quantities Contracts are occasionally used where speed is important and design data is incomplete at the tendering stage. A Standard Form of Contract has been prepared by the Joint Contracts Tribunal (JCT) for use with bills of approximate quantities. The main problems lie with post-contract administration when full remeasurement will be required.

Bills of Quantities Contracts are still a very commonly used contractual arrangement for building projects of all but the smallest in extent, where the quantities of the bulk of the work can be ascertained with reasonable accuracy before the work is commenced. A bill of quantities gives as accurately as possible the quantities of work to be executed and the contractor enters a unit rate against each item of work. The extended totals are added together to give the total cost of the project as described in chapter 9.

In the absence of a bill of quantities each contractor tendering will have to assess the amount of work involved and this will normally have to be undertaken in a very short period of time, in among other assignments. Under these circumstances a contractor, unless he is extremely short of work, is almost bound to price high in order to allow himself a sufficient margin of cover for any items which he may have missed. Furthermore, there is no really satisfactory method of assessing the cost of variations and the contractor may feel obliged to make allowance for this factor also, when building up his contract price. Bills of quantities thus assist in keeping tender figures to a minimum.

(4) *Contracts based on Drawings and Specification*

These are often described as 'lump sum' contracts although they may be subject to adjustment in certain instances. They form a useful type of contract where the work is limited in extent and reasonably certain in its scope and are frequently used for works of alteration and conversion. They have on occasions been used where the works are uncertain in character and extent, and by entering into a lump sum contract the employer hoped to place the onus on the contractor for deciding the full extent of the works and the responsibility for covering any additional costs which could not be foreseen before the works were commenced. The employer would then pay a fixed sum for the works, regardless of their actual cost, and this constitutes an undesirable practice from the contractor's point of view. A RICS publication[4] has advocated the use of quantified specifications (specifications containing quantities) for projects consisting principally of extensive minor alterations.

(5) Design and Build Contracts

These constitute a specialised form of contractual relationship in which responsibility for design as well as construction is entrusted to the contractor. The less developed the design, the less detailed the specification and hence the less precise must be the calculation of the price. Contingencies must be included to provide for the unknown. Design and build (package deal) contracts have been used for local authority housing, often incorporating heavy industrialised systems; hospitals; defence installations and factories.

With housing schemes up to six contractors may be invited by a local authority to submit a complete development scheme for a large site. The contractors may use their own design teams or private architects to prepare schemes within the specified requirements of densities and costs, and the successful contractor will subsequently be required to collaborate with the authority's architect. This form of approach is particularly favoured where special factors exist such as the use of building systems, although the post-war record of the latter has been far from satisfactory.

Contractor-sponsored systems make it essential for the contractor to be brought into the design team. It has also become necessary to formulate procedures which will allow competition between contractors offering different systems. This has involved two separate stages:

(i) a competition to select the contractor who can best satisfy the functional, aesthetic and economic needs; and

(ii) a period of negotiation with the selected contractor using data, especially prices, derived from the first stage.

Public Sector Maintenance Contracts

Where there is a concentration of maintenance work in a comparatively restricted geographical area, the Property Services Agency frequently makes use of *term contracts*. These are contracts under which the builder undertakes to carry out a specific type of work within a range of contract values for a specified period in a particular area at agreed rates. By this means, projects which are appropriate to the contract can be ordered without the extensive preliminary work which would be necessary, both by PSA and the contractors, if each were the subject of lump sum tendering; at the same time the method preserves the principle of competitive tendering necessary to ensure that all approved firms have an equal chance of obtaining work, and that the agency pays the lowest reasonable price. These contracts can take the form of measured, daywork or specialist term contracts.

Measured term contracts are used particularly for maintenance and minor new works in the middle band of contract sizes, although the upper and lower limits of cost vary with the type of work. Each is based on a priced schedule of rates[5] which includes a specification, and tenders are submitted in the form of a percentage addition or deduction from the rates as a whole. To assist contractors in computing the appropriate percentage, they are supplied with extensive information by PSA, including details of buildings and the estimated annual value of work. Tenderers are expected to visit the sites. This type of contract is used for building and civil engineering work, painting and decorating, roadworks, ground maintenance, electrical work and heating, hot water and ventilating installations. Their prime disadvantage is the time taken in measuring each project in accordance with the priced schedule of works.

Daywork term contracts are used in areas where the volume of work is too small and too spasmodic to justify the use of measured term contracts. Tenders are sought for daywork term contracts on the basis of a percentage addition but in this case two figures are required, one for labour and the other for materials, with quotations for transport costs. These percentages are the additions required by the tenderer to cover profit, expenses and overheads in addition to the refund of his actual labour and material costs. PSA normally restricts this type of contract to small projects and smaller catchment areas, where smaller firms can provide personal supervision and a good and economical service.

Specialist term contracts are used where services are required regularly over a period of time. The type of work appropriate to this contract includes gully and window cleaning, boiler descaling, lift inspection and maintenance, cleaning bulk fuel tanks, inspection of pressure vessels, masts and towers, and the maintenance of small isolated gardens and grounds. A pre-priced schedule of rates with percentage additions is not usually appropriate, but a schedule may be prepared for pricing by tenderers or a quotation obtained in the form of a price per visit or operation.

Term contracts have the advantage of enabling the contractors to become familiar with the buildings to be maintained and they assist in the planning of an annual maintenance programme. This is reflected in lower prices and a better service than could be obtained from separate orders for each project on an *ad hoc* basis. PSA gains from a saving in staff time and the contractor benefits from his ability to forecast his resource requirements and to programme them efficiently, although it is not easy to plan a programme of maintenance work well in advance and foresee all contingencies.

Another alternative is to use fixed price maintenance contracts, whereby a lump sum is agreed with an existing contractor for a wide range of recurring jobs of a similar kind in a specified group of buildings over a specified period. Minor items of maintenance not covered by term contracts may be the subject of lump sum contracts against a specification and drawings.

National Schedules

A number of local authorities have developed their own schedules of rates for building maintenance work and there are several national schedules of which probably the widest used is the *National Schedule of Rates* published jointly by the Building Employers Confederation (BEC) and the Society of Chief Quantity Surveyors in Local Government (SCQSLG).[6] A fundamental requirement for schedules of rates for general use is that they be soundly based.[7] This was, in fact, the basic aim of the BEC/SCQSLG document,[6] although the Association of Metropolitan Authorities (AMA)[8] came to the conclusion that a common standard schedule of rates for maintenance was not a practical proposition and that schedules should be individually prepared to suit the needs of each local authority.

Tendering Arrangements

Conventional tendering procedures have been criticised on the grounds that they fail to take full advantage of modern techniques and do incorporate unsatisfactory features. All tendering procedures aim at selecting a suitable contractor and obtaining from him at an appropriate time an acceptable offer, or tender, upon which a contract can be let.

The Simon Committee in 1944[9] drew attention to the fact that low prices resulting from indiscriminate tendering result in bad building and that resources are wasted when many firms tender for the same project. In 1964 the Banwell Report[10] suggested that invitations to tender should be limited to a realistic number of firms, all of whom were capable of executing the work to a recognised standard of competence. The Banwell Committee appeared to favour the general use of standing approved lists of contractors and that *ad hoc* lists should be used mainly when the work was of a specialist or one-off nature. The Committee further recommended that the period allowed for tendering should be adequate for the type of project and welcomed 'firm price' contracts (contracts without a price fluctuations clause). The former Ministry of Housing and Local Government issued revised model standing orders to local authorities in 1964 to facilitate the wider use of selective tendering procedures, and in 1965 the Ministry gave guidance to local authorities on the operation of selective tendering.

In 1965 a working party was established by the Economic Development Committee for Building to examine the Banwell Report and its implementation and it submitted its report in 1967[11]. The working party considered that insufficient attention was paid to the importance of time and its proper use and that clients seldom define their requirements in sufficient detail at the start of negotiations. It favoured the main contractor joining the design team at an early stage.

The working party urged the wider adoption of the practices which are well detailed in *Code of Procedure for Single Stage Tendering*,[12] although they recognised that in the public sector this would require a more flexible approach to satisfy standards of accountability. Although the working party saw merit in 'firm price' contracts, they stressed the difficulties involved in producing firm tenders in a market where materials prices tend to fluctuate and contractors are often invited to tender on incomplete documentation. This problem has now been largely solved by the introduction of price adjustment formulae described later in this chapter.

Summing up, tendering arrangements can be broadly classified into three main groups

(1) advertising for competitive tenders;
(2) inviting tenders from selected contractors; and
(3) negotiating a contract with a selected contractor.

Method 1 ensures maximum competition but there is the disadvantage that tenders may be received from firms who have neither the necessary financial resources nor adequate technical knowledge and experience of the class of work involved. Hillebrandt[13] lists two principal objections to open tendering

(1) No guarantee that the contractor who submits the lowest tender is technically, managerially or financially capable of carrying out the project. If he is not, the costs of remedial work may far outweigh the gain by the lower initial cost.

(2) The costs of tendering to the would-be contractor are high and have to be recouped so that they will be reflected in higher overhead costs.

Method 2 is becoming increasingly popular as it provides for the use of a list of contractors of proven competence for work of the size and nature contemplated and retains the competitive element. Method 3 is often restricted to special circumstances, as for instance when the contractor is already engaged on the same site and space is very restricted. Negotiation might also offer advantages where it is required to make an early start with the work or where the contractor in question has exceptional experience of the class of work covered by the particular contract.

The number of firms invited to tender will depend on the size and type of contract, ranging from a maximum of five for contracts up to £50 000 in value, six between £50 000 and £250 000, and eight between £250 000 and £1 million. A period of four weeks is normally allowed between the receipt of tender documents by contractors and delivery of the actual tenders.

Contract Documents

There are five contract documents which are often used in connection with building contracts. With small contracts it is likely that a bill of quantities will be omitted. In cases where a bill of quantities is issued the specification is not a contract document unless the contract expressly provides for it. The following can constitute contract documents.

(1) Articles of Agreement
(2) Conditions of Contract
(3) Specification
(4) Bill of Quantities
(5) Contract Drawings

The nature and uses of each of these documents are as follows.

(1) *Articles of Agreement.*   These constitute the formal agreement between the employer and the contractor for the execution of the work in accordance with the other contract documents for the contract sum. The contractor

covenants to construct, complete and maintain the works in accordance with the contract, and the employer covenants to pay the contractor at the times and in the manner prescribed by the contract.

(2) *Conditions of Contract.*   These define the terms under which the work is to be undertaken; the relationship between the employer, architect, quantity surveyor and contractor; the powers of the architect and the terms of payment. The normal standard set of conditions used for most building contracts is that issued by the Joint Contracts Tribunal and is generally known as the JCT Conditions, currently JCT80. There are four separate sets for use on public or private contracts and with or without quantities in each case.[2] Practice notes are issued from time to time to clarify doubtful points. Where a contract is of very limited extent and the use of the standard comprehensive set of conditions is not really justified, an abbreviated set of conditions may be used. Another and popular alternative is to use the *Agreement for Minor Building Works,*[14] which is well suited for works of alteration and improvement, where no bills of quantities are prepared, and is described in some detail later in this chapter. On government contracts the general conditions (GC/Wks/1)[15] are frequently used, while the Form GC/Works/2 is particularly designed for minor works[16] and there is also a set of conditions for small works.[17]

(3) *Specification*   The specification performs a vital role in any building contract but in recent years there has been a tendency to include it in the bill of quantities in the form of preamble clauses, or even more effectively in annotated bills. Where a bill of quantities has been prepared, the specification will not constitute a contract document unless it is a requirement of the particular contract, and the contract documents have precedence over documents of lesser standing. Where there is no bill of quantities, the specification constitutes a contract document. As described in chapter 8 the specification amplifies the information given in the contract drawings and bill of quantities. It describes in detail the work to be executed under the contract and the nature and quality of the materials and workmanship.

Details of any special responsibilities to be borne by the contractor, apart from those listed in the conditions of contract, are often incorporated in this document. It may also contain clauses specifying the order in which various sections of the work are to be performed, the methods to be adopted in the execution of the work, and details of any special facilities that are to be afforded to other contractors or sub-contractors.

An excellent arrangement for a building specification is to commence with any special conditions relating to the contract and the extent of the work, then to follow with a list of contract drawings, details of the programme, description of access to site, supply of electricity and water, offices and mess facilities, and statements regarding suspension of work during frost and bad weather, damage to existing services, details of borings if any, groundwater levels and similar general matters.

This section could conveniently be followed by detailed clauses covering the various sections of the work, commencing with materials in each case and then proceeding with workmanship and other clauses. The specification constitutes a schedule of instructions to the contractor with particular reference to the way in which the work is to be undertaken.

(4) *Bill of Quantities.*   This consists of a schedule of the items of work to be carried out under the contract with quantities entered against each item, the quantities normally being prepared in accordance with the Standard Method of Measurement of Building Works or the Code for the Measurement of Building Works in Small Dwellings. The bill of quantities provides a uniform basis on which tenders can be obtained and, when these are priced, they provide a means of comparing the tenders received and of pricing the work on site as executed. The unit rates entered by the contractor against each measured item in the bill of quantities normally include an allowance for general overheads and profit, as described in chapter 9.

(5) *Contract Drawings.*   These depict the nature and scope of the work to be carried out under the contract. They must be prepared to a suitable scale and be in sufficient detail to permit a contractor to price the bill of quantities and to carry out the work satisfactorily. For instance site plans will normally be drawn to a scale of 1:200 or 1:500, working drawings of buildings probably 1:100, assembly drawings 1:20 or 1:10 and details to 1:10 to 1:5.

All available information on the topography of the site, the nature of the ground and the groundwater level, should be made available to contractors tendering for a project. Existing and proposed work must be clearly distinguished on the drawings. For instance, old and new drains and other services are often depicted in different colours or by different types of line. With alterations to buildings it is often preferable to prepare separate plans of old and new work.

All drawings should contan an abundance of descriptive and explanatory notes which should be clearly legible and free from abbreviations. Ample figured dimensions should be inserted on the drawings to ensure maximum accuracy in taking off quantities and in setting out the constructional work on site.

Parties Involved in Contracts

A number of persons are involved in the execution of a building contract operating under a quite complicated set of interrelationships.

Employer.   The employer, building owner or client is the person, firm or body on whose behalf building work is undertaken and who is responsible for paying for the work as executed and certified in architects' or surveyors' certificates. There is little contact between the employer and most of the parties to the contract as he deals mainly through his agent—the architect or surveyor for the contract. It assists in the smooth running of a contract if the employer is reasonably clear about his requirements, refrains from changing his mind or interfering with the work on site, and honours the certificates within the prescribed period. There is no contractual relationship between the employer and any nominated sub-contractor.

Architect.   The architect acts as agent and technical adviser to the employer and prepares all the contract particulars. He ascertains the employer's requirements, selects any necessary consultants and a quantity surveyor if required, prepares preliminary drawings and obtains approximate estimates, selects sub-contractors, prepares detailed drawings and specification, obtains a bill of quantities where required and invites and advises on tenders. He subsequently selects a tender for the employer's approval, supervises the work, certifies payments to the contractor, issues variation orders as necessary, decides how provisional sums are to be spent, secures the remedying of defects at the end of the defects liability period and certifies the final account. On smaller maintenance, alteration and conversion contracts a surveyor may act as agent and plan and supervise the work on behalf of the employer.

Engineer.   Engineers may be employed on large contracts to advise on specialist work such as structural steelwork and heating and ventilating installations. The engineers will prepare designs and specifications for the specialist work, obtain quotations and submit reports to the architect. Subsequently they will supervise the specialist work on the site.

Quantity Surveyor.   The quantity surveyor is concerned with the cost and measurement aspects of building contracts. He advises the architect on the cost implications of design decisions, prepares approximate estimates and often cost plans, and later tender particulars. He values the work on site, assesses the effect of variations and finally prepares the final account on the basis of which the architect certifies final payment.

Contractor.   The contractor undertakes to construct and complete the building work in accordance with the contract documents. He must have sole responsibility for all executive work on the site. He is generally required to proceed regularly and diligently with the work and to complete it by a specified date, otherwise he may be liable for the payment of liquidated damages. The contractor is required to comply with all statutory regulations affecting the works, to give all necessary notices and to insure the works. He receives his instructions from the architect or surveyor for the contract.

Sub-contractors.   There are a variety of sub-contractors engaged on most large building contracts. They specialise in a wide range of crafts, structural work, mechanical and other equipment, and decorative and other finishes. They are frequently able to produce cheaper and yet higher-quality work than the main contractor. Sub-contractors have to look to the main contractor for instructions and payment; there is usually a sub-contract regulating the relationships between the two parties.

Contractor's Supervisory Staff.   The contractor is required under the

Standard Form of Contract to keep upon the works a competent person-in-charge, generally a foreman on smaller contracts, to whom the architect may issue instructions; all verbal instructions being subsequently confirmed in writing. The foreman is generally responsible for organising the labour employed on the site, securing efficient use of plant and materials, setting out the works, liaising with the clerk of works, programming and progressing the work and recording daywork and other data. In the case of very large contracts an agent may be employed to take over some of these responsibilities.

Clerk of Works. The clerk of works is appointed by and acts as an inspector on behalf of the employer, but he carries out his inspection duties and issues instructions only under the directions of the architect, who must confirm them in writing within two working days. He is usually a craft operative with a wide experience of building work and his main function is to ensure that all the work on the site is in accordance with good construction practice and that it complies with the contract documents. He often records details of drainage and other work which will be hidden by subsequent building operations and agrees details with the contractor's foreman. Although the architect may delegate supervision of the work to the clerk of works, he remains nevertheless responsible. In the case of *Leicester Board of Guardians* v. *Trollope* (1911), the architect was held negligent for a structural defect (absence of damp-proof membrane), despite collusion between the contractor and the clerk of works.

Building Control Officer. Building control officers are employed by local authorities and examine submitted plans under the Building Regulations and inspect the work under construction to ensure compliance with the current regulations and the submitted and approved plans. A more detailed consideration of the building control process is given in chapter 11.

Conditions of Contract for Minor Works

The full JCT Conditions are sometimes considered unduly complicated and extensive for works of building maintenance, alteration and conversion, often referred to as 'minor works'. Abridged or condensed formats have been prepared by the Joint Contracts Tribunal[14] and the Department of the Environment.[16,17] A description of the principal provisions of the JCT Agreement for Minor Building Works[14] follows.

General Principles

The JCT Agreement for Minor Building Works has the attractions of brevity and simplicity. The simplicity has been achieved by the omission of many of the clauses of JCT 80, the reduction of others to basic requirements and the omission of detailed legal provisions and administrative procedures.[18]

This Form of Agreement is designed for use where minor building works are to be carried out for an agreed lump sum and where an Architect/Supervising Officer has been appointed on behalf of the Employer. The

Form is not for use for works for which bills of quantities have been prepared, or where the duration is such that full labour and materials fluctuations provisions are required; nor for works of a complex nature which involve complicated services or require more than a short period of time for their execution. Furthermore, no sub-contractors or suppliers are to be nominated. It is however possible to incorporate a fluctuations clause similar to clause 39 of JCT 80. Readers requiring more detailed information on the operation of the Agreement are referred to Wood.[19]

Recitals

The Agreement contains four Recitals which explain and set out the purpose of a contract and the facts upon which the contract is based.

Recital 1 describes the contract works, names the architect and defines the contract documents which together with the conditions may be one of the following:

(1) the contract drawings, the reference numbers of which have to be stated, together with the contract specification and schedules;
(2) the contract drawings and the contract specification;
(3) the contract drawings and schedules;
(4) the contract drawings;
(5) the contract specification and schedules;
(6) the contract specification;
(7) the schedules.

Recital 2 states that the contractor has priced either the specification or the schedules, both of which are contract documents, or has provided a schedule of rates. Presumably the contractor will have to provide a schedule of rates if the contract documents consist only of the contract drawings and the conditions.

Recital 3 states that the contract documents have been signed by or on behalf of both parties.

Recital 4 provides for the appointment of a quantity surveyor should one be required in connection with the contract, although it is difficult to envisage how this can arise.

Articles

There are four Articles in the Agreement.

Article 1 states that the contractor will carry out and complete the works in return for the payment of the contract sum.

Article 2 sets out the contract sum and states that the sum is exclusive of Value Added Tax (VAT).

Article 3 names the person appointed as architect for the purposes of the contract and gives the employer the right to nominate some other person to act as architect should the named architect cease for any reason to be the architect.

Article 4 follows the procedure in JCT 80 and makes the arbitration clause part of the Articles of Agreement. Any disputes or differences concerning the contract that may arise between the employer or architect and the contractor, shall be referred to arbitration at any time, whether or not practical completion of the works has been achieved.

Contract Clauses

Intentions of the Parties (clause 1.0). The Contractor is to carry out and complete the Works with due diligence and in a good and workmanlike manner, using materials and workmanship of the prescribed quality and, where appropriate, to the reasonable satisfaction of the Architect/ Supervising Officer. The Architect/Supervising Officer shall supply any further information as necessary, issue all certificates and confirm all instructions in writing.

Commencement and Completion (clause 2.0). Dates are inserted for commencement and completion. If it becomes apparent that the Works will not be completed by the agreed completion date for reasons beyond the Contractor's control, the Contractor shall notify the Architect/Supervising Officer, who shall make, in writing, such extension of time for completion as may be reasonable. If the Works are not completed by the prescribed or extended completion date, the Contractor shall pay to the Employer the appropriate weekly rate of liquidated damages for the period during which the Works remain uncompleted. The Architect/Supervising Officer shall certify the date when the Works have reached practical completion.

Any defects, excessive shrinkage or other faults which appear within three months of the date of practical completion and are due to materials or workmanship not in accordance with the Contract or frost occurring before practical completion, shall be made good by the Contractor at his own cost unless the Architect/Supervising Officer instructs otherwise. The Architect/ Supervising officer shall certify the date when the Contractor has discharged all his obligations.

Control of the Works (clause 3.0). The Contract shall not be assigned by the Employer or the Contractor without the written consent of the other. The Contractor shall not sub-contract any part of the Works without the written consent of the Architect/Supervising Officer which shall not unreasonably be withheld.

The Contractor shall at all reasonable times keep upon the Works a competent person in charge and any instructions given to him by the Architect/Supervising Officer shall be deemed to have been issued to the Contractor. The Architect/Supervising Officer may (but not unreasonably or vexatiously) issue instructions requiring the exclusion from the Works of any person employed upon them.

The Contractor shall carry out any written instructions issued by the Architect/Supervising Officer, and oral instructions require confirmation within two days. If the Contractor fails to comply with a written notice requiring compliance with an instruction within 7 days after receipt, the Employer may employ other persons to carry out the work and deduct the costs from monies due to the Contractor or they shall be recoverable as a debt.

The Architect/Supervising Officer may vary the contract by ordering additions, omissions or other changes, and value them on a fair and reasonable basis, using where relevant prices in the priced specification, schedules or schedule of rates. Alternatively, the price may be agreed between the Architect/Supervising Officer and Contractor before the Contractor carries out the instruction.

The Architect/Supervising Officer shall issue instructions as to how provisional sums are to be spent and the basis of valuation shall be the same as for variations.

Payment (clause 4.0). Any inconsistency between the Contract Drawings and Contract Specification and schedules shall be corrected and treated as a variation. Nothing contained in these documents shall override, modify or affect the application or interpretation of the Contract Conditions.

The Architect/Supervising Officer shall if requested by the Contractor, at not less than four weekly intervals, certify progress payments to the Contractor in respect of the value of the Works properly executed, and the value of any materials and goods which have been reasonably and properly brought on to the site and which are adequately stored and protected, less a retention of 5 per cent and less any previous payments made by the Employer. The Employer shall pay the amount certified within 14 days of the date of the certificate.

The Architect/Supervising Officer shall within 14 days after the date of practical completion certify payment to the Contractor of $97\frac{1}{2}$ per cent of the total ascertainable amount due to the Contractor, less previous progress payments, and the Employer shall pay the Contractor within 14 days of the certificate. The Contractor shall supply within three months of the date of practical completion all documentation required for final certification by the Architect/Supervising Officer, who shall within 28 days issue a final certificate certifying the amount remaining due to the Contractor or the Employer, and this sum shall constitute a debt payable from 14 days after the date of the final certificate.

Statutory Obligations (clause 5.0). The Contractor shall comply with, and give all notices required by, any relevant statute, regulation, by-law and the like, and shall pay all fees and charges legally recoverable. If the Contractor finds any divergences between the statutory requirements and the contract documents and Architect's instructions, he shall immediately notify the Architect/Supervising Officer in writing. Subject to this latter obligation, the Contractor shall not be liable if the Works do not comply with statutory requirements.

The Employer shall pay to the Contractor any value added tax properly chargeable on the supply to the Employer of any goods and services by the Contractor under the Contract. The Contractor shall in respect of all persons employed by him comply with the Fair Wages Resolution.

The Employer shall be entitled to cancel the contract and to recover any consequent loss from the Contractor, if the Contractor shall have offered or given or agreed to give to any person any gift or any consideration of any kind, or if he shall have committed any offence under the Prevention of Corruption Acts 1889 to 1916, or shall have given any fee or reward the receipt of which is an offence under section 117 of the Local Government Act 1972 or any subsequent re-enactment.

Injury, Damage and Insurance (clause 6.0). The Contractor shall be liable for and shall indemnify the Employer against any expense, liability, loss, claim or proceedings arising under any statute or at common law in respect of personal injury to or death of any person arising out of the Works, unless due to any act of the Employer or of any person for whom he is responsible. Without prejudice to his liability to indemnify the Employer, the Contractor shall maintain and shall cause any sub-contractor to maintain the necessary insurance to cover the liability of the Contractor or sub-contractor. Similar provisions also apply in respect of property.

The Contractor shall in the joint names of the Employer and the Contractor insure against loss and damage by fire, lightning, explosion, storm, tempest, flood, bursting or overflowing of water tanks, apparatus or pipes, earthquakes, aircraft and other aerial devices dropped from them, riot and civil commotion, for the full value, plus a prescribed percentage to cover professional fees, of all work executed and all unfixed materials and goods intended for the Works, but excluding temporary buildings, plant, tools, and equipment owned or hired by the Contractor and his sub-contractors. Upon acceptance of any claim under the insurance, the Contractor shall with due diligence restore or replace work or materials damaged and dispose of any debris and proceed with and complete the Works.

In the case of works to existing structures, the Employer shall maintain adequate insurance encompassing all the matters previously listed. The Contractor or sub-contractor and Employer shall produce evidence of insurance as and when required by the other party.

Determination (clause 7.0). The Employer may, but not unreasonably or vexatiously, by notice by registered post or recorded delivery to the Contractor, determine the employment of the Contractor, if the Contractor shall make default in one or more of the following respects:

(1) if the Contractor without reasonable cause fails to progress diligently with the Works or wholly suspends the carrying out of the Works before completion;

(2) if the Contractor becomes bankrupt or makes any composition or arrangement with his creditors or has a winding up order made.

In the event of the Employer determining the employment of the Contractor, the Contractor shall immediately give up possession of the site of the Works and the Employer shall not be bound to make any further payment to the Contractor until after the completion of the Works, but this shall be without prejudice to any other rights or remedies which the Employer may have.

The Contractor may similarly determine his employment if the Employer shall make default in any one or more of the following respects:

(1) if the Employer fails to make any progress payment under the Contract within 14 days of such payment being due;

(2) if the Employer or any person for whom he is responsible interferes with or obstructs the carrying out of the Works or fails to make the premises available for the Contractor;

(3) if the Employer suspends the carrying out of the Works for a continuous period of at least one month;

(4) if the Employer becomes bankrupt or makes a composition or arrangement with his creditors or has a winding up order made.

Under sub-clauses (1), (2) and (3), determination does not become effective unless the Employer continues the default for seven days after receipt of the notice.

In the event of the Contractor determining his employment, the Employer shall pay to the Contractor, after taking into account amounts previously paid, such sum as shall be fair and reasonable for the value of work begun and executed, materials on site and the removal of all temporary buildings, plant, tools and equipment, but without prejudice to any other rights or remedies which the Contractor may possess.

Contract Procedures

Placing the Contract

In the case of a bill of quantities contract, where no serious errors have been found in the priced bills of the lowest tenderer, the architects's report to the employer will normally recommend acceptance of that tender, particularly where selective tendering operates. As soon as the employer has made his decision, all contractors who tendered should be notified. If priced bills have been submitted with tenders, these should be returned to unsuccessful contractors unopened. Letters to unsuccessful tenderers should include a list of tenders in ascending order and a list of tenderers in alphabetical order.

If the quantity surveyor finds serious errors in pricing, the contractor should be advised and given the opportunity of withdrawing or standing by his tender. Where the errors are significant, an adjusting lump sum will be added to or deducted from the corrected total of the summary. This adjustment can be applied as a percentage to any billed rates subsequently used for valuing variations.

All contract documents must be signed by both parties and deletions or alterations initialled by them. The contractor is then supplied with the appropriate number of copies of documents and insurance cover obtained and agreed.

Variations

As the contract proceeds, the architect will almost invariably issue further drawings, details and instructions. All architect's instructions must be in writing and it is advisable to use standard forms covering both instructions and variations. Provisional quantities may be inserted for work which is uncertain in extent such as foundations; the provisional quantities are omitted in the variation account, and the actual work done is measured and included as an addition. Alterations in design and finishings are frequent subjects for variation orders.

Variation orders are generally issued in triplicate; one copy each for the contractor, architect and quantity surveyor. If the contractor considers that any work constitutes a variation, he should at once draw the architect's attention to it. Subsequently, the quantity surveyor and the contractor's representative meet to agree the amount of work omitted from the contract and to measure the additional work. Items measured in the office are usually entered on dimensions paper and those on the site recorded in dimension books. After the measurements have been agreed the quantity surveyor will work them up into a variation account, containing a bill of omissions and a bill of additions, when suitable rates for the various items will be agreed.

Certificates

As the contractor is continually financing the works in progress, it is imperative that he should receive interim certificates at the appropriate time and for the full amount due on all but the smallest contracts. It is equally important that he should be paid promptly on the certificates, especially since he is expected to pay sums due to sub-contractors whether he receives payment or not. It will also be appreciated that no interest is paid on sums outstanding which arise from incorrect certification.

Certificates are normally issued monthly by the architect; these are based on the quantity surveyor's valuation of the work done and of unfixed materials properly and not prematurely brought upon the site and also of materials in workshops subject to suitable safeguards. The valuations will include a proportion of preliminaries and the cost of any recoverables and daywork which have been certified. The amount of detail in a valuation varies according to circumstances and a percentage of the total of each work sectional bill is often considered adequate. In some cases the contractor's surveyor prepares a statement containing his estimate of the sum due for checking by the quantity surveyor. Previous payments and the retention percentage will be deducted from the total valuation to give the sum due to the contractor.

Daywork

Until 1966 the National Schedule of Daywork Charges, with its fixed percentage additions, provided the basis for the payment of daywork. Since that date, fixed percentages cannot be directed or authorised by any professional body or trade federation. Every contractor must be free to decide what percentage addition he deems necessary to cover overheads and profit and to vary such additions from contract to contract if he deems fit.

Some contractors welcome daywork as with an adequate percentage there is no risk of loss. Employers and their professional advisers tend to view daywork with suspicion, believing that there is no incentive to reduce costs or increase efficiency, and that the contractor may be tempted to use his slowest men on this work and to regard it as a standby or 'hospital' job. On occasions, daywork can be disruptive of the smooth operation of a contract.

Daywork should be restricted to work which cannot satisfactorily be measured and valued at billed rates. Suitable examples are the repair and reinstatement of work stemming from a bad outbreak of dry rot, exploratory work and underpinning of a settled building, and opening up and reinstatement of defective drains. The surveyor must carefully examine the daywork sheets and rigorously check the quantities of labour and materials listed. Milne[20] has amplified the methods of controlling daywork.

Claims

Variations which cause loss or expense, not being due to underpricing in the tender or the contractor's inefficiency, may be the subject of a loss and/or expense claim, and these must be submitted in writing within a reasonable time of the event. Where a claim is appropriate the amount of loss and/or expense is to be ascertained by the architect, or quantity surveyor if so instructed, but the contractor will need to submit the necessary information to support his claim. If a variation has caused delay an extension of time may be granted.

The contractor's site management needs to keep adequate records and take other appropriate action as follows to support any claims for disruption and delay:

    (1) ensure that appropriate written notice, application or confirmation of instruction is made;

    (2) use best endeavours to prevent delay;

    (3) keep a programme showing the dates on which it is planned to carry out the various operations making up the works;

    (4) record the effect on the programme of any events which disrupt or delay the regular progress of the works;

    (5) record any loss of productivity resulting from these events or any waiting time;

    (6) if the character or conditions under which work is to be carried out have changed as a result of the architect's instructions, record the item as daywork.

Price Adjustment Formula

The easiest and quickest method of assessing price fluctuations of labour and materials is to apply the NEDO price adjustment formula.[21] Where this formula is used building works are allocated between the indices for each of the work categories, such as *in situ* concrete, brickwork, blockwork and asphalt work. The formula permits a regular assessment of price fluctuations as valuations are made of work executed for each work category during the period covered by the interim valuation.

Work category index numbers are calculated monthly by the Property Services Agency (PSA) and published in HMSO Monthly Bulletin Construction Indices. Each work category index measures the price level changes which occur in respect of certain items of work chosen as being representative of that work category. This is done monthly by revaluing the resources required (labour, plant and materials) at current prices.

When the formula method of price adjustment is to be used, the base month must be stated in the tender documents and an appropriate clause inserted in the contract conditions. The bills of quantities will contain a schedule indicating the work category into which each item in the bills of quantities falls. The value of preliminaries will be totalled and spread proportionately over the value of the main contractor's work which is subject to price adjustment. Where no bill of quantities is provided and the work tendered for is on a lump sum basis, the full amount of the tender subject to price adjustment will be allocated to work categories.

The value of each item of work executed and included in an interim valuation will be allocated to the appropriate work category. The formula will then be applied to the value of work performed and included in the interim certificate for each separate work category. The indices from which all fluctuations are calculated are the final index numbers for the base month and the month of valuation. The formula is as follows

$$C = V\left[\frac{I_v - I_0}{I_0}\right]$$

where $C =$ the amount of the price adjustment for the work category or group

$V =$ the value of the work executed in the work category/group during the valuation period
$I_v =$ the work category/group index number at the mid-point of the month of valuation
$I_0 =$ the work category/group index number for the base month.

The principal advantages of the price adjustment formula method are:

(1) the contractor secures the benefit of good buying;
(2) fluctuations are dealt with monthly, thus improving cash flow;
(3) there can be no arguments about how labour is employed as the labour element is recoverable no matter how the work is done;
(4) time sheets and invoices need no longer be produced with all the consequent work in checking and cross-checking.

A typical example is given and comparisons made with the orthodox analysis method in *Quantity Surveying Practice*.[22]

Final Account

In order that the architect may issue the final certificate, he requires from the quantity surveyor the final account for the contract. The adjustment of the contract sum in the final account falls under the following headings:

(1) variations
(2) remeasurement of provisional quantities
(3) nominated sub-contractors' accounts
(4) nominated suppliers' accounts
(5) loss and expense caused by disturbance of regular progress of the works
(6) fluctuations (where applicable).

Delays in the settlement of the final account are a cost to the contractor and the employer is usually anxious to know his ultimate financial commitment. The architect and quantity surveyor both have a contractual responsibility to adhere to the date stipulated in the contract, and the contractor should provide every assistance in the prompt provision of sub-contractors' and suppliers' accounts, agreement of measurements and prices and similar matters. Where no quantity surveyor is employed, the responsibility lies with the contractor to supply all relevant cost particulars to the architect.

References

1   I. H. Seeley. *Civil Engineering Quantities*. Macmillan (1987)
2   Joint Contracts Tribunal for the Standard Form of Building Contract. *Standard Form of Building Contract* (1980)
3   The Aqua Group. *Tenders and Contracts for Building*. Granada (1982)
4   Royal Institution of Chartered Surveyors. *Refurbishment and Alteration Work: Quantity Surveying Documentation* (1982)
5   DOE. *Schedule of Rates for Building Works*. HMSO (1984); *Schedule of Rates for Decoration Work*. HMSO (1981); *Schedule of Rates for Minor Works and Maintenance of Roads and Pavings*. HMSO (1981); *Schedule of Rates for Ground Maintenance*. HMSO (1981); *Schedule of Rates for Electrical Installations*. HMSO (1980); *Schedule of Rates for Heating, Hot Water and Ventilating Installations*. HMSO (1983); *Schedule of Rates for Electrical Distribution Systems External to Buildings*. HMSO (1978)
6   Building Employers Confederation and Society of Chief Quantity Surveyors in Local Government. *National Schedule of Rates*. (1982)
7   Chartered Institute of Public Finance and Accountancy and Institution of Municipal Engineers. *Local Authority Maintenance Work: Interim Users' Guide to Schedules of Rates* (1982)
8   Association of Metropolitan Authorities. *Guidance Notes on Preparation of Term Contracts using Schedules of Rates* (1981)
9   Simon Committee. *The Placing and Management of Building Contracts*. HMSO (1944)

10   Ministry of Public Buildings and Works (Banwell Report). *The Placing and Management of Contracts for Building and Civil Engineering Work.* HMSO (1964)

11   Economic Development Council for Building. *Action on the Banwell Report.* HMSO (1967)

12   National Joint Consultative Committee for Building. *Code of Procedure for Single Stage Tendering.* (1977)

13   P. M. Hillebrandt. *Analysis of the British Construction Industry.* Macmillan (1984)

14   Joint Contracts Tribunal for the Standard Form of Building Contract. *Agreement for Minor Building Works* (1985)

15   Form GC/Works/1. *General Conditions of Government Contracts for Building and Civil Engineering Works.* HMSO (1977)

16   Form GC/Works/2. *General Conditions of Government Contracts for Building and Civil Engineering Minor Works.* HMSO (1980)

17   Form C1001. *General Conditions of Government Contracts for Building, Civil Engineering, Mechanical and Electrical Small Works.* HMSO (1982)

18   J. M. Audus. *A Builder's Guide to the Agreement for Minor Building Works—January 1980 Edition.* CIOB (1981)

19   R. D. Wood. *The JCT Agreement for Minor Building Works.* Estates Gazette (1983)

20   R. D. Milne. *Building Estate Maintenance Administration.* Spon (1985)

21   NEDO. *Price Adjustment Formulae for Building Contracts: Guide to application/Description of the indices.* HMSO (1979)

22   I. H. Seeley. *Quantity Surveying Practice.* Macmillan (1984)

11 BUILDING CONTROL

It is doubtful whether any other industry is subject to so much legislative control as the construction industry. A government department or local authority cannot exercise any control over building work unless it can show that an Act of Parliament confers this power upon it. Acts of Parliament also confer powers to make regulations, orders and by-laws which have the force of law. This type of legislation is often called delegated or subordinate legislation; it has the same validity and effect as if it formed part of the law which authorised it. For example, much of the control of building work is exercised through Statutory Instruments, particularly in the field of town and country planning. Up to 1966, local authorities were authorised to make building by-laws under the Public Health Act 1936, but the 1961 Act deprived them of this power and conferred upon the Minister the power to make regulations covering the same field as the former by-laws. The currently operative Building Regulations 1985, were made under the provisions of the Building Act 1984, and apply to England and Wales, including London.

Building Regulations

Building Regulations[1] may apply to alteration and improvement works and to buildings which are to undergo a material change of use, such as the conversion of a house designed for occupation by a single family to multiple occupation. The Building Regulations[1] cover the construction of and materials used in building, including resistance to moisture, cavity insulation, means of escape and fire resistance, resistance to passage of sound, ventilation, hygiene, drainage and waste disposal, heat producing appliances, conservation of energy and facilities for disabled people. Approval of alteration and improvement work under the Building Regulations will be required in the following circumstances in particular.

(1) Addition to a building.
(2) Any work involving structural alterations, which includes breaking out an opening in a loadbearing wall and inserting a lintel.

317

(3) Any alteration to an existing drainage system.

(4) Installation of new sanitary conveniences and certain heating appliances.

Even when maintenance work is exempt from the provision of the Building Regulations, it is advisable to adhere to them, particularly in relation to structural fire precautions. There is a danger that constructional methods and components, such as fire check doors, which originally satisfied the Building Regulations, could as part of maintenance work be replaced with new components which fail to comply.

Exempted Buildings

Certain buildings are exempt by the Building Act 1984 from the operation of the Building Regulations; these are as follows.

(1) Buildings exempt by direction of the Secretary of State.

(2) Schools and other educational establishments to be erected in accordance with plans approved by the Secretary of State for Education and Science or the Secretary of State for Wales.

(3) Buildings of statutory undertakers, the United Kingdom Atomic Energy Authority, the British Airports Authority or the Civil Aviation Authority, to be used for the purpose of the undertaking, but excluding houses and hotels or buildings used as offices or showrooms.

It also exempts public bodies from the procedural requirements of the Building Regulations.

A number of exemptions are contained in the Regulations themselves (Schedule 3) and comprise the following.

(1) Buildings under the Explosives Acts 1875 and 1923, licensed under the Nuclear Installations Act 1965 or scheduled under the Ancient Monuments and Archaeological Areas Act 1979.

(2) Buildings not frequented by people, such as those housing fixed plant or machinery.

(3) Greenhouses or agricultural buildings, subject to certain conditions.

(4) Temporary buildings and mobile homes.

(5) Ancillary buildings on housing sites or in connection with mines and quarries.

(6) Small detached buildings with a floor area not exceeding 30 m^2, with no sleeping accommodation or to shelter people from the effects of nuclear, chemical or conventional weapons; subject to certain conditions.

(7) Extensions at ground level to a building by a greenhouse, conservatory, porch, covered yard or covered way, or a carport open on at least two sides, with a floor area not exceeding 30 m^2.

Relaxation of Regulations

The Building Act 1984 empowered the Minister (Secretary of State for the Environment) to issue a directive dispensing with or relaxing a requirement of the Regulations where, in any particular case, he considers its operation would be unreasonable, but he must consult the local authority before so doing. Under Part III of the Building Regulations 1985, the Secretary of State's powers to dispense with or relax any requirement in the regulations shall be exercisable by the local authority.

Applications for relaxation must be made on the prescribed forms and shall contain such particulars as may be required. Prior to authorising relaxation, the Minister or local authority shall give notice of the application in a local newspaper, and state that representations on grounds of public health or safety may be made by a specified date, not less than 21 days from the date of the notice. The applicant may be required to pay the cost of publication of the notice. If it appears that the relaxation will only affect adjoining premises, it is necessary to notify only the owner and occupier of those premises. No publicity is needed for relaxations confined to internal work. If, after receiving representations, the local authority refuses an application and an appeal is brought against its refusal, the local authority shall transmit to the Secretary of State copies of the representations.

Implementation of the Building Regulations means that the appropriate local authority has to be notified and the building work will have to comply with the Regulations. The main purpose of the Regulations is to ensure the health and safety of people in or about the building, and they are also concerned with energy conservation and access to buildings for the disabled. The client may choose either the local authority or a private approved inspector to supervise the work.

Where the client opts for local authority supervision, he has a further choice of depositing full plans or submitting a much less detailed building notice, and a fee is payable to the local authority. The local authority can prosecute if work is started before either course of action has been taken. Where a private approved inspector is selected, the client and the inspector must jointly give the local authority an initial notice accompanied by a site plan. Work must not be commenced before the notice has been accepted by the local authority. The inspector's fee is negotiable.

The two alternative procedures are now considered in more detail.

(1) *Local authority control*

Where the proposal encompasses the erection of offices or shops, full plans must be deposited. Where full plans are deposited, the local authority may pass or reject them within 5 weeks, or 2 months if the client agrees. The plans must be accompanied by a certificate that the plans show compliance with the structural stability and/or energy conservation requirements of the Regulations. The local authority has to consult the fire authority about proposed means of escape in the case of certain factories, offices, shops, railway premises, hotels and boarding houses.

The local authority may pass plans subject to either or both of the following conditions:

(1) modifications in the deposited plans;
(2) the depositing of further plans.

Work may begin at any time after the submission of a building notice or deposited plans, provided the local authority is given 48 hours' notice. If the local authority considers that any work contravenes the requirements of the Regulations, it may serve a notice requiring the demolition or alteration of such work within 28 days.

(2) *Supervision by approved inspector*

The approved inspector and the client should jointly give the local authority an initial notice together with a declaration that an approved scheme of insurance applies to the work, which must be signed by the insurer. The initial notice must contain a description of the work and, in the case of a new building or extension, a site plan and information about drainage.

The local authority has 10 working days in which to consider the notice and may only reject it on prescribed grounds. On acceptance the local authority may impose conditions. It is a contravention of the Regulations to start work before the notice has been accepted. As a general rule the approved inspector must be independent of the designer or builder, but he need not be if the work consists of alterations or extensions to one or two storey houses. The National House-Building Council is a major provider of private building control services, alongside practising professionals.

Where a client wishes to have detailed plans of work certified as complying with the Building Regulations, he should ask the approved inspector to supply a plans certificate and the local authority also receives a copy. When the work is complete the approved inspector should give the client and the local authority a final certificate.

Unlike a local authority, an approved inspector has no direct power to enforce the Building Regulations. He is, however, required to inform the client if he believes that any work being carried out under his supervision contravenes the Regulations. If the client fails to remedy the alleged contravention within 3 months he is obliged to cancel the initial notice. He must also inform the local authority of the contravention, unless a second approved inspector is taking over responsibility.[2]

Building Notice Procedure

A person intending to carry out building work may give a building notice to the local authority unless the building requires a means of escape and is designated under the Fire Precautions Act 1971, such as certain hotels and boarding houses, and prescribed factories, offices, shops and railway premises. Full plans must also be deposited for dwellinghouses of three or more storeys and buildings of three or more storeys containing a flat.

There is no prescribed form of building notice. It must, however, be signed by the person intending to carry out the work or on his behalf, and must contain or be accompanied by the following information.

(1) The name and address of the person intending to carry out the work.

(2) A statement that it is given in accordance with Building Regulation 11 (1)(a).

(3) A description of the location of the building and the use or intended use of the building.

(4) If it relates to the erection or extension of a building, it must be supported by a plan to a scale of not less than 1:1250, showing its size and position in relation to streets and adjoining buildings on the same site, the number of storeys, and details of the drainage.

(5) If section 18 of the Building Act 1984 (building over a sewer) or section 24 (provision of exits) applies, the necessary particulars must be given.

(6) Where cavity wall insulation is to be inserted, information must be supplied as to the insulating material to be used and whether or not it has an Agrément certificate or conforms to British Standards, and whether or not the installer has a BSI Certificate of Registration or has been approved by the Agrément Board.

(7) If the work includes the provision of a hot water storage system with a capacity of 16 litres or more, details of the system and whether or not the system and its installer are approved.

The local authority is not required to approve or reject the building notice and has no power to do so. It is, however, entitled to request any plans that it considers necessary to enable it to discharge its building control functions and may give a time limit for their provision. Plans in this context include drawings, specifications, structural calculations and other information. The building notice and plans are not to be treated as having been *deposited* in accordance with the Building Regulations (regulation 12 (6)).

Deposit of Full Plans

The traditional method of building control by which full plans are deposited with the local authority is covered under section 16 of the Building Act 1984, as supplemented by regulation 13 of the Building Regulations. The local authority must give notice of approval or rejection of plans within five weeks unless the period is extended by written agreement.

The full plans required under the deposit procedure are the same as those required by the building notice approach, together with such other plans as are necessary to show that the work will comply with the Building Regulations. The plans must be deposited in duplicate; the local authority retains one set of plans and returns the other set to the applicant. They must be accompanied by a statement that they are deposited in accordance with regulation 11 (1)(b) of the Building Regulations 1985. Since the reason for depositing full plans will normally be to have them passed, a person who is in doubt would be wise to provide more information rather than less, to avoid the possibility of the plans being rejected on the grounds that they are incomplete.

The advantage of the deposit of full plans method is that the work has to be carried out in conformity with the plans as approved by the local authority. Furthermore, the work will be supervised by a building control officer of the local authority.[3] A typical approval notice from a local authority follows.

PENDLETON DISTRICT COUNCIL

Date: 4 November 1987
Reference: 217/BR/87

Messrs Simmons and Peters
10 High Street
Newville

Dear Sirs

Building Regulations

Your Building Regulation application dated 21 October 1987, relating to the conversion of a bedroom to bathroom and the provision of a new WC at 65 Thorney Lane, Bedrock for Mr E. J. Robinson, has been approved.
The necessary notices for completion and submission to this office at the various stages of the work are attached.

C. A. Smith
Chief Building Control Officer

Planning Control

External maintenance, improvement or alteration, when they are works which materially affect the external appearance of a building, require planning permission. While acknowledging that painting is exempt from planning control, there is a need to keep the external painted surfaces regularly painted for both protective and decorative reasons. Planning control is unlikely for instance to prevent the spoliation of the elevations of buildings, particularly old ones, by incompatible pointing of brickwork, rendering attractive brickwork and replacing with modern unsuitable windows. This can only be secured on the wider front by raising the standards of aesthetic appreciation and sympathy with older buildings of both clients and builders alike and recognition of the importance of harmony, scale, composition, texture and colour.

Improvement works do not usually require consent under the Town and Country Planning Acts unless the size or appearance of the building is altered; all conversions schemes require consent to cover the change of use.

Buildings of special architectural and historic interest are protected under the Town and Country Planning Act 1971 and the Civic Amenities Act 1967. There are restrictions on the demolition, alteration and extension of these buildings. All scheduled buildings require separate planning applications and press and site advertisement notices must be displayed. There are a number of advisory bodies whose function it is to promote public interest in and assist in preserving older buildings; a selection of these bodies is given in table 11.1[4]

Table 11.1 Advisory bodies for repair and preservation work

Bodies	*Main Objectives*
Ancient Monuments Society	Advise on sources of finance, new uses for redundant buildings and dating and value of historic buildings
Historic Buildings Councils (England, Scotland and Wales)	Advise on grants for repair and maintenance of buildings of architectural and historic interest
The Building Conservation Trust	Promote proper conservation, maintenance, alteration and uses of buildings of all types and ages
Civic Trust (Various regional offices)	Encourage protection and improvement of the environment
Georgian Group	Promote public interest and give advice on repair, adaptation and preservation of Georgian buildings
National Trust	Promote preservation of buildings of architectural and historic interest
Victorian Society	Promote preservation of 19th century architecture
Pilgrim Trust	Make grants for repair of ancient buildings
Society for the Protection of Ancient Buildings	Advise on problems of conservation repair and arrange lectures and training courses

Planning Permission

The law relating to town and country planning is mainly contained in the Town and Country Planning Act 1971, which consolidated in a single Act most of the previous law on the subject. The local planning authorities are primarily the county councils but the development control functions are shared with district councils in varying ways. Steps have been taken to accelerate and simplify planning procedures by giving executive powers to local planning committees to decide the less complex applications, without reference to higher committees or full council, and the delegation of a wide range of decisions to planning officers. In addition, increased consultation has been developed between applicants and planning officers and many local authorities have compiled comprehensive guides to their development control policies as a code of practice for the public and for developers. For its part, the Department of the Environment has reduced the number of applications referred to the Department and have suitable appeals against planning decisions settled by the written representation method to avoid delays in arranging formal public inquiries. At the time of preparing the second edition, the government was considering further measures to streamline and simplify the development control arrangements.

Planning permission is required for 'development' as defined by the Town and Country Planning Act, namely the carrying out of building, engineering, mining or other operations in, on, over or under land, or the making of any material change in the use of buildings or other land. The Act lists a number of operations or uses which do not constitute development. These include road improvement works within the boundaries of a road and repairs to and renewal of services by local authorities and statutory undertakers.

The Use Classes Order 1972 classifies buildings with similar uses into 18 different use classes. For instance, class I embraces shops of all kinds but excludes garages, fun fairs and petrol filling stations. It does however include hairdressers' and undertakers' establishments, laundries and dry cleaning receiving offices. Class II consists of offices, including banks, while class III comprises light industrial buildings, which can be located in a residential area without detriment by reason of noise, smell, vibration, fumes, soot, ash or grit.

The Order provides that if the use of a building is changed but it still remains in the same class, then the change will not constitute development and planning permission will not be required. Hence permission will not be needed to change a newsagent's into a hardware shop, or a butcher's into a greengrocer's shop, as these do not constitute a material change of use. On the other hand if it is proposed to change the use of a building from one class to another, such as from a shop to an office, then permission will be required.[5]

It will be observed that the change in use must be *material* to constitute development; trivial changes will be disregarded. Whether a change of use is material is often a matter of degree. For example, an architect may use one room of his house as an office without changing the predominantly residential character of the dwelling, but if he converts a number of rooms to drawing offices, print rooms, plan stores and similar uses, the essentially residential use may cease, involving a material change of use and the need for planning permission.

The 1971 Act empowered the Minister to make *development orders* which grant planning permission for specified types of development. It is unnecessary to make application for planning permission for development coming within one of the twenty-three classes of permitted development listed in the Town and Country Planning General Development Order 1973. The two most common forms of permitted development which are particularly appropriate to maintenance and improvement work follow.

Class I—Extension of a Dwelling House

The cubic capacity may be increased by up to 70 m^3 or 15 per cent of the original capacity (50 m^3 and 1/10 in the case of a terrace house), whichever is the greater up to a limit of 115 m^3. The extension must not increase the height of the building nor must it project beyond the front of the original building. The erection of a garage in the curtilage of a dwelling house is treated as an enlargement of the house. Permitted development under class 1 includes the erection of a porch outside an external door of a dwelling, subject to the general requirements that its height must not exceed 4 m and no part must be within 2 m of any boundary of the curtilage of the dwelling house.

Other developments permitted under class 1 comprise the construction within the curtilage of a dwelling house of a hardstanding for vehicles or the erection of a storage tank for oil for domestic heating. The tank must not have a capacity exceeding 3500 litres and is also subject to other conditions.

Class II—Sundry Minor Operations

Permitted development within class II includes the erection of gates, fences and other means of enclosure, provided the height does not exceed 1 m where abutting on a road used by vehicular traffic, and 2 m in other cases. The painting of the exteriors of buildings, except for purposes of advertisement, is also permitted development within class II.

Development Plans

Local planning authorities (county councils) are required to institute a comprehensive survey of their area and, among other matters, must examine and keep under review such matters as the principal physical and economic characteristics of that area and adjacent areas likely to influence it; the size, composition and distribution of the population; communications and transport; and any changes in these matters already projected which are likely to have any bearing on the development of the area.

The local planning authority must then prepare and send to the Minister for his approval a structure plan for its area.

The *structure plan* consists of a written statement which formulates the local planning authority's proposals for the development and use of land in its area and relates these proposals to those for neighbouring areas. The plan must be accompanied by diagrams, illustrations and such descriptive matter as the authority considers necessary for explaining its proposals, but not a map. In preparing a structure plan, the authority must take steps to secure adequate publicity for the report on which it is based and the matters which it proposes to include in the plan and the proposed content of the explanatory memorandum. It must make copies of the plan available and ensure that persons interested are given adequate opportunity to make representations. It must also see that each copy of the plan is accompanied by a statement of the time within which objections can be made to the Minister.

When submitting a structure plan, the authority must inform the Minister of the steps taken to publicise it and of the consultations with, and consideration of the views of, interested parties. If the Minister is satisfied that the appropriate steps have been taken he may then consider the plan and, if he does not reject it, he must consider any objections which have been made. He must give objectors an opportunity of appearing before a person appointed by him and, if he decides upon a local inquiry, he must give a like opportunity to the planning authority and such other persons as he thinks fit.

In addition to the structure plan, a local planning authority may, and if required by the Minister must, prepare a detailed plan for the development of any part of its area, called a *local plan*. The proposals it contains must conform generally to the structure plan. The proposals must be given adequate publicity and the steps to secure this follow closely those required for a structure plan. Objections to a local plan are considered at a local inquiry arranged by the authority and presided over by a person appointed by the Minister or in some cases by the authority itself. After considering any objections, the authority by resolution may adopt the plan as originally prepared or as modified to take account of objections. The express approval of the Minister is not required unless he has, before adoption, directed that the plan shall not come into effect unless approved by him.

Planning Applications

If a person wishes to undertake development, for which planning permission is not granted by the current General Development Order, then he must make application to the local planning authority for planning permission. Where in doubt he may apply to the authority for a decision as to whether an application is necessary. The application for planning permission is made on the appropriate form obtained from the local planning authority or the local authority with whom the application is to be lodged (district council), and accompanied by the appropriate fee. The application must be accompanied by a plan sufficient to identify the land and such other plans and drawings as are necessary to describe the development. The local planning authority may require additional copies of the plan(s) and drawings, not exceeding three, and such further information as may be needed to determine the application. The plans consist of a 1:2500 site plan and normally a 1:500 block plan showing the site, its boundaries and adjoining buildings. Other drawings should normally be to a scale of not less than 1:100 showing existing features of the site, and accesses thereto, the appearance of any proposed buildings and, where change of use of part of a building is proposed, floor plans indicating the extent of the new use.[6]

In practice the following particulars are generally required on a planning application: name and address of applicant; applicant's interest in land (owner, lessee, prospective purchaser, etc.); name and address of agent (if any); address or location of site; present use; vehicular access requirements; total area of site; area of site covered by existing and proposed buildings; cubic content of existing and proposed buildings; description of processes in industrial buildings, together with floor area, provision for loading and unloading vehicles and means of disposal of trade refuse and effluents. In addition they may require an indication as to whether it is an outline application and a brief description of the development, including materials to be used in roofs and external walls of buildings if it is a detailed application.

Under the 1971 Act (section 26) the classes of development that follow, sometimes referred to as 'bad neighbours', must be publicised in the prescribed manner (by a notice clearly visible on the land and by advertisement in the local press describing the applicant's intentions and indicating where the application may be inspected).

(1) Construction of buildings as public conveniences.

(2) Construction of buildings and use of land for disposal of refuse or waste, or as a scrap yard or coal yard or for winning or working of metals.

(3) Construction of buildings or use of land for sewage treatment or disposal.

(4) Construction of buildings to height exceeding 20 m.

(5) Construction of buildings or use of land as slaughter house or knacker's yard or for killing or plucking poultry.

(6) Construction and use of buildings as a casino, fun fair, bingo hall, theatre, cinema, music hall, dance hall, skating rink, swimming bath or gymnasium (not being part of an educational establishment), or a Turkish, vapour or foam bath.

(7) Construction and use of buildings or land as a zoo or for the breeding or boarding of cats or dogs.

(8) Construction of buildings and use of land for motor car or motor cycle racing.

(9) Use of land as a cemetery.

Outline Applications

A person may be considering the purchase of land for a particular purpose but can be uncertain as to whether he can obtain planning permission for that purpose. In these circumstances he will be unwilling to purchase the land or to prepare detailed plans for its development if there is the possibility of planning permission being refused. He is however able to make application for outline permission, as, for example, to erect a certain number of houses on a particular plot of land. The local planning authority may approve the proposals in principle but at the same time will specify certain matters, known as reserved matters, to be subsequently submitted for approval before development can proceed. These matters usually include the siting, design and external appearance of buildings and means of access. Permission is generally available for a specified period.

Certificates

The 1971 Act (section 27) requires every application for planning permission to be acompanied by a certificate relating to the applicant's interest in the land which is the subject of the application, otherwise an application for permission to develop land might be made without the owner's knowledge. There are four forms of certificate to be used according to circumstances.

Certificate A   is used when the applicant is either the owner or tenant of all the land in the application.

Certificate B   states that the applicant has notified all owners of any part of the land he wishes to develop.

Certificate C   states that the applicant has notified some of the owners but has failed to notify others because he does not know their names and addresses. It also states that he has advertised his application in a local newspaper.

Certificate D   is used when the applicant cannot discover the names and addresses of any of the owners. This certificate must also certify advertisement in a local newspaper.

Planning Register

On receipt by the local planning authority, or possibly a district council, an application will be date-stamped and entered in the planning register. A copy of the application is usually submitted to the highway authority when the development affects a road for which the authority is responsible. A district council normally retains one copy of the application and forwards two copies to the local planning authority (county council). The arrangements for the determination of applications may vary from one county to another. Part of a typical planning register with two entries is shown in table 11.2. Planning registers are to be open to public inspection.

Notices

When considering a planning application the planning officer will have regard to the Development Plan (Structure and Local Plans). He will also consider the effect of proposals on adjoining developments, amenities of the area and other relevant factors. In general, the decision of the local planning authority must be notified to the applicant within eight weeks or such extended period as may be agreed by the authority and the applicant. Notice must be in writing and, where planning permission is refused or granted subject to conditions, the authority must state the reasons for its decision. An aggrieved applicant has a right of appeal against a planning decision to the Minister.

Table 11.2 Planning register

Application Nr	Date of Application	Brief description of proposed development	Location	Name and address of applicant	Name and address of agent (if any)	Date considered by Committee	Committee's decision	Date of appeal	Date of local inquiry	Decision on appeal
GP/223/87	15/10/87	Outline application for residential development at a density of 20 houses per hectare	South Bilney, Castle Road OS plot 364 (1.22 ha)	Mr P. J. Lewis, 15 St. Brelade Walk, Jacksville	D and E Carruthers, 8 Aubrey Avenue, Chatsford	3/11/87	Permission refused as development would be a substantial departure from the County Development Plan, which envisages the retention of the existing agricultural use. Development in the Bilney Valley must be restricted because of deficiencies in water supply and drainage			
GP/224/87	20/10/87	New bungalow and garage	St Watcham, Peters Road	Mr R. S. Shipley, 20 Beaufort Close, Bradwell Church End	—	3/11/87	Permission granted subject to vehicular access being constructed and maintained to the approval of the County Council, to minimise danger, obstruction and inconvenience to users of the highway and the premises			

329

A typical planning approval subject to conditions follows.

CRINGLEFORD DISTRICT COUNCIL

Date   5 November 1987
Reference   GP/220/87

Herbert S. Scribbington
6 Hawthorn Avenue
Tupton-on-Marsh

Dear Sir

Town and Country Planning Acts

Your planning application dated 13 October 1987, relating to the erection of a new parsonage house for the sequestrators of the Benefice of Newley, in part of the field adjoining the Bewley rectory garden is hereby approved subject to the following condition:

> Both vehicular and pedestrian access to serve the proposed parsonage house shall be obtained through the combined vehicular and pedestrian access serving the existing rectory in the manner indicated on the deposited plan.

The reason for this condition is to minimise danger, obstruction and inconvenience to users of the highway and of the premises.

D. F. Marshall
Chief Planning Officer

Procedure at Public Inquiries

Public inquiries are held before inspectors of the Department of the Environment, in connection with appeals against decisions on planning applications, building applications and other matters. The procedure is similar to that adopted in a Court of Law, although a greater measure of flexibility and informality is permitted. There is, for instance, some relaxation of the rules of evidence, and it is usual for witnesses to read their evidence after circulating copies to the inspector and the opposing party. The procedure at a public inquiry dealing with an appeal on a planning or building application is as follows

(1) The applicant or his representative makes an opening statement outlining his case.

(2) The appellant calls witnesses to give evidence and these are cross-examined by the local authority's representative and re-examined, if necessary, by the appellant. The inspector may also ask questions if he wishes to do so.

(3) The local authority's representative makes an opening statement and calls his witnesses in the same manner as the appellant.

(4) The local authority's representative can make a closing statement if he wishes.

(5) The appellant or his representative makes a closing statement.

(6) Any other interested parties are invited to make statements and may be cross-examined.

(7) The inspector may visit the site accompanied by a representative of each party, who can identify objects mentioned at the inquiry.

The inspector subsequently prepares a report for submission to the Minister. The Minister's decision and reasons must be given in writing to the parties and be accompanied either by a copy of the inspector's report or a summary of his conclusions and recommendations. Since July 1981, appeals relating to planning decisions, enforcement notices and related matters are heard and determined by an Inspector appointed by the Secretary of State (Determination of Appeals by Appointed Persons (Prescribed Classes) Regulations 1981).

Other Statutory Requirements

Highway Matters

The public have a right to pass and repass over a highway and it is an offence to obstruct a highway in the exercise of this right. At common law, any encroachment on the highway is a public nuisance and a member of the public can accordingly bring an action for damages if he can show that he has suffered 'damage' in excess of that suffered by the public generally. It is also a nuisance to render the use of the highway unsafe for the public as by making excavations adjoining it and leaving them unfenced.

The contractor, with the consent of the local authority, may deposit temporarily building materials, equipment and rubbish in the street or make temporary excavation there. Should the authority refuse consent, the contractor has right of appeal to the Magistrates' Court. Where consent is given, the contractor must ensure that the obstruction or excavation is properly fenced and lighted during hours of darkness. He must remove the obstruction or fill the excavation when required to do so by the highway authority and in any event must not allow the excavation or obstruction to remain any longer than necessary (Highways Act 1980 s. 171). A local authority may remove things unlawfully deposited on a highway which are a danger to users, without giving notice or obtaining a court order, and can recover expenses (Highways Act 1980 s. 149). There are also extensive conditions prescribed in respect of builders' skips located on highways, and the builder requires the highway authority's consent before placing the skip (Highways Act 1980, ss. 139 and 149).

The Highways Act 1980 s. 172 prescribes that a contractor who is about to erect, demolish, alter or repair the outside of a building in a street, before commencing work must erect a close boarded hoarding or fence separating the building from the street to the satisfaction of the local authority. This obligation may be dispensed with if the local authority agrees. The builder, if the local authority so requires, must make a convenient covered platform with a handrail to serve as a footway for pedestrians. This must be kept in good condition and, if the authority so requires, must be lighted during hours of darkness. The hoarding must be removed when required by the authority.

The Highways Act 1980 also empowers the highway authority to recover the expenses of repairing a road where damage has been caused by 'extraordinary traffic'. Extraordinary traffic is traffic which by reason of its weight, nature, extent or mode of operation is likely to cause damage to the highway in excess of that caused by traffic normally carried by it.

Building and Improvement Lines

The siting of new buildings and extensions to existing buildings is often affected by *building lines* under the Highways Act 1980 s. 74. A building line may be prescribed by the highway authority on either one or both sides of a public highway, and no new building, other than a boundary wall or fence, shall be erected and no permanent excavation below the level of the highway shall be made, nearer the centre line of the highway than the building line, except with the consent of the highway authority, who may impose conditions. The frontage line of existing buildings may in some cases form the building line. The main purpose of the building line is to ensure that all buildings are kept a reasonable distance from the highway to preserve amenity and obtain good sight lines.

Under section 73 of the same Act, a highway authority may prescribe *improvement lines* on one or both sides of a street and they represent the lines to which the street will eventually be widened. This procedure is normally adopted when a public highway is narrow or inconvenient without any sufficiently regular boundary line, or it is necessary or desirable that it should be widened. Similar prohibitions apply to building work in advance of an improvement line as those applicable to building lines, except that boundary walls and fences are not excluded.

Amenity and Safety

The Building Act 1984 contains several provisions regarding the safety and appearance of buildings and sites. If a structure is in a dangerous condition the local authority may apply to a Magistrates' Court, and the court may order the owner to carry out remedial works or, if he so elects, to demolish the building and remove the rubbish. If the owner defaults, the authority may carry out the work and recover its reasonable expenses from the owner. In addition, the owner is liable to a penalty (section 77). If it appears to an authority that immediate action should be taken, it may take appropriate action to remove the danger, recovering the cost from the owner through the Magistrates' Court. However, the authority must first, if it is reasonably practicable, give notice to the owner and occupier (section 78).

Local authorities may exercise considerable control over demolition works. With minor exceptions, a person who intends to demolish a building must notify the local authority, and the authority may require action to be taken in various ways, such as shoring up adjacent buildings, weatherproofing exposed surfaces, removing rubbish and sealing or removing drains and other pipes. The recipient of the notice may appeal to the Magistrates' Court where the required action could result in problems concerning the adjoining building or its owner. Furthermore, the local authority may undertake the work if the recipient defaults and also seek a fine, additionally or alternatively (sections 81–83).

Local authorities also have powers to deal with ruinous buildings and vacant sites used as rubbish dumps. If a building, because of its ruinous or dilapidated condition is seriously detrimental to the amenities of the neighbourhood, the local authority may by notice require works of repair or restoration or, if the owner chooses, its demolition, and the clearance of the resultant rubbish. In case of default the authority may do what is necessary and can seek a fine. Before taking direct action the authority must serve a notice on the owner or occupier stating what it proposes to do. The recipient may serve a counter-notice stating that he will do the necessary work or he may appeal to the Magistrates' Court on the ground that the authority's action is not justified. If a counter-notice is served, the authority must wait to see if the necessary steps are taken within a reasonable time, or, if work is begun, whether satisfactory progress is made (section 79).[7]

Other Statutory Provisions

In addition to the provisions of the Building Act 1984 concerning the compulsory repair of dilapidated property, the *Housing Acts* also contain provisions for the compulsory repair of houses unfit for human habitation.

The *Factories Act 1961* makes provision for securing satisfactory means of escape in case of fire in the majority of factories and for the issue of certificates by the fire authority. The same Act prescribes certain minimum standards for space, heating, ventilation, lighting, cleanliness, sanitary conveniences and welfare provisions. This Act is supplemented by the *Health and Safety at Work Act 1974.*

The storage of petroleum in bulk requires a licence from the local authority to which conditions may be attached under the *Petroleum (Consolidation) Act 1928.* The conditions may cover the mode of storage, the nature and situation of the premises, the nature of any goods stored with it, facilities for testing and other safeguards.

The *Clean Air Act 1956* aims at the ultimate elimination of smoke and other forms of atmospheric pollution arising from the operation of industrial furnaces and domestic heating appliances. The Act makes it an offence to emit dark smoke from a chimney but the Minister may by regulation exempt periodic emissions. All new furnaces must be smokeless and minimise the emission of grit and dust as far as practicable. This provision does not apply to domestic furnaces with a heating capacity not in excess of 55 000 BTU/h. Furthermore, a local authority may, by order confirmed by the Minister, declare the whole or any part of its area to be a smoke control area, when it will be an offence to emit smoke within that area. If the Minister confirms a smoke control order, anyone with an interest in a dwelling situated in the area who spends money on an installation designed to prevent a breach of the order is entitled to claim 70 per cent of his expenditure from the local authority. The local authority may also pay all or some of the remaining 30 per cent.

The *Offices, Shops and Railway Premises Act 1963* is designed to give protection to office and shop workers comparable to that given to industrial workers by the Factories Act. The most important regulations made under the Act relate to dangerous machines, washing facilities and sanitary conveniences. The Act also prescribes minimum standards in respect of cleanliness, overcrowding, temperature, lighting, accommodation for clothing, construction and maintenance of floors, passages and stairs, and fire precautions. Hotels and restaurants are examined under the Food Hygiene Regulations.

Other regulations affecting building work include *water by-laws* administered by the water authorities and covering water supply installations and the use of water.

The *IEE Regulations* issued by the Institution of Electrical Engineers lay down minimum requirements for electrical work. Finally, as described in chapter 1, the *Defective Premises Act 1973* places a statutory duty on any person responsible for the provision of a dwelling, whether by building, conversion or enlargement, to use proper materials and for the work to be carried out in a professional or workmanlike manner, providing a dwelling fit for habitation.

The *Fire Precautions Act 1971* ensures a minimum standard of fire precautionary measures for certain buildings such as institutions for entertainment, teaching, treatment or care and most purposes entailing access by the public, and the issue of certificates specifying the means of escape, fire fighting equipment and fire alarms.

Control of Work in Progress

Under the Building Regulations,[1] a contractor carrying out building work which is subject to the regulations and to supervision by the local authority, is to give the local authority not less than 48 hours' notice in writing of when work will be commenced; 24 hours' written notice before the covering up of any excavation for a foundation; any foundation, damp-proof course; or concrete or other material laid over a site; before any drain or private sewer is haunched or covered; and not more than 7 days after the work of laying any drain or private sewer.

If the contractor neglects to give any of these notices, he may be required to cut into, lay open or pull down so much of the building work as prevents the local authority from ascertaining whether any of the regulations have been contravened. Finally, the contractor shall give notice in writing to the local authority not more than 7 days after completion of the building work which is subject to the regulations.

Building work in progress will also be inspected as part of development control under the Town and Country Planning Acts, to ensure that development proceeds in accordance with the approved applications and any conditions that may have been imposed. Where development has been carried out without planning permission or conditions attached to permission have not been satisfied, then there is a breach of planning control and the local planning authority may serve on the offender an enforcement notice requiring him to remedy the breach. If the breach consists of carrying out, without permission, building, mining, engineering or other operations, or the failure to comply with conditions relating to these operations, then the enforcement notice must be served within four years from the date of the breach.

The local planning authority is also empowered to serve a 'stop notice' on any person having an interest in land and who is carrying out operations on it. The authority must serve an enforcement notice and then, at any time before the notice takes effect, may serve a stop notice which will prohibit continuance of the operations specified in the enforcement notice. The stop notice must specify the date upon which it is to take effect, which must not be less than 3 or more than 28 days after service, and the stop notice will cease to have effect when the enforcement notice is withdrawn or quashed; or the period for compliance with the enforcement notice expires; or the stop notice is withdrawn.

Easements

Difficulties may sometimes arise in carrying out building extensions through the existence of easements, in which the owner of the dominant tenement secures a right over another property (the servient tenement). The tenements must be in different ownerships, the right must be capable of being granted, the servient owner must not be involved in any expenditure in complying with the easement and it must not involve the removal of anything other than water from the servient tenement. A legal easement is made by a grant from the owner of the servient land or by prescription (long use of the privilege by the dominant owner under certain conditions), and is binding on all persons who occupy the servient tenement.

The more common easements affecting building development are as follows

(1) Right of light: the right of light to a building becomes legally protected after it has been enjoyed for a period of twenty years. A right of

light prevents the owner or occupier of land from creating buildings on it which will obstruct the light passing on to the other person's land.

(2) Right of support: the ownership of land carries with it the right to support from the adjoining land but not for the support of any buildings subsequently erected on it.

(3) Right of way: a private right of way may be presumed if it has been enjoyed for twenty years and becomes absolute if enjoyed for a period of forty years. Public rights of way are not easements.

(4) Right of drainage or pipe easements: the right to drain across the property of another. An easement which permits the dominant owner to lay pipes across the land of the servient owner, whether for purposes of drainage or water supply, may sterilise the strip of land through which it passes and provision will have to be made for access to and inspection of the pipes.[8]

Disputes over Site Boundaries

On occasions disputes arise as to the ownership of land and the boundaries between adjoining plots. The difficulties may stem from one of a number of causes and the more common ones follow.

(1) Fences are normally erected by persons with limited knowledge of setting out.

(2) Many fences are constructed of materials which are subject to decay.

(3) Deeds often fail to specify the position of the boundary line in relation to the fence—whether on the centre line or one face of the fence.

A common approach is to assume that the smooth side of a fence faces the adjoining owner. Unfortunately this rule is not infallible, as for instance where an owner has renewed his fence without being permitted access to his neighbour's land. This results in the owner being obliged to nail the pales on his own side owing to the physical impossibility of working overhand. Again it might be that a previous owner preferred to see the more attractive side of the fence and hence erected the posts on the boundary and the boarding on the inside face. Aldridge[9] has pointed out that a wall or boarded fence may be set back slightly from the boundary to avoid trespassing on erection or maintenance. Reference should always be made to the appropriate deeds.

In the case of a bank and a ditch, it is often presumed that these belong to the person whose land is on the bank side of the ditch. The owner would normally dig the ditch up to the edge of his land, throw the earth on to his land to form the bank and then often plant a hedge on the bank. This presumption can apply only to artificial or man-made ditches.[10]

References

1   DOE. *Manual to the Building Regulations*. HMSO (1985)
2   I. H. Seeley. *Building Technology*. Macmillan (1986)

3 V. Powell Smith and M. J. Billington. *The Building Regulations: Explained and Illustrated*. Collins. (1986)
4 E. D. Mills (Ed.). *Building Maintenance and Preservation*. Butterworths (1980)
5 D. Heap. *An Outline of Planning Law*. Sweet and Maxwell (1987)
6 J. Stephenson. *Planning Procedures*. Northwood Books (1982)
7 C. Cross and S. Bailey. *Cross on Local Government Law*. Sweet and Maxwell (1986)
8 I. H. Seeley. *Building Economics*. Macmillan (1983)
9 J. M. Aldridge. *Boundaries, Walls and Fences,* Longman (1982)
10 I. H. Seeley. *Building Surveys, Reports and Dilapidations*. Macmillan (1985)

12 PLANNING AND FINANCING MAINTENANCE WORK

Planning, budgeting and controlling the cost of maintenance work are essential operations if buildings are to be maintained effectively within available funds. This leads automatically to regular inspections and the implementation of programmes of planned maintenance. These activities need to be backed up with adequate data and particularly a full awareness of maintenance and operating costs.

Planning, Budgeting and Controlling the Cost of Maintenance Work

Planning

Planning and budgeting are two highly interrelated activities which must proceed simultaneously. It is not possible to plan maintenance work without knowledge of the costs involved, or to budget for the work in the absence of an effective programme. Nevertheless, effective building maintenance is dependent upon making the correct decisions and satisfactorily implementing them.

Building maintenance should be regarded by management as part of the total operating strategy; that far from being a 'make do and mend' service, it should be viewed as a property conserving activity contributing significantly to the success and well-being of the operations and occupants within it. With manufacturing organisations adopting carefully structured procedures, building maintenance claims can be justified on financial grounds in comparison and in competition with other direct profit making projects,[1] showing dividends as good as or better than other activities.

A study of local authority housing maintenance[2] showed the general standard of maintenance on certain post-war housing estates to be reasonably high, but the greater part of expenditure was made on basic fabric maintenance and on satisfying tenants' requests. Some areas of maintenance were neglected and there was a long waiting list of requests requiring attention, showing the need for a positive programme of maintenance. Closer examination revealed that building maintenance policy was influenced by four criteria which could on occasions be conflicting, namely:

338

(1) social—to provide a quick service to high standards of quality;

(2) financial—to invest funds in activities in the most efficient manner with due regard to the effects on debt charges, subsidies and rents;

(3) technical—to maintain property at a level deemed necessary after a thorough and regular technical survey;

(4) to provide continuous employment for certain operatives within a fixed budget.

Policy Formulation

The ability to formulate a long-term maintenance strategy and prepare budgetary forecasts is one of the benefits of having a maintenance policy.[3] The following five factors deserve consideration when formulating a maintenance policy for a manufacturing organisation.

(1) The aims of the organisation—the nature of the end product and how it is produced and the requirements in buildings and services.

(2) The standards required—influenced by aims of the organisation but may vary between different buildings.

(3) Legal liability—compliance with statutory requirements.

(4) Method of execution—such as direct labour or outside contractors, with particular attention paid to the effect on production.

(5) Cost and method of financing—with decisions supported by cost–benefit analyses where appropriate showing that the previous criteria (1–4) have been considered and optimum solutions proposed. All this information can then be translated into maintenance, cleaning and operating profiles for use not only in management of property but also for guidance of designers of new buildings so that total cost concept may be used.

In the detailed formulation of maintenance policy for a specific property, the following approach has much to commend it.

(1) Analysis of present condition of buildings, their nature and use, and estimated life cycle, as described by Smith.[4]

(2) Outline programme of work necessary to put and keep the buildings in satisfactory condition.

(3) Determine the method of implementing the programme.

(4) Calculate the approximate costs—total and annual. In most cases two assessments will be needed: first, for the period while the buildings are put in repair—including the routine repair in this phase—and then the assessment of the cost of keeping them in that state.[5]

It may be helpful at this stage to apply these principles to a specific situation, in this case a major property holding consisting of a range of large and small buildings of diverse age, construction and use. There is a large maintenance staff of poor calibre with greater experience of plant than fabric, and they are often diverted on to minor improvement work. The buildings reflect an absence of regular routine maintenance and finance is limited.

An *investigation* shows that there is no planned routine maintenance and that the owners are unaware of the serious and expensive results of neglect. The roofs are in particularly poor condition. Although finance is limited, it is still ill-used and often diverted to improvements at the expense of maintenance. There is no central control of maintenance expenditure or priorities, and the works managers of individual buildings undertake their own repairs.

The *outline policy* recommends the following action.

(1) Centralisation of administrative and technical control of maintenance.

(2) Listing of priorities and drawing up a programme for remedying the serious backlog, with emphasis on financial consequences of further *ad hoc* arrangements.

(3) Assessing the finance required over a five-year period and justifying the need for a larger allocation of funds, while ensuring that the programme is sufficiently flexible to fit into the existing allocation if no further funds are forthcoming.

(4) Ensuring that the neglected structural repairs are carried out and that external painting is undertaken at the same time to make the best use of scaffolding.

(5) Reorganisation of maintenance staff securing a suitable balance between engineering and building operatives.

(6) Newer buildings to be subject to planned preventive maintenance programme to prevent deterioration.[5]

Long-term policy concentrates on putting buildings into a satisfactory state of repair, because of the prevalence of *ad hoc* maintenance based on breakdown maintenance. The earlier years of the programme may be relatively expensive (the cost of neglect) but succeeding years should, assuming the correct diagnosis and decisions, result in a lower and steadier level of expenditure.

Sometimes the cost of putting in good repair so greatly exceeds current annual maintenance as to disturb management, hence the case must be argued soundly and well. The most convincing argument for spending more is an assessment of the eventual cost of deferment. If inadequate funds are available, the rectification period will have to be extended. Periodic inspection—preferably annually—is the best method of ensuring that the right policy has been devised and is being implemented and adapted, if necessary, to meet changing conditions.[5]

The Department of Health and Social Security has emphasised the need for maintenance managers to prepare costed long-term and annual programmes of work which distinguish between work of a periodical nature, work of an irregular nature, planned preventive type maintenance and provision for day-to-day requisitions and emergencies. Provision must also be made for minor improvements. It is intended that long-term programmes should be on a rolling basis being reviewed and moved forward as each annual programme is prepared.[6]

Bushell[7], in assessing maintenance priorities in the National Health Service, proposed the following approximate order of priority: safety; essential service; statutory requirements; security; initial cost; revenue saving; spares availability; alternative source of supply; delivery time; manpower; public relations.

Standards of Maintenance

If insufficient maintenance is carried out, the fabric of buildings first become unattractive, then unacceptable to the occupants and finally dangerous and uninhabitable. The maintenance manager has to decide the optimum level of maintenance work required on the fabric to preserve an acceptable environment in the buildings under his care. He has for instance to decide whether a building should be patched temporarily and replaced later or be replaced immediately. To determine the best course of action he needs to consider the use and condition of the building, the comparative cost and effectiveness of different types of repair, the expected future life of the building, acceptable standards of maintenance and similar matters.

The first step is to determine reasonable standards of maintenance for the various building elements, such as paintwork, rainwater goods, and windows and paths. These usually fall into two categories:

(1) The smaller number where standards can be related directly to cost. For example, it is evident that a roof should not be permitted to deteriorate until it leaks as this will give rise to higher future maintenance costs.

(2) The majority where maintenance costs do not increase appreciably as the condition deteriorates. For instance it costs little more to repaint internal wall surfaces after seven years than after five, although the appearance has worsened. Hence discussions are needed with management and occupants to agree appropriate standards.[8]

After establshing reasonable standards it is necessary to estimate the deterioration rate of each element, so that changes in its condition can be related to its age. This rate is influenced by a number of factors such as aspect, age and location. The maintenance manager should supplement published data with his own information on the history of elements.

The next step is to decide the maintenance policy to be implemented for each element, determining also the method and materials to be used. The costs of maintaining each element can then be estimated over a period of time—possibly 20 or 30 years. Over this time scale most elements will need replacing or repairing and average annual maintenance costs can be computed.

Finally the average annual costs of implementing the maintenance policy can be assessed, by summing the average annual costs of all the elements. If the total cost can be met from available resources, the maintenance manager can prepare a programme for work, but if the resources are insufficient then lower standards of fabric maintenance will have to be set which are consistent with available funds.[8]

In general, most local authority maintenance sections are organised on hierarchical lines often at three or four levels. Attempts to overcome the problem of achieving consistent standards throughout county council administrative areas include meetings between the chief officer and divisional staff, when guidelines will be discussed and periodically reviewed, and occasionally officers from one area inspecting and reporting on properties in another area to identify differences in assessment of needs. Difficulty may also arise from the fact that there is some improvement element in much maintenance work.[9]

Budgeting

A budget has been defined as 'a financial and/or quantitative statement prepared prior to a defined period of time of the policy to be pursued during that period for the purpose of obtaining a given objective'. The budget limits will be established after inspections, critical analyses and estimates have provided the essential supporting data as previously described. Budgetary control is an important management function aimed at planning and controlling the use of its resources in order to achieve its objectives. In practice this is not always achieved, as indicated by the Committee on Hospital Building Maintenance[10] which reported "there is no evidence to show that the existing financial allocation for building maintenance was either sufficient or that it was being distributed on the most equitable basis relative to need and priority." All too frequently a maintenance budget is based on the previous year's allocation plus a percentage. Admittedly, there is no standard method of budgeting, but the skills and empiricism of building surveyors with their wide ranging experience of the construction, use, performance and cost of repair of buildings, can help to provide a sound base for budgeting.

The budget as a plan stipulates the use of the organisation's available funds over the projected time span towards the various objectives and opportunities within the total plan. It is thus the basis of control—the monitoring, evaluation and provision of a basis for decision taking upon ongoing operations and future plans.[11]

To devise an effective budgetary control system the following criteria must be satisfied.

(1) A clear understanding of objectives and their order of priority.

(2) A systematic analysis and evaluation of the needs and demands stemming from the objectives.

(3) A rational balancing of these demands against the desired objectives within the known constraints of labour, materials, time, managerial skills and funds.

(4) The avoidance of waste of financial resources.

(5) The development of a control system based upon identification of needs, adequate measurement of resource requirements, setting of work standards, measurement of performance, evaluation of significant deviations from standards, and the control of present and future opportunities through this knowledge.[11]

From these principles, implementation of the budgetary control system will require: organisational plans for a specific period; definition of resources required to accomplish the allotted tasks, establishing operating standards and targets to be achieved; information systems to generate, collate and evaluate data for decision taking and control purposes; detailed plans and programmes for achieving objectives; a framework for decision taking and a basis for measuring efficiency, effectiveness and profitability.[11]

Cost Structure Analysis and Budgeting

Budgets are based upon work to be done and are expressed in terms of financial expenditures. The costs are built up from labour, material and other expenses contributing to the maintenance work. The costs can be classified under various headings.

Bedrock costs are incurred to maintain the assets in a serviceable condition, regardless of whether or not the property is used productively.

Programme costs are incurred by specific decisions of management with a view to improving the level of activity, technology of operations, quality of the environment and/or public relations image.

Operating costs are incurred by decisions of management in fulfilling the operating role of the enterprise; they may be *variable* when geared to the level of business activity, or *fixed* as in the case of heating plant.

Committed costs stem from past decisions over which present management has little or no control, while management has some discretion about the level of *managed costs*.

Engineered costs arise where for a given future activity the optimum amount of work and hence of cost can be measured.[11]

Budgets and Building Maintenance

There are often technical difficulties in assessing the quantity, problems in execution and costs of building maintenance work, but overruns and underestimates frequently result from failure by management to recognise the value and need for realistic budgets. In many enterprises the cost of building maintenance is such a small proportion of total expenditure that it is not accorded a very high priority. Hence the process of compiling and appraising the budget is conducted on a level of additions to last year's expenditure rather than upon current costs or future needs.[11]

The building maintenance budget is easily pruned if pressure is exerted for funds for other purposes. It is comparatively easy to defer the incidence of spending on maintenance because the impact of so doing is seldom obvious, and the insidious nature of the needs is not fully recognised. Few organisations regard building maintenance as the preservation of the value of the asset as a functioning property. Thus budgeting for the total upkeep of the property is rarely conducted in full knowledge of all relevant facts. Various aspects of property upkeep are often tabulated under differing budget heads; accountability is thus diffused and control becomes difficult to exercise.

Constant or bedrock costs represent an inescapable minimum expenditure. General and specific programmes may impose additional maintenance responsibilities which can be separately identified and costed. The identification of the fixed and variable elements of operating expenditure related to changes in the level of activity should assist in avoiding illogical budget pruning.[11]

A basic aim of management is to minimise the discretionary or contingency amounts in a budget, and wherever possible to determine the proper amount for a given level of activity. In establishing the budget for building maintenance too much attention is often paid to the previous year's expenditure and insufficient to technical requirements.

In preparing a maintenance budget a distinction can be made between current foreseeable work such as painting, clearing gutters and cleaning, and non-recurrent foreseeable work such as repairs to floors, roofs and the like. In the absence of a technical survey, the contingency sum for unforeseen work is likely to be high and it becomes more difficult to plan and control efficiently.

Many county councils in preparing their budgets, identify three main components in their annual planned programmes:

(1) non-recurring or irregularly occurring needs of individual properties which have been accorded the highest priority in the year in question;

(2) cyclical work, stemming from a predetermined policy to renew or refurbish certain elements at fixed intervals (normally staggered, to even out annual workloads);

(3) service contracts.[9]

Relationship between Budget and Finance

Finance is provided from either capital or revenue sources. Capital is normally provided for new works only and revenue for running costs, but this seemingly straightforward procedure is upset by one factor—the expenditure on renewal or replacement. These latter processes often involve large sums that increase the revenue budget by a varying and substantial extent. In theory 'depreciation' should provide the financial aid required for renewal or replacement—the non-recurrent maintenance costs—but too often it is merely a 'book' figure and the cash is not available. There is a tendency to invest the capital in new works only and to pay insufficient or no regard to a replacement policy.

There is a tendency to pay for renewals and replacements by reducing expenditure on routine maintenance. Furthermore, variations in the cash flow available, such as from profits, can cause a serious disruption to renewal and replacement programmes. Delayed expenditure in these areas means higher future costs, owing to increased prices and possibly higher operating costs. Larger items of machinery are generally financed out of capital whereas items like roof renewals or replacement of building services are often treated unsatisfactorily as write-off or revenue expenditures.

To overcome these difficulties, it is good practice for financial authorisations for new assets to include the capital cost of the project, depreciation and running costs, so that it is considered as a whole from initial conception. With existing assets, the condition should be assessed and monies set aside for planned renewal and replacement.

Controlling Cost

The management process of control should incorporate the following activities:

(1) setting performance standards at the appropriate level to achieve a given objective;

(2) measuring actual performance and comparing it with the standard;

(3) taking appropriate action in the event of actual performance deviating from standard.

To achieve effective total cost control, the following criteria need to be satisfied:

(1) A sound knowledge of the relationship between budgeting and finance.

(2) A logical breakdown of the budget into specific sections under capital and revenue, with particular reference to renewals and replacements.

(3) Reasonable assessment of the factors affecting the budget, including a plan for maintenance.

(4) A method of calculating economic assessments of capital, renewal and replacement expenditures, using discounted cash flow analysis.

(5) Evaluation of the results of non-maintenance, such as lost amenities or production due to breakdown, excessive running costs and increased health hazards.

(6) Budgetary control, including a calendar programme for authorisation and implementation of plans for capital, renewal and replacement expenditures.

(7) The use of accurate costing techniques, including cost coding and classification systems, methods of cost collation and investigation, and feedback of relevant cost information for control purposes.

A computer program can be used to ensure effective budgetary control by providing a pattern of regular and frequent printouts giving for each level of management appropriate details of expenditure—actual and committed—and physical progress for each project under each separate budget heading. In this way performance can be compared with forecast in terms of time and resources but particularly in relation to finance. Likely problems and deviations in the budget can be quickly identified and corrective action taken in good time.

The detail into which each separate project is subdivided and pro-grammed will depend on the degree of management control that is required, but it is important that the printout for the lower levels of management should show separate projects, each with its own sub-budget. A computerised system will meet the overriding requirement which is the continuous provision of up-to-date information and a quick reaction at the first sign of deviation from the budget or programme.

Planned Maintenance

Nature of Planned Maintenance

There is growing interest in applying planned preventive maintenance to buildings following its established use in engineering. In engineering the maintenance operative takes with him tools and materials and performs such tasks as lubricating bearings and adjusting tolerances, and will in most cases attend to these items fairly frequently whether or not a defect is apparent. In the context of building maintenance a change of emphasis is needed with a higher proportion of inspections, with treatment only if a defect is detected.[12] It is advisable to keep day-to-day maintenance to a minimum as its nature and extent cannot be forecast and it can frequently be detrimental to longer-term action, in addition to being expensive in terms of both unit costs and staff time.[9]

Planned maintenance of buildings can be subdivided into three main categories:

(1) Preventive *running maintenance*—work which can be done while the facility is in service.

(2) Corrective *shut-down maintenance*—work which can only be done when the facility is, or is taken, out of service.

(3) Corrective *breakdown maintenance*—work which is carried out after a failure, but for which advance provision has been made, in the form of spares, materials, labour and equipment.

The terms 'emergency maintenance', 'condition-based maintenance' and 'scheduled maintenance' are also used in maintenance practice, and these were identified and defined in chapter 1.

A system of planned maintenance consists of two mutually balanced components—planned preventive maintenance and planned corrective maintenance. Both must be organised with forethought, control and records, but their nature is different. In the case of planned preventive maintenance, each item of work is identified some time before failure or a diminution from an acceptable standard of the facility. Planned corrective maintenance differs in that restoration to the acceptable standard is re-quired, and corrective maintenance must have a prior claim on available resources.

In practice, the most common approach to building maintenance is to wait until a defect is reported to the maintenance organisation by the occupants. Often a better approach would be to adopt a policy of periodic inspection of the property and subsequent rectification of observed defects. Observing and rectifying a defect at an early stage is likely to reduce repair costs. Furthermore a large proportion of maintenance work is identified and grouped at discrete points in time. The maintenance organisation is thus able to allocate its resources and rectify the defects in the most efficient manner. In many large building complexes the majority of maintenance repairs are single trade repairs, such as plumbing or joinery work. These trades can operate as autonomous groups within the parent maintenance department and a strong case could be made for inspections on a trade basis, such as plumbing systems every 2 months and slated roofs once a year.

Maintenance programming should ideally be preventive as far as practicable, based on regular inspection at intervals designed to prevent trouble from developing or accumulating. Admittedly not all building defects can be prevented but many can and others will be rectified before they become more expensive. The frequency of inspection is the crux of the matter so that the right balance is struck between the cost of inspection and prevention on the one hand and expenditure on repairs on the other. Dramatic examples can be given of expensive repairs which could by relatively small outlay have been avoided, such as a major outbreak of dry rot stemming from a single long-neglected leaking rainwater pipe. It is not possible to over-emphasise the high cost of the consequences of neglect of historic buildings, not only in hard-to-raise funds, but also in the irretrievable loss of our heritage of ancient materials, craftsmanship and even entire buildings.

Having decided a maintenance policy, the next step is to prepare a maintenance programme. There may be a need to deal with a backlog of general disrepair; to plan major restoration works some years ahead; to deal with year to year painting, decoration and associated repairs; and to operate a system of regular inspection and minor repairs. Very often it entails programmed maintenance within a restricted budget—deploying scarce resources to satisfy many demands—entailing a professional assessment of the overall situation, inspection of specific problem areas, formulation of general strategy for containing or removing critical problems and for reducing the breakdown aspect of maintenance to an acceptable level.

Inspection Cycles

Inspection cycles are an important component of an efficient maintenance service. The facilities which are most subject to wear and tear are the services which contain parts that are affected by friction, heat or dynamic stress. Suppliers normally prescribe inspection cycles for these items which are kept within anticipated endurance limits. Planned maintenance of services is essential to avoid inconvenient and often costly failures. Similarly the fabric of a building must also be inspected at regular intervals and this can be related to the endurance characteristics of a significant component or material. For example, external painted surfaces normally last for about 5 years without attention, although there are variations due to geographical location and the degree of atmospheric pollution, and this could provide the basis for the cyclic inspection of the fabric. Buildings cannot however be left uninspected for 5 years and interim examinations should be carried out at intervals of not less than 12 months. The aim of the intermediate inspection should be to detect defects which would result in progressive deterioration if left unattended until the next cyclic inspection. The linking of inspections of local authority houses with external painting cycles was advocated by the Local Government Operational Research Unit,[13] although the Unit also examined the purpose of inspections—whether they should be confined to individual elements in a large number of houses or to all elements in particular houses, either of the same age and type or on the same estate.

A study of hospital maintenance[14] examined a variety of approaches ranging from *ad hoc* inspections to annual inspections and different inspection frequencies for different elements. Having regard to the dispersed locations of hospitals, the study recommended 2-yearly regular inspections together with interim inspections where necessary. On inspection it was suggested that the following information should be recorded on standard inspection forms and this information provides the starting point for the next inspection:

(1) locality and identity of elements;
(2) type of work (for example, patch or replace);
(3) extent of work (such as area involved);
(4) estimated total cost;
(5) estimated year of treatment.

Stevens[15] devised a classification system for assessing building condition to secure uniformity of approach. He prescribed five categories of condition ranging from class 1 (very good) to class 5 (dangerous). In like manner the levels of maintenance to be achieved will be influenced by the type of building and the use of the part under consideration, and these can also be assessed on a five-point system ranging from level 1 (very high; applicable to board rooms and operating theatres) to class 5 (very low; prior to demolition).

Programming Maintenance

To prepare a programme it is necessary to assess the general condition of the buildings, services and external works and to consider these against the criteria currently adopted. The repair and replacement work is costed and priorities established having regard to any cyclic arrangements.

Maintenance of the very large Shell Centre on London's South Bank is based on a 9-year cycle and covers fabric maintenance, redecoration, major cleaning, and maintenance of plant, services, fixtures and fittings. The building and site are divided into nine zones for maintenance purposes. Workload is controlled through a card index system which permits levelling out work peaks, control of costs against budgets and the forecasting of areas where reductions in the maintenance budget can be accommodated in the short term if spending has to be curtailed. Based on experience the programme adopted is a 9-year redecorating cycle, 3-year cleaning of offices and an 18-monthly cleaning of walls in corridors.

Painting

Painting embraces a variety of activities including washing down; washing down and touching up; washing down and applying one coat of paint; and washing down and applying one undercoat and one finishing coat. Paintwork should be inspected to verify the need to repaint as programmed; for example, the exterior may last 6 years instead of 5 and exposed elevations may require more frequent decoration. Hence on large buildings or groups of buildings some maintenance personnel advocate painting by elevations rather than buildings, although it results in fragmentation of the maintenance work.

If the painting of external woodwork is neglected, the timber may decay prematurely. The effectiveness of paintwork on external joinery is dependent on a wide range of factors—painting cycle; age of building; quality of timber, joinery, paint and painter's work; system of painting; use of building; local climatic conditions; design of building; and degree of exposure. The greatest rate of deterioration occurs on south and west elevations to sills and bottom rails of both upper and lower sashes. There is no universal economic optimal painting cycle.

The former Greater London Council formulated the following effective painting programme for housing estates.

External. 5-year cycle with new property painted in third or fourth year if found necessary; short life property to be painted if life is 2 years or more and it has not been painted for 4 years; extensive repairs prior to painting are to be executed 1 year ahead of the painting programme and minor repairs completed 3 months before painting commences. The author[16] identified the need to repaint externally within two to three years of the initial painting.

Internal. Normally every 5 years.

Choosing the correct paint for the project is vitally important. The difference in initial cost between a satisfactory material and a cheap paint is insignificant compared with the additional labour and disruption costs in having to repaint 1 year earlier, together with possible deterioration of the base material in the meantime.

It is generally necessary to schedule painting of factories to avoid interference with production and this often entails doing the painting during the 2-week annual shut-down period, with any unfinished work completed during evenings and weekends. The problems can be reduced by the use of modern paint systems and methods of application.

In aggressive areas it may be better not to use cycles based on complete repainting but to spot prime and apply one finishing coat as and when necessary. For instance one large factory chimney is repainted every 2 years at a cost of about £3000, whereas if painted on a 5-year cycle the cost of repainting and disruption of production would be in excess of £30 000.[16]

The Scottish Special Housing Association has linked the general maintenance of buildings to a five-year painting maintenance cycle in a very effective way, with plumbing and sanitary fittings in year 1, specialised services and painting defective work in year 2, painting guarantee check and gas and electrical services in year 3, detailed inspection of building fabric, including windows and doors, in year 4, and check on items to be repainted and inspection of external services in year 5.[18] Another alternative approach to the preparation of a 5-year planned maintenance programme is shown in Table 12.1.

Cleaning

The cost of cleaning and maintaining a normal building over a 20 to 30-year period can equal the initial cost of the building. Cleaning can in fact be regarded as part of maintenance in that in part it is a preventive and protective activity. For example, regular washing down of painted wall surfaces reduces subsequent painting costs.

Programming of cleaning work should be preceded by a survey of the various floor and wall finishes with their respective areas and uses. Uses have a bearing on cleaning frequency as for example windows in offices, display rooms and dining areas are usually washed more frequently than those in storage and factory production areas. Generally, 40 to 45 per cent of all cleaning time is devoted to floor surfaces. Work sheets are compiled for daily, weekly, monthly, quarterly and annual operations and these will provide the basis for estimating labour requirements.

Cleaning equipment and products must be carefully selected to secure the best results at an advantageous price. The choice is influenced by such factors as standard of appearance, amount of wear, degree of soiling, type of equipment available and quality of cleaning staff. It is good policy to minimise the number or type of products in use, and the cheapest product is not necessarily the most economical. Supervision and inspection of cleaning work is vitally important, and cleaning equipment must be properly maintained.

Engineering Services

In the planned maintenance of engineering services in buildings, it is vital to define clearly the tasks to be performed and then to programme the work in a logical manner. If properly planned, malfunctioning components will be detected at an early stage, enabling simple corrective action to be taken, rather than waiting for a complete failure and then being faced with an expensive replacement; in addition productivity will be increased by the reduction in travelling and waiting time. Kelly[19] has described how condition monitoring can be applied in three ways: simple inspection; condition checking; and trend monitoring. The frequency of monitoring will depend on the deterioration characteristics of the item and the costs involved.

Table 12.1 5-year planned maintenance programme

YEARS		1	2	3	4	5
EXTERNAL FABRIC	SURVEY					- - - →
	WORK EXECUTION	— →				
PRE-PAINTING MAINTENANCE	SURVEY					
	WORK EXECUTION	— →				
EXTERNAL DECORATION	SURVEY	- - →				
	WORK EXECUTION		— →			
EXTERNAL WORKS	SURVEY	- - →				
	WORK EXECUTION		— →			
PLUMBING	SURVEY		- - →			
	WORK EXECUTION			— →		
INTERNAL FABRIC	SURVEY			- - →		
	WORK EXECUTION				— →	
INTERNAL FINISHES & FITTINGS	SURVEY			- - →		
	WORK EXECUTION				— →	
HEATING INSTALLATION	SURVEY				- - →	
	WORK EXECUTION					— →
ELECTRICAL INSTALLATION	SURVEY				- - →	
	WORK EXECUTION					— →

△
└── SURVEY OF FABRIC & PRE-PAINTING MAINTENANCE HERE PRIOR TO SCHEME'S INTRODUCTION

Furthermore, building owners are also requiring a good service and a better environment.[1]

The majority of engineering planned maintenance schemes have tended to concentrate solely on planning the activities of craft operatives to carry out predetermined tasks at regular frequencies, and this often falls short of the ideal. The operative is inevitably limited in his ability to inspect installations for correct operation. For example, all the component parts of an air conditioning plant may appear to operate satisfactorily but unless the airflows, temperatures and humidity are measured and checked against the design one cannot be certain that the plant's performance is satisfactory. Similar checks should be carried out on boiler efficiency, water treatment, electrical insulation and earthing, heating systems and domestic hot water calorifiers.[17]

Another major shortcoming of frequency-based maintenance for engineering services is the large number of inspections that results. Most of the maintenance tasks are better dealt with by periodic inspections carefully planned and carried out by competent persons, followed by specific corrective action. The person inspecting equipment should be supplied with full diagrams and operating data, any instruments required and information on their method of use. Some preparatory work such as draining down, removing lagging and opening up may be required and this should all be pre-planned.[17]

For example, the method of controlled maintenance adopted by the British Steel Corporation at a major plant resulted in large savings in operating and maintenance costs. Lubrication tasks are carried out on a frequency basis but otherwise all maintenance is done on request only. The plant is inspected regularly and any faults found are assessed for urgency. Some will require immediate attention if only to effect a temporary repair. In these cases the inspector raises a requisition and work is put in hand straight away. In most cases, however, the work requires planning and if detected early enough can wait for a convenient shut-down. The work is planned, spare parts ordered and labour and time requirements assessed, and then all this information is stored in a computer memory until the appropriate time. When convenient the plant is shut down and the carefully pre-planned remedial work is carried out in the shortest possible time. A considerable reduction in the spares inventory has been possible owing to the decreased likelihood of an unexpected breakdown.[17]

Edwards[1] has described how a diverse estate of offices in several central London buildings under individual manual plant control was equipped with a remote control system designed and engineered for the specific purpose required, with greatly improved operation and halving of the manpower requirements.

Computerised Maintenance

Following technical inspections and completion of standard inspection reports, the appropriate data can be fed into the computer which can produce detailed orders based on standard specification items, and these can be incorporated in contract documents if required. As an extension of this process, the computer can print out complete and ready priced maintenance schedules based on the inspection report and schedules of prices submitted by contractors. Responsibility for ordering minor every-day maintenance work can be delegated to caretaking staff, provided the extent of the delegation is clearly defined and the work restricted to certain contractors. Accounts can be checked for accuracy and paid by the computer.

The maintenance manager must keep detailed records of maintenance expenditure and must know what each unit costs to maintain. To establish control figures, maintenance and running costs of each individual building will be recorded and analysed to show costs by age group of property, type of construction, and location and user. Comparison of expenditure with control figures will indicate where further investigations are needed to determine whether a property is being over or under maintained, is being misused, let at an uneconomical rent or is suffering from design deficiencies. These processes could be carried out laboriously by clerical staff but the computer is able to receive, process and print out this information quickly and accurately.

The advantages of computer programming are numerous and include the following:

(1) reduces clerical staff and management costs;
(2) lessens risk of human error and increases reliability;
(3) provides a check on work done;
(4) permits instant updating;
(5) capable of ready adaptation to suit alterations to buildings and additions and variations in estates;
(6) can incorporate various management and administrative functions;
(7) can provide management with cost and other supporting data.

Elliott[20] has shown how computerisation will deal effectively with the provision of information, sorting and quantitative and chronological analysis, but it will not solve policy and people problems. However, accurate information and effective channels of communication can result in the formulation of sound policies and good performance.

The following example helps to illustrate the practical benefits to be gained from the use of a computer. On receipt of the relevant information the computer can be used to perform the following operations relating to the production of works orders with speed and efficiency:

(1) immediate display of all work outstanding, in progress or completed, to avoid duplication;
(2) provision of details of cyclical maintenance such as repainting and rewiring that may affect the response;
(3) provision of information from past records to identify trends in maintenance/repair works and patterns of failures;
(4) identification and arrangements for tasks requiring pre-inspection and appointments, to validate accuracy of original work descriptions;
(5) production of works orders with full descriptions of requirements from basic codes as predetermined and access to schedules of rates;

(6) automatic up-date of records of committed expenditure to assist with budgeting and cost control.

Minicomputers or even microcomputers are likely to have the capacity necessary for a maintenance/repairs system, but investigation of available software should precede any consideration of hardware. Once the software is agreed then computer suppliers can be invited to submit quotations.

There is a growing range of software available, but often coupled as packages with particular types of computer. Software specifications should be checked against the defined maintenance/repair requirements, practical demonstrations held, and existing users contacted to gain knowledge of how well their systems perform in practice and how well needs are satisfied.

Pettitt[21] has provided guidance on the design and operation of a simple computer system and described the scope of a comprehensive system. He has also given a useful code specification of building maintenance operations.

Recording and Dissemination of Maintenance Data

Information Systems

Information and information flows are normally justified by their relevance to decision-taking and the action which has to be initiated. However, the general view of many concerned with building maintenance is that there is already too much information in the form of Codes of Practice, British Standards, Building Regulations, manufacturers' catalogues, research reports, specifications and drawings. An individual is primarily interested in the information that affects his area of decision-taking—he needs to know the source of information, what decisions have to be made and what action initiated. To be effective, information has to be collected, collated, presented, and be easily retrieved and capable of direct application in problem-solving and decision-taking. Designers can benefit from having more knowledge of how structures operate, useful lives of buildings and components, and how much is spent on their maintenance and operating costs. Useful sources of information are DOE (construction feedback series), BRE (digests and other publications) and BMI (Building Maintenance Information Ltd).

Building Records

The first requirement of a maintenance or property manager is to know in detail what he is managing. Without this basic knowledge he will not be in a position to decide his maintenance policy or to prepare estimates of expenditure which go to form his budget. The information he needs includes the geographical location of each property, the constructional details by elements, age and condition, details of services, the superficial area and cubic content, the accommodation available, the current user and any proposals for the area by the local authority which might affect the property.[22]

There are various ways of recording this information ranging from the simple card index to the computer, and the method chosen will depend on the size and resources of the estate. The method of recording is however of secondary importance; the prime requirement is for accurate and relevant information to be available as quickly as possible. It is likely that in the foreseeable future the development of computers will make it possible for the manager of even a small estate to have access to data through his own minicomputer or microcomputer.

There is a risk of accumulating masses of data which can be expensive and of only limited practical value. The maintenance manager must retain a sense of proportion and recognise the limitations as well as the benefits of recorded data. Records cannot for instance detect the leaking drain caused by minor subsidence or indicate the untrue ring of a timber floor affected by rot.

Maintenance Cost Records

There is a vital need for cost information for both overall budgetary control and for making day to day management decisions. There is a need for more information on the consequences of design decisions as they affect maintenance costs and insufficient feedback of maintenance cost data.

Maintenance Cost Records may be kept to fulfil three separate functions.

(1) Budgetary control—to produce the annual or other periodic sum which needs to be set aside to provide for maintenance and operating services.

(2) Management control—to permit the day to day control over maintenance expenditure.

(3) Design cost control—to provide full information concerning causes of failure, types of failure, design faults and similar particulars. Records of roof repairs are of little use unless, for instance, they show the type of tile, quality of tile, method of laying, angle of pitch and degree of exposure.

Some have advocated a hierarchical system of grouping or classification of items for recording maintenance work and costs. A primary grouping would be of value to organisations who do not wish to prepare very detailed records, and would also form the basis for more detailed recording. One possible primary grouping could be as follows:

(1) external painting
(2) internal decorations
(3) main structure
(4) internal structure
(5) finishes and fittings (including kitchen equipment)
(6) plumbing and sanitary services
(7) mechanical services, including heating, ventilating and gas installations

(8) electrical services
(9) external works
(10) miscellaneous and ancillary works.

A secondary grouping of plumbing and sanitary services could be hot water and heating services, and a tertiary grouping of this element could embrace pipes of different materials, and valves, tanks and cylinders of different materials.

In 1970 the Department of Health and Social Security requested hospital authorities to introduce an improved method of recording and analysing expenditure on hospital building and engineering maintenance by the use of fourteen primary and fifty secondary codes.[6]

A detailed coding system for building maintenance has been developed by Holmes[23] with job codes made up of a comprehensive elemental code and a process code. The elemental code consists of a main elemental code and up to five subelements, in a hierarchical structure. For example, 3 is structure; 34 is roofs; 342 is roof covering; 3422 is a flat roof; and 34221 is an asphalt covered flat roof. The process code includes such items as renew, remove, refix, and ease and adjust. Additional codes were developed to further define the specific items. These consisted of a descriptive code to identify materials and components; a block code and a location code to identify locations at different levels; and a reason code to encompass the type of maintenance and the nature and/or cause of the defect.

This coding system provides an effective coding frame for all types of building. Subsequently, two case studies were presented to illustrate the use of the code,[24] and they do show the advantages to be gained by the use of a universal but flexible system of recording maintenance cost data for comparative, analytical and monitoring purposes.

There is a wide variation in the method of recording maintenance costs as betwen different organisations. Some organisations using directly employed labour will include overheads and management costs in varying proportions. These variations in costing procedures make for difficulties when comparing maintenance costs.

The breakdown of expenditure into elements will isolate expensive items which can then be investigated. Where expenditure reveals the failure or short life of an item, an alternative replacement should be considered. The analysis of expenditure may also be subdivided according to use and cause of failure under such headings as vandalism, boisterous use, fair wear and tear, design failure, permanent failure and poor workmanship. This information should also be fed back to the design team to secure improved future designs and consequent lower maintenance costs.

Holmes and Marvin[25] have advocated that any statement of costs and trends will be more meaningful if the basic cost data used as evidence were supplemented by details of the factors which influenced policy within the hierarchical structure, and the extent of discretion exercised by the decision makers. A further policy variation which will affect maintenance costs is the extent of self-help maintenance, whereby tenants often undertake small repairs themselves. Additionally, tenants of some local authorities are permitted to make certain improvements to their homes, such as installing central heating or modern kitchen fitments, with subsequent maintenance implications for the authority on a change of tenancy. It has been suggested that it may be possible to construct a discretion model, in terms of policy weightings, which could be applied to the annual maintenance costs at the same time as applying cost indices.[25]

Need for Clearing House

A vast amount of information flows from many different sources to the designer and his often limited technical library facilities and he tends to become the human computer which sifts, selects, analyses and co-ordinates this information for design purposes. A central clearing house can perform this task far more effectively, as the information needs constantly to be updated, catalogued and cross referenced, supplemented by graphic details and costs recorded on standard forms. Such a centre can disseminate information on performance of materials, formulate standard classification routines for maintenance cost analysis, receive and analyse costs and circulate results. The RICS Building Maintenance Information Ltd (BMI) aims to provide this service.

Building Maintenance Information Ltd (BMI)

Building Maintenance Information Ltd (formerly BMCIS Ltd) is administered by the Royal Institution of Chartered Surveyors to help maintenance managers and property administrators in the maintenance of property, and the service relies upon their support in providing information and in assisting with feedback. BMI supplies general information on cost indices, labour costs, materials and equipment, control techniques, legislation and statistics, together with digests of articles of relevance to property managers.

Feedback of technical information from properties in use to the design team occurs through design/performance data sheets of building failures, which describe the element or component, forecast the likely long-term effect, analyse the cause of failure, and suggest any design or construction correction which might have avoided the situation and the necessary remedial action. Case studies consist of factual reports of typical situations and exemplify current operative practice in both public and private sectors. They can reveal strengths and weaknesses in an organisation and a commentary on each case illustrates the lessons to be learnt.

Probably the most valuable part of BMI is the occupancy cost analyses covering a range of property types. Annual property costs are analysed on the basis of standard instructions and elements on a unit area basis, and these form part of a technical/financial budgetary control system. The analyses provide a unique library of property occupancy data on organisations and costs. Each analysis contains details of the building and a note of the managerial criteria which decide maintenance policy. Examples of BMI occupancy cost analyses are given in appendixes 2 and 3, covering a laboratory and halls of residence. The importance of monitoring and costing energy usage is illustrated in the BMI energy cost analysis of a hospital shown in appendix 4.

BMI centres its operations on the regular distribution of up-to-date data which, when filed in loose leaf binders, provide a comprehensive library of information and costs concerning property occupancy. It collects and disseminates information from many sources but to a large extent the data is derived from the records, operating conditions and experience of its own subscribers. The service encourages better communications between upper management and maintenance management and also between property occupiers and the design team. The service is non-profit making and anyone can become a subscriber.

Maintenance Feedback

Maintenance feedback should be an essential part of any maintenance administration. Feedback may be mainly injected into the system in two ways:

(1) directly to the design team; particularly information on design faults, faulty workmanship and materials failures;

(2) by general discussion within the maintenance team, when solutions to problems should be documented and passed on to all appropriate personnel.

A visual representation of feedback is illustrated in figure 12.1, and this shows some of the major stages in the operation of a maintenance scheme.

(1) management organisation of resources;

(2) work execution;

(3) appraisal of results; and

(4) corrective action through feedback to design and management teams.

To assist in the feedback of information, site defects are suitably recorded showing the symptoms, diagnosis, prognosis (projection of defect performance in time), and the agreed remedy.

Skinner and Kroll[26] believe that through a thorough analysis of past maintenance data, the components in each project need to be identified, together with the nature of the work undertaken, and preferably the position in the building and the cause of failure. Hence, an adequate coding system is required, simple enough for every day use with every job coded.

There is unlikely to be much benefit derived from the detailed feedback on an individual small building, whereas a large stock of buildings could produce meaningful data over a relatively short period. This, together with the fine breakdown required normally necessitates computerised analysis.

For feedback purposes, the information has to be retained after the point where its usefulness for day to day management has ceased, and it should be

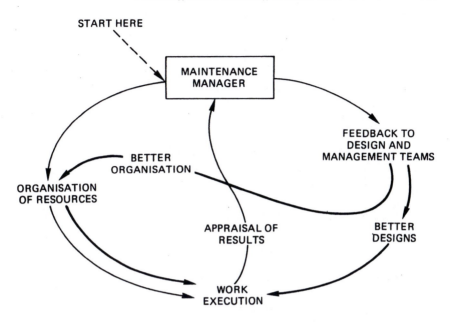

Figure 12.1 Feedback in maintenance supervision

coded on an elemental basis. Facilities for the analysis of work and consequent feedback are normally included in the computer programs of the larger maintenance organisations.[26]

The Princes Risborough Laboratory[27] has found that:

(1) stored maintenance information is sparse and rarely goes back more than a few years;

(2) it is seldom in the form required and often inextricably agglomerated;

(3) it is very time-consuming to retrieve; and

(4) results are highly dependent on specific local factors of specification, design and workmanship.

Furthermore, Holmes[27] found that few local authorities know how much different dwelling types are costing; therefore to budget for maintenance on particular elements the authorities had to depend on site feedback, usually in the form of estimates from area managers. Authorities were unable to plot costs or indices of failure for individual house types, and maintenance budgets were compiled from experience rather than from an analysis of the work needing to be done.

In general terms, maintenance costs rise with age; during the first twenty years costs rise fairly steeply and then settle down to a more gradual increase for the rest of the building life. However, the age/cost profile is not a simple one, and there are a number of factors which affect the shape of the profile, such as policy and type of construction.

Some aspects of dwelling design, in terms of maintenance, are critical, such as flat roofs finished with lightweight membranes and large areas of painted woodwork. It is more difficult to pin-point many of the problems with smaller components. Government pressure on local authorities in the 1980s to provide financial statements of maintenance expenditure could lead to improved maintenance systems and, hopefully, to the production of data on component lives and costs.

Current analyses show that a high percentage of repairs to finishes and fixings result from work to doors and windows. Finishes account for approximately 27 to 30 per cent of maintenance work, plumbing for about 25 per cent and heating and lighting for around 20 per cent. Defective ball valves were the most common plumbing problem. More attention needs to be paid to the design and treatment of external doors and windows with a view to extending their lives.

It is probably unwise for designers to judge a component or system on the basis of annual maintenance costs. Holmes[27] suggests that the ratio of capitalised maintenance costs to initial costs provide a better guide. The main problem is that few designers receive analytical data (cost and performance) from maintenance departments. Most of the feedback to designers appears to come from the results of their own activities or through the experience of other designers, rather than from systematic records of behaviour maintained by the user. There should be a more analytical approach to feedback, which has become increasingly important with the greater use of new materials and components, otherwise their cost benefits may never be soundly evaluated.

Smith[28] has provided extensive information on the estate data base and performance monitoring system employed by the National Health Service.

Maintenance Manuals

Need for Manuals

Most buildings, from the dwelling house to the prestige office block, have increased in complexity in recent years. An increasing number of domestic buildings are being provided with thermostatically controlled heating and hot water systems, while large office blocks will probably be equipped with high-duty heating and air conditioning systems, lifts and other items of sophisticated plant and equipment, all requiring regular and specialist inspection and servicing. The form of construction may involve new techniques for which conventional maintenance practices may be unsuitable. Although the designer may be aware of servicing requirements, these are rarely communicated to those responsible for maintenance of the property. A convenient form of communication is the maintenance manual which should provide clearly and concisely all the information needed to maintain and operate the building satisfactorily.[29]

Maintenance manuals will facilitate building maintenance and there is a growing awareness of the need for them to be prepared for new buildings by the design team, as part of the building contract at an appropriate fee, possibly in the order of 0.5 per cent of the contract sum. A properly prepared and comprehensive manual will serve three principal functions.

(1) It will enable the property manager to organise the repair and maintenance of the building, its services and surrounds effectively and economically.

(2) It will enable the occupier to clean the building and operate its services efficiently, thus reducing loss of time and production.

(3) It will establish a link between the project design team and the client and his maintenance organisation to their mutual benefit.[29]

Contents of Manuals

It has been suggested that a maintenance manual should consist of three basic parts.

(1) A physical record of the building and site, to include materials, services, superficial area and cubic content, all in sufficient detail to assist the manager in looking after property efficiently.

(2) Time-based inspection and maintenance cycles for the various elements, including services, together with detailed check lists, maintenance schedules for engineering services and list of specialist sub-contractors and suppliers.

(3) Information and instructions on maintenance delegated to the occupier.[22]

The type and amount of information in a maintenance manual will vary with the nature and complexity of the property and its services. The detailed contents could include the following matters.

(1) *Contract and legal particulars*, including the design team, contractor and sub-contractors, nature of tenure and details of any easements or other restrictive covenants, contract particulars and statutory consents, and any operative guarantees.

(2) *Housekeeping*—details of surface finishes and decorations both internally and externally, with information concerning cleaning and periodical routine maintenance; this is best prepared in schedule form set against a time scale for the operations or inspections.

(3) *Operation of plant*—means of operating mechanical, electrical or solid fuel plant or fittings, with details of requisite periodical routine maintenance or servicing; the location of meters, recording devices, stop-cocks, valves and the like should be recorded.

(4) *Maintenance and repairs*—full details of materials, components and constructional processes should be given, preferably on an elemental basis in the form given in appendix 5. All hidden features should be described and special items noted such as jointing and replacement techniques, method of fixing cladding components to structure and form of repair or replacement, and the method of dismantling and re-erecting demountable components.

The names and addresses of sub-contractors or suppliers of all fittings, components and plant installed in the initial project shall be given together with catalogue numbers, colours and other relevant information which may ease the task of ordering replacements or spare parts.

(5) *Record of maintenance executed*—provision for a maintenance log to permit constant updating and the inclusion of any changes or additions.

(6) *Plans and drawings*—plans of each floor to a small scale with permissible floor loadings and usable areas, all as built, on size A4 or A3 sheets for convenience, and relevant service layouts.

(7) *Emergency information*—names, addresses and telephone numbers of contacts in the event of fire, theft or burglary; gas, electricity or water failures or leaks; or failure/breakdown of plant; together with location of appropriate equipment.

(8) *Manufacturers' leaflets*—these should be incorporated to give an 'after sales' service, with technical information on cleaning, operating, maintenance and repairing products.[29]

Many local authorities provide their house tenants with a maintenance manual giving information on the cleaning and operating of the dwelling, its services and equipment, together with the locations of stopcocks, valves and other fittings, and any restrictions imposed on the tenant by the nature of the construction, regarding the fixing of any fittings or execution of any alterations or adaptations. The inclusion of emergency information is essential.[29]

The RICS has issued a set of guidance notes for persons preparing manuals, titled *Building Management Manuals*.

Types of Manual

Carnwath[30] has described the criteria which he established when preparing a maintenance manual—absolute simplicity of identification; minimum of coding and maximum of easily identifiable abbreviations; special drawings to show drainage, buried service runs, floor loadings, fire alarm shut off points and like features; and a general attempt to make the building as easy to run and the manual as easy to read as possible. His manual was divided into three volumes as follows.

(1) Building specification with key drawings, names and addresses, paint colours, materials, plant type numbers, light bulb wattages, and the like, to assist anyone requiring detailed information quickly or to replace damaged items.

(2) Maintenance references, describing all cleaning and maintenance requirements and frequencies, and giving areas of particular materials and numbers of plant and equipment. It assists the maintenance manager in planning his workload and contains particulars to send to cleaning contractors as a basis for their quotations.

(3) Current and forecast budgets and programmed replacement dates for plant.

All volumes are contained in loose leaf folders and the maintenance manager is encouraged to update it by inserting comments on performance, revising specifications as new items are fitted and modifying maintenance procedures and frequencies.

Typical manual schedules follow in tables 12.2 and 12.3.

Table 12.2 Maintenance manual materials schedule

Element Number—2 (floors/ceilings)

Item Nr	Item	Location	Description and comments	Manufacturer or supplier
2/1	Acoustic tiles	Basement access corridor	Gold bond fissured solitude tiles, 300 mm square, self-finish on Anderson's 'J-type' suspension, with white semi-gloss stove enamelled edge trim	Supplied and fixed by Anderson Construction, Twickenham
2/2	Asphalt floor	Tank room	Includes skirting 300 mm high	
2/3	Carpet tiles	Offices	'Debron' $^1/_2$ metre square Stipple range, colour 103 'brindle brown'. Arborite skirting.	Supplied and laid by Carpet Tile Co. Berkhamsted. Maker—Carpet Manufacturing Co.

Another form of maintenance manual adopted by the Clwyd and Deeside Hospital Management Committee, contained the following data:

(1) brief history

(2) short specification outlining system of construction, principal materials and finishes;

(3) floor plans as a minimum with sections and elevations wherever possible;

(4) description of renovations, extensions, adaptations and major repairs;

(5) graph showing maintenance expenditure from 1960 to date with provision for continuation for further ten years for the particular hospital and average of all Welsh hospitals, and material and labour costs;

(6) estimate of annual maintenance costs for next ten years subdivided into (i) larger repairs and infrequent items; (ii) periodic repainting, wall washing and road surfacing; and (iii) minor day to day repairs;

(7) schedule of statistics—number of beds and net cubic capacity provide two standards for assessing performance;

(8) schedule of measurements for external repainting;

(9) room schedule.

Table 12.3 Maintenance manual cleaning schedule

Element Number—2 (floors/ceilings)

Item Nr	Item	Maintenance specification	Quantity	Location	Maintenance frequency						
					Twice daily	Day	Week	One month	Three months	Six months	One year or longer
2/1	Acoustic tiles	Vacuum clean with brush attachment	210 m²	Basement access corridor							Vac. once year
2/2	Asphalt floor and skirting	Wash with clean water and wipe dry	25 m²	Tank room		Vac.				Wash	
2/3	Carpet tile	Vacuum shampoo clean with proprietary carpet shampoo and report wear	4900 m²	Offices					Clean		

Other Matters

Information should be provided on methods of jointing and techniques for repair and replacement of units, especially those likely to be the subject of maintenance work. For instance, information could be included on the fixing of cladding panels and how they may be removed and replaced without damage and methods of weatherproofing joints between units. The manual should enable faults to be analysed quickly with a minimum of preliminary investigation, and to permit rectification without causing further damage.

All finishes should be cleaned in accordance with the manufacturers' instructions, as failure to do this may reduce their effective lives. For example, vinyl floor finishes should normally not be cleaned with strong detergents, wax polishes or scouring powders; in like manner steel wool, strong acids and alkalis and abrasive cleaners should not be used on aluminium window frames. The success of a manual is to a considerable extent dependent upon manufacturers providing suitable literature prepared as an after sales service to consumers, giving the means of cleaning, operating and servicing their products.

Costs in Use/Life Cycle Costing

When designing a building the design team should ideally consider the total costs or life cycle costing of alternative designs, embracing both initial and future costs. The costs in use approach enables the way in which a building functions to be expressed in terms of the costs of renewing and repairing the fabric and fittings, of heating, lighting and fire insurance, and of the labour needed to operate the building. The life cycle costing approach is extended to include rates, insurances and other charges, and energy costs feature very highly in the life cycle cost analyses. These costs can be added to the amortised initial cost of the building to give the annual cost in use. The use of this technique makes it possible to combine all the costs of the building and so enables the vast range of factors on which judgement is necessary to be reduced to a comparison of a single cost,[31] as described in chapter 1. A useful guide to life cycle costing is provided by Flanagan and Norman.[32]

Effect of Taxation and Insurance

The incidence of taxation can have a considerable effect on the design economics of buildings. For instance, with industrial buildings some relief can be obtained on the initial cost, through depreciation allowances, investment, initial and cash allowances, their actual form and impact varying from time to time. Amounts spent on maintenance and repairs, heating and lighting, and other running expenses, are classified as business expenses and are deductible from profits in the case of all types of buildings. The exact incidence of taxation varies with the circumstances of the taxpayer. Public authorities do not normally pay tax and so are not affected. Current regulations and levels of taxation tend to favour alternatives with low construction costs and high running costs. The total costs of buildings can thus be influenced considerably by the form of taxation. For example, valued added tax partially offsets the tax advantage previously accruing to running costs, as new construction work was still zero rated in 1987.[31]

There is thus a wide variation of fiscal relief against building expenditure, ranging from the total absence of relief against investment in commercial or residential property, through the general run of investment in industrial property, to the favoured case of a building treated as plant for tax purposes and situated in a development area. The case has often been argued that while maintenance expenditure is wholly allowable against liability to tax, and capital expenditure, subject to the incidence of grants and allowances, is not allowable, then a given volume of maintenance work must be less expensive to the property owner than a corresponding volume of new construction; hence building expenditure is liable to bias against new construction in favour of maintenance, even when maintenance would otherwise be uneconomic. If this is so, the demand for maintenance is increased at the expense of demand for new construction, which would put the same volume of physical resources to more productive and less labour-intensive use. Maintenance-saving investment is also stifled; the use of buildings is prolonged beyond their natural life, existing buildings are put to uneconomic uses, and the quality of the environment deteriorates.[31]

There can however be alternative explanations. Despite every fiscal inducement, it remains genuinely uneconomic to retire apparently obsolete buildings; fiscal policy has yet to realise the full development potential of land; and finally there may be an innate resistance among businessmen to investment in construction unless necessary, or for prestige.

It is also argued that fiscal considerations often have but a marginal influence on investment decisions in new building, and that the harmful effects of fiscal discrimination between new construction and maintenance may thus be less prevalent than is supposed. Furthermore, the argument described earlier is conceptually in error since money saved by building more cheaply initially would be invested elsewhere to produce at least an equal return and consequentially equal tax liability.[31]

The design and layout of buildings may also influence rating valuations and premiums payable for fire insurance. For industrial buildings, floor space is rated according to the level of amenities provided; thus upper floors, and areas which are unheated, have low storey heights or can only carry low loads will be assessed at lower rates. However, an attempt to reduce rateable value by lowering standards may adversely affect efficiency and flexibility. Fire insurance premiums are related to the degree of fire risk and reductions in premiums may be made for design features which are likely to reduce fire spread, such as the use of non-flammable materials or those which resist the spread of fire, and the provision of fire-fighting equipment like sprinklers. It may not pay to install sprinklers or automatic fire alarms where the annual equivalent cost of provision and maintenance is greater than the reductions in fire insurance premiums.[31]

References

1 E. J. Gibson (Ed.). *Developments in Building Maintenance—1.* Applied Science Publishers (1979)
2 Building Maintenance Cost Information Service (BMI). *Maintenance Management of Local Authority Housing* (1973)
3 Chartered Institute of Building. *Managing Building Maintenance* (1985)
4 R. N. B. Smith. *Condition Appraisals and Their Use.* CIOB Technical Information Service No. 52 (1985)
5 B. A. Speight. Formulating maintenance policy. *Chartered Surveyor* (April 1970).
6 H. R. P. Gregson. Putting maintenance into perspective. *DOE Fourth National Building Maintenance Conference.* HMSO (1973)
7 R. J. Bushell. *Assessing Maintenance Priorities—Guidelines based on Health Service Experience.* CIOB Maintenance Information Service No. 17 (1981)
8 R. G. Howell. Making the right decision. *DOE Third National Building Maintenance Conference.* HMSO (1971)
9 DES Architects and Building Group. *Design Note 40: Maintenance and Renewal in Educational Buildings: Needs and Priorities* (1985)
10 Committee on Hospital Building Maintenance. *Report of Committee, 1968–70.* HMSO (1970)
11 Building Maintenance Cost Information Service (BMI). Occasional Paper No. 5. *Budgetary Control and Building Maintenance* (1973)
12 Local Government Operational Research Unit. *Report D9: Building Maintenance How Can OR Help?* HMSO (1972)
13 Local Government Operational Research Unit. *Report C188: Evaluating Alternative Housing Maintenance Strategies: Proposals for a Study to Develop a New Method.* HMSO (1973)
14 Local Government Operational Research Unit. *Report C144: Hospital Building Maintenance: Can Decision Making be Improved?* HMSO (1972)
15 R. F. Stevens. *BRE Current Paper 55/74: Maintenance Standards and Costs* (1974)
16 I. H. Seeley. *Blight on Britain's Buildings: A Survey of Paint and Maintenance Practice.* Paintmakers Association (1984)
17 B. P. Holloway. Considering planned maintenance. *DOE Fourth National Building Maintenance Conference.* HMSO (1973)
18 R. D. Milne. *Building Estate Maintenance Administration.* Spon (1985)
19 A. Kelly. *Maintenance Planning and Control.* Butterworth (1984)
20 D. A. Elliott. *Computerisation of Maintenance/Repairs Systems.* CIOB Technical Information Service No. 54 (1985)
21 R. Pettitt. *Computer Aids to Housing Management.* HMSO (1981)
22 Department of the Environment. *Practice in Property Maintenance Management.* HMSO (1970)
23 R. Holmes. *A Coding System for Building Maintenance.* CIOB Technical Information Service No. 47 (1985)

24 R. Holmes. *Maintenance Coding and Monitoring: Two Case Studies*. CIOB Technical Information Service No. 53 (1985)
25 R. Holmes and H. Marvin. Maintenance Costs and Policy.*Housing* (January 1980)
26 N. P. Skinner and M. E. Kroll. *Maintenance Feedback*. CIOB Maintenance Information Service No. 18 (1981)
27 Building Maintenance Cost Information Service (BMI). Occasional Paper No. 123. *Feedback from Housing Maintenance* (1983)
28 R. N. B. Smith. Estate maintenance monitoring and appraisal. *DOE Ninth National Building Maintenance Conference*. HMSO (1986)
29 Department of the Environment. *Maintenance Manuals for Buildings*. HMSO (1970)
30 D. Carnwath. Design responsibility and maintenance manuals. *DOE Third National Building Maintenance Conference*. HMSO (1972)
31 I. H. Seeley. *Building Economics*. Macmillan (1983)
32 R. Flanagan and G. Norman. *Life Cycle Costing for Construction*. RICS (1983)

13 EXECUTION OF MAINTENANCE WORK

Maintenance work can be undertaken by contractors, direct labour organisations or a combination of both systems, and the decision will be based on a number of criteria. The structures of maintenance organisations are examined together with programming and operational activities. Finally the training of maintenance staff and the operation of incentive schemes are considered.

Choice Between Direct and Contract Labour

General Background

Building maintenance is not a single industry and is executed by contractors working for profit and by commercial enterprises and public authorities as an adjunct to other functions. Contractors in fact accounted for less than half of all estimated expenditure on building maintenance in the early 1980s; nevertheless they have in the past constituted the more flexible part in that they are subject to direct economic pressures and can respond fairly readily to changes in the pattern of demand. Direct labour organisations are rather more rigid in structure and the previous absence of a profit motive required the substitution of other forms of motivation and objective. In all cases the services should be effective and the total cost economic.

It is possible that the form of construction, use of the building or maintenance policy may determine whether directly employed labour, or contractors, or both will be the most advantageous. There are no general recommendations that will provide the correct proportions of direct and contract labour for all maintenance departments; each has to be treated separately. In reaching a decision, the maintenance manager in any organisation should compare the costs and services provided by contractors with his own directly employed labour force, taking into account the availability of labour and the type and location of buildings to be maintained. The type of maintenance organisation serving a single large complex of hospital buildings in continuous use will be ill-suited to maintain the dispersed properties of a large county council. Direct labour forces are particularly well placed to cope with emergency repairs to large commercial and public buildings, while small contractors can provide a good service to house-holders.

The cost of directly employed labour is made up of wages and materials; consumable stores; administrative overheads such as labour oncosts, and associated clerical, travelling and supervisory costs; and depot costs. The cost of employing a contractor consists of the contractor's charges plus administrative overheads, such as inviting and comparing tenders, drawing up contracts, work supervision and checking invoices.

Application in Practice

The main advantages claimed for the use of directly employed labour are:

(1) it allows full control of activities of operatives, permitting reasonable flexibility, a more rapid response and direct quality control;

(2) it should ensure a good standard of workmanship by craft operatives who enjoy continuity of employment and are suitably trained, although recruitment may be a problem;

(3) there is a standard complement of labour available;

(4) the maintenance manager can introduce and operate incentive schemes, with resultant improved productivity;

(5) it is particularly well suited for the execution of emergency repairs, as the labour force is familiar with the location of stopcocks, switches, manholes and the like, and for operational services and services requiring particular or unusual skills for which employees can be trained;

(6) there is little opportunity for trade demarcation in small maintenance organisations.

On the other hand the establishment of a direct labour maintenance force will require the provision and administration of supporting facilities such as stores, workshops, transport and accounting services; a high standard of supervision and control to ensure economic programming, and good productivity and quality of work; and experienced and efficient management to provide effective labour relations and communications.[1] With direct labour work the difference between estimated and actual cost can be considerable; a report by one local authority on schemes undertaken by its works department showed that some jobs cost two to three times the original estimate, while claims that direct labour yields a higher standard of work are often difficult to substantiate. Accounting procedures of direct labour organisations have varied considerably in the past and some rationalisation would facilitate statistical analysis to general advantage. Some criticism of local authority direct labour expenditure is founded on the assertion that some costs are hidden and one local authority auditor questioned how it could be established that a direct labour organisation was effectively competing with contractors, when it operated depots for which the costs were unknown. Direct labour had no bad debts, no costs of tendering in competition but, on the other hand, often had a high rate of sickness around 6 per cent, and its higher ratio of staff to operatives could exceed the contractor's profit element.

Contractors play an important part in maintenance work, both for putting buildings in repair and for the larger periodic works. Successful results depend on exacting and well-detailed specifications and close supervision. Special contractors are indispensable for maintenance of lifts and other sophisticated plant and for specialist trades such as asphalt and terrazzo work.[2]

Maintenance work loads tend to fluctuate, particularly with redecorations where external work is seasonal. With certain buildings such as universities, polytechnics and colleges, some work can be undertaken only during vacations, resulting in heavy demands in the summer. Direct labour gangs cannot be built up to cope with such diverse seasonal loads, otherwise they will be underemployed at other times of the year. Milne[3] has emphasised the desirability of keeping the workforce to such a size that full employment can be found for them irrespective of the economic climate, by identifying the essential and irreducable maintenance operational requirements.

Some have argued that it would be better to confine direct labour to little more than emergency and scheduled maintenance, and to use contractors for the seasonal, major and specialist work, although many efficient direct labour organisations would quarrel with this approach. Contractors generally need long-term contracts to give the employer good service on advantageous terms.

Effect of Local Government Planning and Land Act 1980

The Local Government Planning and Land Act 1980 was designed to provide greater flexibility, better accountability and firmer financial disciplines for local government. Under Part III of the Act, the Secretary of State was empowered to introduce regulations so that where local authorities continued to operate direct labour organisations they must compete for work with private contractors, keep their accounts on a trading basis and earn a specified rate of return, to ensure that they would be cost-effective in operation. Failure to achieve the prescribed rate of return could result in the DLO being ordered to cease its operations. Although Part III of the Act does not apply to small maintenance organisations employing fewer than 30 manual workers.

The *Direct Labour Organisations (Rate of Return on Capital) Directions 1981* set the rate of return at 5 per cent, calculated on a current cost accounting basis. The new legislation was aimed at bringing to an end the unfair advantages that DLOs were considered to enjoy as a result of their non-profit making financial structure supported by subsidies from the rates. It did, however, go much further by imposing a profit margin which private contractors were unable to achieve.[4]

Edwards[5] has described how DLOs have been placed at an unfair advantage when comparing with the private sector. For example, there are limits on the types of work for which they can tender; they are confined to certain geographical areas; are required to achieve a prescribed percentage return on capital; must make a profit in each of the specified work categories; and must bear their share of the authority's central operating expenses.

The main implications of the new legislation in the operation of DLOs is now examined.

(1) *Increased accountability*: DLOs now have to make a profit to survive while continuing to satisfy the demands for service.

(2) *Potential reduction in workload*: in order to achieve previous levels of workload a high success rate in tendering would have to be achieved.

(3) *Charging to clients*: work can no longer be charged at cost; it must be based on a tender price or estimate and any excess costs will be shown as losses against the particular contract or account.

(4) *Allocation of overheads*: many of the overheads could remain in the event of a reduction in workload, thus adversely affecting the chances of gaining new work.

(5) *Local authority responsibilities*:   various aspects of national agreements relating to pay and service conditions could put DLOs in an adverse position as compared with private contractors. Examples are the arrangements for absences due to sickness and industrial injury and the requirement for all full-time employees to participate in the local government superannuation scheme.

(6) *Monitoring of expenditure*: need to have more rapid access to up-to-date financial information on works in progress.

(7) *Additional work/overheads*: significant increase in the surveying and estimating workload with a consequent increase in overheads with a potential diminishing workload.

(8) *Local issues and policies*: difficulties in operation within locally determined policies and strategies, such as a non-redundancy agreement or a 'buy British' policy, restrictions on the use of suppliers of materials or services, or any local betterment of basic pay or conditions of service.

(9) *Problems with tendering*: much of building maintenance work is in the under £10 000 category and day to day repairs, which is difficult to put out to tender as the nature and amount of the work involved is not known prior to going out to tender.[5]

Edwards[5] has advocated the implementation of the following actions to overcome the problems previously outlined:

(1) improve efficiency in operations and organisation;
(2) review assets used to improve the rate of return;
(3) review overheads to effect reductions wherever possible;
(4) maximise work available as by right;
(5) review contract conditions to ensure that wherever possible contractors and DLOs compete on equal terms;
(6) review the range of operational procedures.

Much of the action centres around improving efficiency and cost effectiveness, and Edwards[5] has aptly listed the main areas where this could be achieved.

(1) reduction or elimination of wasted time such as that spent waiting for materials, plant or transport, travelling to and from and between jobs, and abortive calls;

(2) effective organisation and deployment of labour by trade and area to give satisfactory service at minimal cost;

(3) monitoring of incentive payment schemes;

(4) improvements in the planning and scheduling of the larger maintenance schemes;

(5) rationalisation of plant and transport and depot location and usage to give the best service possible;

(6) review of working methods to maximise flexibility and minimise labour input and materials usage for each task.

Building Maintenance Departmental Structures and Arrangements

Hierarchical responsibility

The pattern of responsibility for directing and undertaking repairs and maintenance varies considerably from one local authority housing department to another, and is influenced by such factors as geographical spread of the authority's administrative area and the size of the housing stock. The primary functions will consist of determining general maintenance policy, assessing funding requirements, preparing work programmes, executing maintenance work, progressing the work, monitoring costs and implementing feedback procedures.

Authorities owning less than 5000 dwellings and occupying a compact geographical area are likely to manage repairs and maintenance from the departmental headquarters. Authorities which have more houses or a large administrative area will probably have the work organised on a district basis. The latter arrangement will be examined, although it will be appreciated that the functions performed by district officers will be carried out by headquarters staff in smaller authorities.

Irrespective of the size of the authority, all housing matters will normally be administered by a director of housing, who will be responsible to the Housing Committee for the work of the department. The director will be concerned with the initiation of policy and is the link with the Council. An assistant director is often appointed to liaise with the principal headquarters officers overseeing the districts; these could, for example, be a principal technical officer and a principal surveyor, although on occasions these posts might be combined.

The housing team within each district will be headed by a district officer, with a district technical officer and possibly a district surveyor controlling repairs and maintenance. The likely maintenance personnel and their main functions are now described.

District officers

District technical officer: reports to the principal technical officer at head-quarters and exercises financial control of maintenance work, deals with non-routine matters, arranges payments to contractors and may prepare estimates in conjunction with the district surveyor.

Assistant technical officers: report to the district technical officer, carry out building inspections and compile work specifications, possibly in co-operation with an assistant district surveyor. They will also monitor maintenance costs by overseeing work undertaken by the authority's direct labour organisation.

District administrative staff: report to the assistant technical officers, taking work requests, issuing work dockets to the direct labour organisation or contractors and maintaining adequate records.

District surveyor: reports to the principal surveyor and supplies data obtained from site inspection surveys. He also undertakes structural surveys, organises demolition work and oversees alteration, conversion and refurbishment work, including the preparation of cost estimates and specifications and valuation of payments to contractors.

Assistant district surveyors: assist the district surveyor in carrying out his various duties.

Clerks of works: carry out normal inspection duties to ensure satisfactory quality standards in materials and workmanship.

Works personnel: direct labour organisations require sufficient supervisory staff, craft operatives and labourers to carry out the maintenance work allocated to them. They often operate as mobile teams.

An alternative arrangement consists of area maintenance units operating under the control of an area manager, responsible for carrying out pro-grammed and contingency maintenance works in his area within the prescribed budgetary provisions, and assisted by a supervisor, technicians, work study officer and clerical staff. A central planning and control group prepares routine planned maintenance programmes for each area, and its work includes the allocation of manpower and other resources, preparation of area budget estimates, monitoring the performance of area maintenance units and co-ordination of their activities.

Central administration

The central administration will monitor the progress made by the districts in fulfilling the authority's repair and maintenance objectives, and the effectiveness of its actions. In addition, it will keep comprehensive records of the authority's properties, and of the repairs, maintenance and improvement work carried out to them.

Property details will be stored on a property register, which often ranks property in terms of age, height and size of accommodation. Although property maintenance records will be kept by each district office, information will also be forwarded to the central administration to enable work to be monitored. Information will be received on instructions issued, including their date of issue, brief descriptions of the work, when it was completed and whether by contractor or direct labour organisation. This enables the central administration to examine the effectiveness of the work and to identify responsibility for substandard performance.

Case studies

The organisational structures for maintenance of dwellings of a large city housing department and a much smaller housing association are briefly described to show the different arrangements.

Nottingham City Council owned over 45 000 dwellings in 1987 and the Department of Housing Services controls all aspects of maintenance and management of the properties. The Director of Housing is supported by two deputy directors, one of whom is responsible for the direct labour organisation of about 500 employees. The administrative centre of the DLO contains various trade shops and a minor works section which deals with adaptations and pre-painting repairs. Jobbing repairs are organised through six depots on the basis of orders raised by technical housing officers located in branch and neighbourhood offices.

Repairs are categorised according to priority. Thus 'E' orders relate to emergency items, such as burst pipes, blocked drains, broken windows and serious electrical faults and are rectified within 24 hours. 'A' orders are dealt with within 10 days and include ill-fitting external doors and holed roofs, while 'B' orders are rectified within 10 weeks and include all non-urgent repairs such as defective plaster and ill-fitting windows.

The area covered by each depot is subdivided into cyclical districts each covering about 2000 dwellings. They are visited every three to four weeks, one or two portakabins are set up and craft operatives sent out from these to deal with 'B' orders. The nearest arrangement to an estate-based repair team are the on-site joiners, plumbers and electricians for the flat complexes.

A surveying and estimating section of 30 technical staff deals with work not undertaken by the DLO, such as rewiring and central heating installations and work exceeding £2000 in value. Specifications are prepared and the work is put out to tender. This section is also responsible for the planned programme of repainting, each contract comprising several hundred houses, and for which the DLO is usually invited to tender.

In 1987 the department was in the process of computerising its system for costing and ordering repairs. Each branch and neighbourhood office will have a terminal and printer. When a disrepair is reported by a tenant, the details are immediately keyed in and stored under the appropriate property reference. The printer then automatically produces an acknowledgement for the tenant and a copy is passed to a technical housing officer to assess the work. He codes the work which is then typed back in and an order automatically raised with the appropriate depot which has its own VDU and printer. When the work is completed the depot clerk books off the repair. Managers are able to ascertain the extent and location of outstanding repairs and detailed information on works completed, costs and many related matters.

The technical aspects of modernisation work are undertaken by the Department of Technical Services, although the improvements section of the Department of Housing Services liaises with tenants, arranges decent accommodation and related matters. In 1985/86 Nottingham City Council spent £250 per dwelling on repairs and improvements compared with the national average of £375.

By comparison, *Nottingham Community Housing Association Ltd* managed 1575 properties in 1987 through three area offices. Maintenance work is organised from the Association's main office, which is staffed by a maintenance officer, two maintenance assistants and administrative support staff. The properties are predominantly refurbished dwellings with a small number of new buildings. Unlike a local authority, properties are rarely located in close proximity to one another.

Most of the maintenance work is complaint-orientated and each defect is recorded on a repair acknowledgement form. Most of the works are pre-inspected, recorded on a card with a priority code and a confirmation order sent to a contractor. If works are estimated to exceed £100 to £200, competitive prices are obtained from selected contractors.

Properties which have not been visited previously are inspected at six-monthly intervals. Cyclical maintenance is carried out every four years. Decorations, gas and electrical installations are inspected on relets and appropriate action taken. Where possible, contractors submit invoices on a weekly basis and these are carefully scrutinised, with details entered on to individual house cards and into an expenditure record book. Each month a statement of expenditure and outstanding works is prepared as a means of budget control. In 1985/86 the Association spent £253 per dwelling on repairs and maintenance.

Maintenance Depots

Area Depots

Craft operatives normally operate from depots which should include facilities for changing clothes, washing, and drawing stores and tools. The economics of providing a central depot as against a number of smaller depots dispersed throughout the administrative area needs to be examined. The optimum number, size and location of depots is an operational research problem that requires for its solution a detailed examination of the total costs of running the depots, geographical distribution of jobs, their urgency and how long they take to complete.

Depots provide craft operatives with the materials they require and the maintenance manager has to determine the correct stock levels. He needs to consider how frequently each item is required, how long it takes to replenish stocks when they run low, the relative seriousness of stocks running out, the cost of purchasing stock in different quantities and the storage costs associated with bulk purchasing.

Mobile Depots

Mobile depots can often be used to advantage with housing maintenance, particularly in rural areas. Their use needs to be carefully planned to provide an efficient service and the following arrangements introduced by a contractor have proved satisfactory in practice.

Site instructions for repairs are as far as possible grouped into areas or streets, whereby the team has sufficient work for at least two days ahead. To overcome the common problem of non-accessibility, the team foreman calls at all houses which it is proposed to visit the following day. He makes the call at the end of the working day by which time there is usually someone at home; arrangements can then be made for someone to be available the next day or for a key to be left with a neighbour.

The problem of stores availability is almost completely overcome by a wide selection of articles carried in the mobile workshop, including standard doors and windows and plumbing spares. It is seldom that the mobile workshop has to return to base during the day. Stores are replenished at the end of each day's work. There is considerable advantage in having a mobile workshop readily accessible to the repair work. Even the most straightforward repairs, such as easing doors and windows present cleaning problems when carried out in the house. They can now be done, whatever the weather, close at hand in the mobile workshop. This particular mobile unit was used for the maintenance of 2600 dwellings with a team consisting of a foreman, 2 carpenters, 1 plumber, 1 bricklayer–handyman and 1 painter–glazier.

A useful type of mobile workshop was based on a diesel-engined 1.75 t vehicle. On one side of the workshop was fitted a joiner's bench and the other side was racked to assist in the identification of small stores. There was an electric circular saw which could be used on the bench or outside; although not large it catered for almost all the requirements of small maintenance jobs.

A mobile workshop could take the form of a motorised workshop or be a caravan type workshop towed by a Land Rover or lorry. In making a choice the following factors should be considered.

(1) How long is the mobile workshop likely to remain in any one place?
(2) Could the towing vehicle be used for other duties when not required for moving the workshop?

(3) The consequential gain or loss from Road Fund licensing costs in each case.

(4) The cost of replating a motorised workshop every 18 months and the effect on overall costs.

Programming of Maintenance Work

Planning the Workload

In planning maintenance work, the maintenance manager aims to match the available resources with the workload. Effective forward planning is difficult with this category of work because of the large number of uncertain factors involved. Hence it is advisable to adopt a two stage planning system.

(1) An annual plan is prepared showing the expected workload for the coming year and how this can be matched effectively with available resources. Jobs are best grouped according to their work content, urgency, amount of notice required and restrictions on timing. A simple classification is jobbing repairs, fabric maintenance and modernisations. Using a classification of this type, a detailed plan can be formulated of the work to be done by each trade during the coming year.

(2) The completed plan is translated into a weekly work schedule, with a rolling programme prescribing the workload for about 6 weeks ahead. This programme is updated regularly to take into account the most recent information on the content and timing of specific jobs and availability of resources. In this way the maintenance manager retains close control of the maintenance work and when events diverge from the plan, it allows him to take corrective action at the earliest possible time.

The majority of local authorities find it helpful to employ their own direct labour organisation extensively in order to control the order of priorities and to build up comprehensive information on the nature, condition and requirements of their own properties. A direct labour organisation generally caters for most routine jobs, while specialist work such as electrical or heating may be let out to private contractors. It is usually advisable to put out to tender any abnormal increases in workload, including possibly part of the painting work to avoid employing too many painters in the winter. Furthermore, obtaining tenders from contractors provides a useful indication of the cost effectiveness of a direct labour organisation, and in many cases the local authority has to tender in competition with private contractors to satisfy the requirements of the Local Government Planning and Land Act 1980. On occasions the extent of a maintenance programme could be too rigidly constrained by the size of the direct labour organisation rather than the needs of an effective maintenance service.

It is surprising that some local authorities organise their maintenance service without attempting to match expected demands and resources. It is possible to estimate annual and seasonal variations in workload and the proportion of the labour force likely to be available. Admittedly, some maintenance jobs can be unpredictable in work content, but without a forecast of probable future work the effects of abnormal delays and the necessary remedial measures cannot be predicted. Even with formal planning, work does not arise in a uniform flow. There are inevitably peaks and troughs for most trades, dependent upon climatic conditions and changes in local authority policies and attitudes.

Many local authorities endeavour to restrict severely the repairs undertaken for tenants, often to the extent of issuing conditions of tenancy of doubtful legality, such as requiring the tenant to protect from frost badly designed and very vulnerable fittings and services. The primary objective of this policy is to limit maintenance expenditure and to avoid as far as possible undertaking small jobs which can be proportionately very expensive. Internal decorations are normally made the responsibility of the tenant and replacing broken glass arising from wilful damage or gross negligence on the part of the tenant. Most local authorities exempt old age pensioners from many repair obligations, and some concessions may be granted to single parent families with children of school age and registered disabled persons living alone. Tenants must be fully aware of their own repair obligations and the procedure for notifying the landlord of repairs for which he is responsible under the tenancy agreement.

Use of Programmes

Frequency-based programmes can be prepared for internal and external painting and for washing down. These programmes can be based on manuals drawn up for each building recording all relevant data such as painting specifications, superficial areas and a cost index based on the actual cost in a particular year. A programme of preventive maintenance such as drain clearing, gutter clearing, tap re-washering and oiling and adjusting door furniture can also be prepared for each building.[6]

The main part of any maintenance programme is usually a series of pre-planned inspections. An annual inspection by a structural engineer ensures the stability of the structure. A checklist ensures effective inspections at suitable intervals. The programme of inspections can be fed into a computer which prints out a slip requesting the inspection on a specific date. The report of defects prepared on inspection is then assessed for urgency. Where these are urgent a requisition can be prepared and corrective work put in hand. Less urgent jobs can be planned and fitted into a work programme provided it can be kept within the budget. The information can then be returned to the computer for updating the historical record.[6]

One effective method is to operate a planned maintenance system based on a 5-year painting cycle, and a typical 5-year planned maintenance programme is illustrated in table 12.1. The first priority is the replacement or repair of any item to be subsequently painted. Alongside this is work urgently needed for reasons of safety or hygiene, such as rewiring, replacement of sanitary ware or making good defective structural work. Effort is made to ensure that the work would not be quickly overtaken by more comprehensive schemes of maintenance or modernisation. More comprehensive improvements can be classified under three main headings, as follows.

(1) Internal improvements: improved insulation, central heating, remodelling of kitchens and upgrading of bathroom fitments.

(2) External improvements: general facelift to property not being part of maintenance paintwork cycle, such as tiled roofs, fencing and footpaths.

(3) Site improvements: car parks, children's play areas, landscaping and general environmental improvement.

Speight[7] has described several approaches to programming or a combination of them, such as determining priorities over a 1 to 2 year period to dealing with emergency work, deciding on a rolling programme covering 5 to 7 years, and identifying the requirements for routine day to day maintenance and servicing. He asserts correctly that sound maintenance is largely dependent upon doing the right thing at the right time.

A typical painting schedule would show the locations of dwellings on the left-hand side and the columns would be headed with years. In addition the number of dwellings, type of dwelling (for example, pre-war three bedroom) and possibly the date of erection may be inserted. The schedule will show the painting programme for each group of dwellings and whether this has been implemented in the past. Schedules can also be prepared to cover other major items of repair or replacement such as re-roofing, replacement of paths and fences, rewiring and replacement of sanitary fitments.

Graham[8] has formulated the following maintenance standards as a basis for programming of housing maintenance work.

> external painting: every 5 years
> electrical re-wiring: every 25 to 30 years
> renewal of gas water heaters: every 25 years
> servicing of gas appliances: annually
> servicing of electric appliances: annually
> cleaning of ventilation ducts: annually
> servicing floor springs to swing doors: annually
> cleaning gullies: annually
> cleaning of water storage tanks: annually
> cleaning refuse hoppers on staircases and balconies: three times each
>   year
> cleaning dust chute chambers: clean three times each year and paint
>   internally or wash down as required
> grass cutting: generally every 9 working days, reducing to 7 days in peak
>   growing periods and increasing to 10 days in slow growing periods
> hedge cutting: twice each year (April/May and August/September)
> hedge splitting: one fifth each year (October/March)

Housing authorities often rank repairs and maintenance work in a stated order of priority to indicate the urgency with which operatives should seek to deal with them. The following represents a typical priority ranking.

> emergencies: make safe or complete within 24 hours of notification
> bricklaying: within 5 weeks

carpentry: within 4 weeks
plumbing: within 3 weeks
glazing: within 2 weeks
electrical work: within 1 to 2 weeks
space heating: within 3 days

Stevens,[9] while accepting that many building components deteriorate gradually, believes that consideration of the physical needs of the building might indicate that different speeds of service would be appropriate for different types of repairs. He suggests the following categories of priority.

(1) several hours: major water leaks, gas leaks and dangerous wiring
(2) several days: minor water leaks, reglazing, repairs to external doors and heating repairs in winter
(3) several weeks: major plastering repairs, and blocked rainwater pipes and gutters
(4) several months: renewal of gutters, easing of windows and doors, and repointing brickwork.

It is possible to estimate approximate times for a range of cleaning tasks calculated on a unit of 100 m^2 of floor area but these will vary according to the type of building and its use. Some of the more common cleaning tasks follow with their likely times.

Activity	*Time in minutes per 100 m^2 of floor area*
Office and toilet cleaning	60–70
Impregnated mop sweeping	10–15
Damp mopping	18–25
Wet mopping, rinsing and drying	60–70
Hand scrubbing and drying	180–200
Machine scrubbing	12–45
Machine drying (suction)	25–30
Stripping and re-waxing	160–180
Vacuum sweeping	30–35
Hand sweeping	20–25
Applying wax polish manually	85–100
Applying liquid polish manually	45–50
Applying liquid polish with applicator	30–40

Organisation of Maintenance Work

Recording of information concerning properties and repairs is best undertaken on a card system providing ready means of communication by means of a computer. The records are sometimes described as a property register. The computer can be programmed to give either fixed format reports providing answers to individual questions or an analysis across the full range of properties. In one such system the information is held on standard 80 column punched cards, using a very simple code to identify each street. On

Table 13.1 Control card

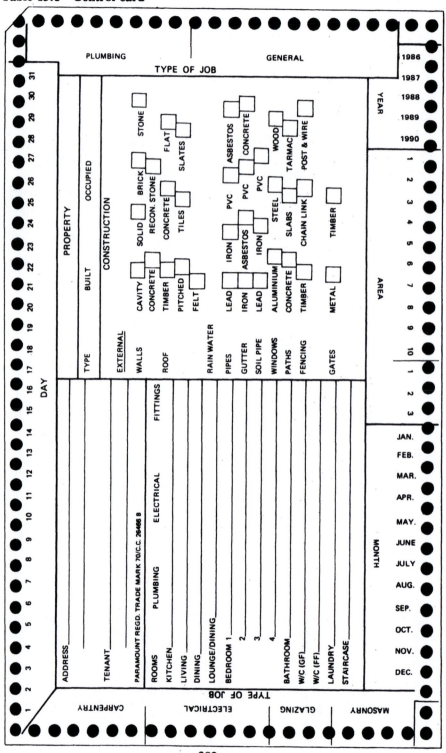

being given a cycle number the computer can provide a printout of street names and house numbers in the sequence of inspectors' areas, to instigate the planned maintenance inspection process. The inspector will be told which houses and elements to inspect.

In this way a centrally updated and accurate information service is secured. It also provides an effective management tool capable of very fast pre-inspection assessment of problems and realistic projections.

The primary functions of control cards are to contain details of property and maintenance work. A typical control card is shown in table 13.1.

With private properties there may be agreements between landlord and tenant which restrict access to the property. Records should show tenancy details and arrangements. Repairs to roofs, or electrical or heating services may be chargeable to a number of tenants. Furthermore, tenants may require notice of intention to inspect.

Notification of Defects

Defects are notified in a variety of different ways, which are now listed roughly in descending order of use.

    telephone call from tenant
    tenant returning pre-paid complaint card
    letter from tenant
    officer of local authority finding defect
    tenant notifying defect in person at a depot or housing office
    tenant notifying complaint to officer of local authority on site

It has been found that requests for work are sometimes mislaid because of the many and varied procedures that exist for notifying repairs, and in the absence of a record card for each property, it is difficult to trace some enquiries. Hence it is better for all complaints to pass through maintenance control, where a maintenance complaint card can be completed, and a property card index system established. A typical tenants' request card is shown in table 13.2 on which is entered the tenant's name and address and the nature of the complaint. An alternative form of card is illustrated in table 13.3, which comprises three NCR (no carbon required) copies in different colours, serving various purposes (request, action and acknowledgement to tenant).

Thurley[10] on a survey of services tenants on married quarters estates found that the percentage of requests for each main category of repair was structural: 2; fixtures and fittings: 33; plastering and tiling: 6; plumbing: 25; electrical: 14; domestic services: 13; and external work: 7.

General Maintenance Procedure

A typical procedure covering the identification of a defective component and the associated remedial work is shown diagrammatically in figure 13.1, and a completed maintenance feedback report is provided in table 13.4.

Table 13.2 Tenant's request card

HOUSING MAINTENANCE REQUEST

BLOCK LETTERS PLEASE

Name.. **Tel No.**........................

Address ....

Keys at.................................... At home a.m./p.m. (Not Sat.) Date..........................

Please give precise nature of fault and, where possible, location (indicate by tick in box)

Location	Nature of Fault

External ☐   Internal ☐

Room: Kitchen ☐
       Living ☐
       Dining ☐
       Bedroom ☐
       Bathroom ☐
       W.C. ☐

NOTES: Please use this card to request repairs. Urgent calls Telephone Newtown 53844
Unless otherwise notified repairs may normally be expected to be carried out
within seven days.

Table 13.3 Housing repair request/acknowledgement

N.C.H.A. 14 Pelham Road, Sherwood Rise, Nottm. Tel: 622531 39 Sneinton Hermitage, Sneinton, Nottm. Tel: 504145 303 Castle Boulevard, Lenton, Nottm. Tel: 411654	DATE:-
HOUSING REPAIR REQUEST ACKNOWLEDGMENT	
ADDRESS:-	POST CODE:-

ITEM	WORKS DESCRIPTION	PRIORITY CODE	ACCESS DETAILS	
			CONTRACTOR CONTACTED	ITEM
			INSPECTION REQUIRED/ADDITIONAL DETAILS	
SIGNATURE:-				

384

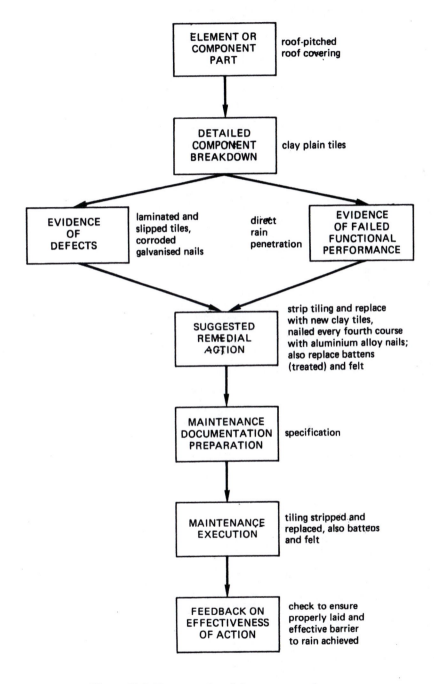

Figure 13.1 Component maintenance procedure

ITEM : Floor finish: clay tiles	AREA : Bransome Estate ADDRESS : 15 St. John's Walk TYPE : C 22 Semi-detached
DEFECT	
SYMPTOMS	Approximately one-third of clay floor tiles in kitchen arching and lifting (approximately one year after laying)
DIAGNOSIS	Shrinkage of cement: sand screed, resulting in breaking of bond between tiles and screed
PROGNOSIS	Progressive damage occurring: recommend immediate action to prevent injury to elderly tenants
REMEDY / RECOMMENDTN.	Take up tiles, remove bedding grout, lay polythene sheet over screed and relay tiles to whole floor, with expansion joint around the perimeter of the room.
Authorisation:	DATE SURVEY : 20 October 1987 COMPILED BY : Peter Johnson

Table 13.4   Maintenance feedback report

These illustrate the need for a systematic approach to maintenance work to secure effective and economical solutions to the defects that occur.

Execution of Maintenance Work by Maintenance Departments

Even with jobbing repairs such as repairs to dripping taps, broken windows or faulty door latches, the maintenance department has to decide on the method of execution and has a number of choices.

(1) Whether to use its own operatives or contractors, or to make the tenants responsible for certain repairs.

(2) Decide on urgency of jobs—whether to put into effect or to postpone until the next planned maintenance cycle.

(3) Decide on response time to tenants' requests for non-emergency work.

Relet repairs undertaken prior to reletting a house to a new tenant often include painting and decorating as well as general maintenance work, such as replacing cracked sinks or relaying badly damaged floors. The extent of the work can be affected by the policy on planned maintenance and jobbing repairs. Once again some types of non-emergency work could be postponed until the next planned maintenance cycle.

The majority of direct labour maintenance systems operate on a job ticket, request or order arrangement, whereby operatives only do work contained on a job form. One such system embraces forms of four colours used for housing maintenance.

Red forms are used for emergency work in response to tenants' requests and must receive attention within 24 hours.

White forms are used for normal tenants' requests and may be prepared by rent collectors. No guarantee is given as to when the job will be completed but the tenant is asked to give access times.

Blue forms are used for tenancy changes where a foreman inspects all properties after the outgoing tenant has vacated, and all work necessary prior to reletting is entered on blue tickets for the appropriate trades. Three weeks are normally allowed between tenancies for such work.

Green forms are used for work involving several trades. The initial form (red, white or blue) is passed to the first trade which after completion arranges for the chargehand to make out a green form for the next trade.

The white forms are prepared from descriptions supplied by tenants and require some degree of pre-inspection depending upon the nature of the job description. On average each man in a gang completes six jobs per day and , on this basis, the foreman sends selected job forms to the chargehand at least three working days prior to the scheduled work visit. This enables the chargehand to make any necessary pre-inspections and to obtain materials and plant in time for the scheduled visit.

```
            ┌─────────────────────────────────┐
 ──────────┘         NEWTOWN CORPORATION       └──────────
               HOUSING MAINTENANCE REQUEST
```

Name...........................	Address...
	....
Keys at............................	At home a.m./p.m. Date...............................

Trade	Location	Repair	Craft operative
☐ Carpenter ☐ Electrician ☐ General ☐ Painter ☐ Plumber ☐ Glazier ☐ Mason	☐ Internal ☐ External ☐ Kitchen ☐ Living ☐ Dining ☐ Bedroom ☐ Bathroom ☐ W.C.		Materials Used Time Taken:

Special Instructions/Comments _____

Other Trades Necessary _____

_____   Signature Date ....
_____   Craft operative ....
                                Foreman ....

Table 13.5   Job order form

A typical job order form is shown in table 13.5 to provide foremen and craft operatives with all requisite information, including location, operatives and time taken and materials used. Once again the colours of forms have special significance—pink for emergencies to be done in hours, green for urgent work to be done in days, and white for routine work which could extend over weeks.

Yet another works order system uses different colours according to the personnel for whom they are intended—pink: record, white: operative, yellow: foreman, buff: plant/transport, and blue: work study/bonus.

With some local authorities the housing maintenance order form is produced on A5 size paper with two different coloured copies. When operatives find no occupants at home, they leave call cards which ask tenants to specify a convenient time for the operatives to call. Tenants who do not return the prepaid call card receive a follow up card which states 'if no reply is received within seven days we shall assume you no longer wish to have the matter attended to'. For jobs delayed beyond two months, maintenance control estimates the probable date of execution and sends a two month delay card containing this information to the tenant.

In addition a check is made on jobs outstanding at the end of each week for each trade and depot. This is multiplied by the appropriate average work time (3.50 hours) to determine the likely delays in execution on the basis of the available labour force.

An alternative housing repair instruction/order form is illustrated in table 13.6. This is made up of four copies of NCR paper each in distinctive colours.

The planned maintenance and operation system used by one government department embraces:

(1) providing operatives with information on work to be done (job sheets);

(2) overall planning of the work (planning charts);

(3) issuing orders for work to be done (work dockets, contractors' orders);

(4) keeping records of work done (by books).

Another system used in Australia and termed the plant information management system has the following routine inputs:

(1) budgets (annually);

(2) work orders or requests or maintenance jobs to be done;

(3) timesheets, on which craft operatives record the hours worked on each job;

(4) stores warrants, recording the value of stores items issued for each job;

(5) purchase orders for each job;

(6) invoices to match the purchase orders;

(7) lost production records;

(8) job completion sheets, outlining what was actually done.

Execution of Maintenance Work by Contractors

The maintenance contractor must be able to process efficiently all enquiries and orders by submitting quotations or issuing works orders. Effective communication is important so that building owners are kept informed and the contractor's own staff are aware of their involvement. Operatives have to be deployed to cope effectively with the workload, adequately backed with materials and plant. An easily operated and yet foolproof booking and costing system is needed to provide quotations and accounts, together with the necessary feedback of cost data, including the analysis of oncosts.

On maintenance work, operatives are required to work more on their own, are more scattered and have less supervision, often restricted to a travelling foreman, surveyor or works manager. On larger projects there may be a foreman resident on the site, although he will probably be a working foreman.

Yard stocks of all regularly used materials should be maintained, as purchasing materials in small quantities is expensive. Purchasing in bulk where practicable is advantageous, when materials will be distributed from

NOTTINGHAM COMMUNITY HOUSING ASSOCIATION LIMITED
14 PELHAM ROAD, SHERWOOD RISE, NOTTINGHAM

HOUSING REPAIR INSTRUCTION/ORDER

CONTRACTOR:—

REPLY TO:—

WORKS ADDRESS

ACCESS INSTRUCTIONS

DATE

INSTRUCTIONS

CODE

CERTIFICATE OF COMPLETION BY CONTRACTOR/DATE

JOB NO.

CONDITIONS OF WORKS ORDER

1. This constitutes a contract between N.C.H.A. and the Contractor named.
2. All specified works and materials shall be complied with and no other works or materials are to be used without prior approval of a N.C.H.A. Officer.
3. The works listed shall be completed within the period set out below.
   Code 1 - 48 hours   Code 2 - Ten working days   Code 3 - Three months
4. The Contractor shall have at all times a current Tax Exemption Certificate and Public Liability Insurance to a minimum sum of £250,000.
5. All damage to the property and tenants effects shall be made good by the Contractor.
6. Any defects which occur from the works shall be made good by the Contractor.
7. All works shall conform to current Health and Safety Regulations.
8. All works shall conform to statutory Requirements.

Table 13.6 Housing repair instruction/order

stock to each job. Bulky materials such as aggregates, bricks and hardcore are expensive to transport and deliveries should be so arranged to reduce double handling to a minimum. Some maintenance jobs may require very small quantities of materials such as 50 kg of cement and a wheelbarrow load of sand, which should be drawn from a bulk supply where possible.

Management of Maintenance Work

The majority of repairs to local authority dwellings are completed within a reasonable time and often it is executed to a higher standard than that expected by private tenants or owner-occupiers. Much of the criticism stems from shortcomings of management, by the failure to inform tenants why a repair cannot be executed or has to be delayed. Easy repairs may be executed promptly while more difficult ones are postponed and sometimes overlooked. There is a good case for issuing more information to tenants on housing maintenance policies and the channels available for securing service and making complaints.

Some of the main problems in housing management have been identified as follows:

(1) communication problems between tenants and management and vice versa;

(2) maintenance standards are seldom clearly defined or generally recognised;

(3) maintenance is susceptible to financial stringency and cut-backs in resources and future plans, with problems in assessing priorities;

(4) difficulties in obtaining adequate control data for maintenance management.

The views of a housing tenants' association are useful. Tenants tend to feel they are regarded as irresponsible as they are given no opportunity to involve themselves responsibly in the decisions affecting their environment. They have little or no say in the choice of colour schemes, in the timing and frequency of repairs, and in the design, provision and management of amenities. London tenants believed they were the recipients of decreasing services through the withdrawal of residential caretakers and substitution of mobile arrangements, reduced rent collection facilities and difference in level of service provided between older and newer estates. They complained about the delays in carrying out repairs, the amount of chasing needed to obtain action and the extent of bad workmanship.

Some of the main needs associated with housing maintenance management are now listed.

(1) All significant repairs, except those resulting from deliberate maltreatment by the tenant, should ideally be the responsibility of the housing authority, to ensure good standards of maintenance. Most non-urgent repairs can wait for execution under a planned maintenance cycle if the delay does not exceed one year; this should be explained to the tenants so that they can make temporary arrangements or even do the repairs themselves.

(2) Many housewives go out to work and access can often be obtained only by appointment.

(3) More time should be devoted to routine inspections to ensure that attention is paid to repairs not reported by tenants. One approach is to use mobile gangs of mixed trades, with foremen more concerned with inspection, management and tenant relations than with supervision of craft operatives, on the assumption that operatives have adequate incentives through bonuses as described later in the chapter.

(4) It is advisable to determine common acceptable standards of maintenance as guidelines for housing authorities. Costs of direct labour

forces and contractors' prices for similar work should be compared to provide comparative yardsticks for efficiency.

(5) More use could be made of local authority consortia to obtain bulk quotations for materials for housing maintenance.

(6) Tenants should be encouraged to take more interest in the maintenance of their dwellings, and in some areas tenants' associations could be consulted more frequently so that they could for instance help to decide priorities of expenditure, treatment of common areas and colour schemes.

Thurley[10] examined the conflicting goals in housing maintenance systems. He described how maintenance engineers or surveyors are anxious to maintain the value of the property and usually favour a system of systematic planned maintenance, rather than relying on tenants' requests as a way of initiating work. They are likely to emphasise structural and preventive repairs rather than, for example, frequent redecoration. Foremen, chargehands and operatives involved with the actual execution of the work are more concerned with maintaining a sufficient programme of work which will utilise their own skills. Hence they are likely to resist pressure from tenants as they have to work under the close scrutiny of the occupants.

Costing and Accounts

The clerical and accounting processes supporting a direct labour organisation can be simplified and accelerated by the use of the computer. Staff records can be filed and quickly retrieved, time sheets can be processed to produce wage make-ups and salary cheques, stores accounting and ordering can be programmed so that stocks are recorded and checked and materials and components ordered at pre-set levels, and the cost of labour and materials can be coded and charged against budgets for control purposes. Time sheets for plant and transport can be recorded, costed and charged to budgets and running times and travel distances recorded against planned maintenance schedules. Pettitt[11] has described how computers can be used to produce detailed financial analyses of maintenance expenditure as an aid to achieving strict budgetary control, and to assist in minimising future breakdown by aiding the preparation of planned preventive maintenance programmes and feedback to designers.

Where work is carried out by contractors, it may be the subject of a daywork order with the account rendered on a time and material basis with percentage additions for overheads and profit; the certified account could be checked by the computer for arithmetical accuracy and payment made direct to the contractor's account. Alternatively it may be by order describing the work, with the account settled on the basis of measurement against an agreed schedule of prices or by lump sum offer based on specification or bill of quantities and drawings, from selected tenderers; the account and measurements could be certified and priced against the contractor's agreed schedule of prices to provide payment or the account certified and checked for arithmetical accuracy by the computer.

Contractors will decide the costing system that best suits their needs and which gives up-to-date information quickly. The basic cost elements of maintenance work are as follows:

(1) labour costs: from weekly time sheets;
(2) material costs: from merchants' or suppliers' invoices and stock issues from depot on costing dockets;
(3) transport costs: from drivers' log sheets or time sheets;
(4) expenses: from claims and petty cash vouchers;
(5) fees or charges: such as statutory undertakers;
(6) plant costs: from hire charges or internal costs.

A typical property maintenance record card is illustrated in table 13.7, to show how works of repair are allocated to a particular property, together with the costs and method of execution. In this way a comparison of the costs of similar dwellings can easily be made. Computerisation of this information simplifies and speeds up the recording, retrieval and preparation of comparative analyses.

Tools and Plant

The maintenance contractor or department needs to have the use of appropriate power tools and plant, the range of which has increased extensively in recent years. The contractor has to decide whether to purchase outright or to hire the plant and there is generally a case for both approaches. Ideally the most regularly used items should be purchased and less popular items hired, but having regard to finance, and storage and maintenance facilities. Plant must be used effectively and reasonably frequently to be viable.

Carrying out maintenance and jobbing work inevitably means a considerable amount of travelling from one job to another, also more materials have to be collected from stores or merchants' or suppliers' stocks because of the need to secure prompt delivery on site. It is not always possible to predict precisely all material requirements until work is exposed opened up or inspected. For instance, the repair of a leaking drain may require bends or other fittings which have to be collected as required.

Light vans and pick-up trucks are very popular with maintenance contractors. It is generally more convenient to have a large number of small vehicles rather than a smaller number of large ones. Small vehicles have various advantages such as ease of handling in confined spaces and ability to load and unload on domestic drives.

Maintenance and cleaning contractors may be faced with maintenance and cleaning of buildings at different angles and at heights of 30 m or more. One useful item of equipment is a hydraulic platform to lift men and tools to overhead working positions for cleaning and renovating buildings, for lighting maintenance, painting, welding, property repairs and general overhead maintenance work. An alternative is multiple staging towers, based on a scaffolding principle but far less costly and involving less trouble and time in erection. There are many companies offering plant hire facilities

Maintenance Work		Address: _____ _____		District code: _____ Property code: _____				
Date	Work Docket Nr.	Work	DLO / Contr.	Cost Est. £	Cost Act. £	Invoice	A/c Code
17/9/86	53217	Repairs to front guttering	Harrison Builders Ltd	45	45	19732	2
18/2/87	53996	Repoint chimney stack	Banham and chalkley	115	115	20011	1
28/8/87	54689	External repainting	DLO	250	237	—	4

Table 13.7 **Property maintenance record card**

but special care is needed in selecting the best type of plant for the particular job.

Ordering Materials and Stock Control

Maintenance operatives need to be backed by a steady flow of materials and components of the right type and at the right time, as waiting time is expensive. Most building maintenance departments have at least one small store to hold those items that are in frequent demand or difficult to obtain at short notice. A number of fundamental questions have to be answered.

(1) What items should be kept in store and in what quantity?
(2) At what level should they be re-ordered and in what quantities?
(3) How to balance the discounts available for bulk purchase against the cost of storage?
(4) Should stores be kept in sub-depots or a central depot?
(5) Which items could be more economically delivered direct to site by suppliers?

Operational research has provided a methodology for tackling these questions which can lead to significant reductions in purchasing costs and store holdings.[1]

Where stores are held against issue it is important that satisfactory storage arrangements are provided to prevent the deterioration of stock, which could offset the savings accruing from bulk purchases. Ironmongery, mastics, emulsions, fire parts and many other materials and components deteriorate rapidly and may become unusable if stored in damp premises. Cement and plasters should be rotated so that old stock is used before new. It is also important to employ a competent storekeeper of high integrity, with a good knowledge of building materials, their worth and how they should be handled and stored, and the ability to keep accurate and neat records.

When operatives require materials from the stores, a stores requisition is normally completed listing the job reference. This requisition will desirably have the workman's signature for receiving the materials, a supervisor's signature authorising the use of the materials, and the storekeeper's signature covering their issue. These forms will pass to the treasurer's department or other appropriate section for costing together with completed time sheets. Where materials have been allocated to a job and not used they can be returned to stock and a credit note issued to ensure that stock records are correct.

It is advisable to keep a catalogue of the stores with each item allocated a code number. The catalogue should ideally be subdivided into categories of stores items for ease of reference and to assist in planning the stores layout. It is also advisable to stipulate the maximum and minimum stock level for each item, together with the current unit price. The computer can print out at prescribed intervals, a list of items which do not conform to the stipulated stock levels. Annually or at other suitable times a complete inventory of stores can be printed, giving the average unit price per item based on invoices of the preceding year, the quantity of each item in stock, the monetary value of stock per item, and the total monetary value of stock. As an additional aid to stock control, attention can be drawn to those items which have remained unchanged throughout the year. Milne[3] has emphasised the importance of computerised storekeeping in order to maintain an acceptable standard of financial control.

The purchasing of materials and components entails investigating and selecting sources of supply, obtaining quotations and placing orders, interviewing merchants and suppliers' representatives, keeping catalogues and other trade literature suitably indexed and being familiar with market conditions and trends. On delivery of materials, the delivery note should be checked against the original quotation for price and the quantity and quality of materials checked before the delivery note is signed. The materials should be recorded in a stock book or on a tally card and delivery notes cross referenced to the order number. Invoice particulars are often recorded in a purchases day book with the costs of materials entered on cost sheets, and the invoices filed for future reference.

Programming and Progressing Maintenance Work

To maintain regular progress of building maintenance work, a progress chart should be prepared before the work is commenced, usually in the form of a bar chart, with bar lines representing the time period allocated to each operation. The timescale, usually related to a calendar, is shown horizontally and the activities listed vertically.[12] The availability of men, materials and plant are the key factors in determining the time required to complete the work. The first step is to determine the various operations involved and which are to be performed by the contractor and which by sub-contractors. The estimated time required for each operation is then logged on the chart in the sequence in which it will be carried out on site, commencing with such items as excavating for and concreting foundations which must be completed before the walling can proceed. Storage capacity on the site may affect the speed at which the work can be carried out. The contractor must then consult with any sub-contractors to ensure that they can work to the programme. The progress chart is used throughout the contract to record the actual progress and any variations require discussion and adjustment of labour and materials to bring the work back on to target.

One commonly used method for planning and controlling building work is network analysis by which analysis of the operations is recorded in a diagrammatic form enabling each fundamental problem to be investigated separately. It employs a network to show the jobs (activities) to be done and their interrelationship. From the statement provided by this diagrammatic method and the estimates of times taken to do individual jobs, it calculates the duration of the project, the critical path (that sequence of jobs which defines project duration) and the float (or permissible delay) for non-critical jobs. A schedule or programme for the whole project is thus devised. Computer-aided network analysis offers the advantages of speed, accuracy and optimisation of resource utilisation.[12]

The prime advantages of network analysis are as follows.

(1) It separates planning the sequence of jobs from scheduling times for the jobs.

(2) It shows the interrelationship of jobs and enables people to see not only the overall plan but also the ways in which their own activities depend upon, or influence, those of others.

(3) By setting out the complete plan it is easier to assess its soundness and so prevent unrealistic or superficial planning.

(4) The effect on the project of alternative methods or individual job times can be examined at the outset.

(5) The total requirements of manpower and plant can be readily calculated; there is a vital need to ensure the use of balanced gangs and to reduce to a minimum the time when plant is standing idle.

(6) If the completion date has to be advanced, attention can be concentrated on speeding up the relatively few critical jobs and avoiding wasting money on accelerating non-critical jobs.

(7) Schedules may be based on consideration of costs so as to complete projects in a given time at minimum expense.

Other programming techniques include elemental trend analysis, precedence network diagrams and partly linked bar charts, but these are too sophisticated for most maintenance and repair projects.[13]

Maintenance Management of Condominiums

Many large condominium housing developments took place overseas in the 1970s and 1980s and in Singapore in particular. These consist mainly of large numbers of flats or apartments, constructed to a high standard of finish, and usually provided with extensive landscaped areas, car parking facilities and recreational features such as swimming pools.

Lim[14] has described how a condominium is a legal device by which a separate interest and an undivided interest in common in a multi-unit structure can be created. It has been aptly defined as a system of separate ownership of individual units in multi-unit projects. The owner is described as a subsidiary proprietor and his interest may be freehold or leasehold.

In Singapore, condominiums are managed by management corporations, whose main duties are to manage and maintain the common property and to keep it in good and serviceable repair; to keep the subdivided building insured; and to comply with all relevant notices and orders. The corporation can recover from any subsidiary proprietor any sum expended by the corporation in respect of the subsidiary proprietor's lot (apartment), and establish a management fund to cover administrative expenses.

The term *common property* includes the following:

(1) the main structure together with stairways, fire escapes, entrances and exits;

(2) car parks, parking areas, storage spaces, recreational and community facilities, and gardens;

(3) central and appurtenant installations for services such as power, lighting, gas, hot and cold water, heating, refrigeration, air conditioning and incinerators.

(4) escalators, lifts, tanks, pumps, motors, fans, compressors, ducts and all other apparatus and installations existing for common use.

The multifarious functions of a management corporation have been fully documented by Goh[15] and the main activities can be categorised as follows.

1. *General administration*
    (a) Overall management of the development, including arranging for fire, public liability and workmen's compensation insurances, and supervision of contract or directly employed staff.
    (b) Preparing annual budgets for the management and maintenance of the buildings, and collection of management fund contributions.
    (c) Payment of accounts for services rendered, keeping proper books of accounts, and preparing annual accounts for external audit.

2. *Security and carparking*
    (a) Ensuring the overall security of the development, including screening all visitors to the buildings.
    (b) Ensuring the observance of the by-laws of the corporation by residents, particularly with regard to recreational facilities.
    (c) Manning of the security counter and the CCTV screens, and operation of the car park.

3. *Cleaning*
    (a) Prepare specification for the regular cleaning of the common property.
    (b) Arrange for cleaning contract for performance of these duties or employ direct staff.
    (c) Supervision of cleaning.
    (d) Purchase of cleaning materials.

4. *Gardening and landscaping*
    (a) Prepare specification and work programme for gardeners.
    (b) General maintenance of gardens and landscaped areas.

5. *M & E Services*
    These will include the lifts, central air conditioning plant, solar hot water system, CCTV and video intercom system, anti-burglar alarm system, fire fighting installations, and electrical installations, including transformers and standby generators. The activities will include:

    (a) Arrangement for a regular maintenance contract with specialist contractors, especially for the lifts and central air conditioning plant.
    (b) Maintenance of service records to ensure that all plant is regularly serviced in accordance with the contract.
    (c) Study plant breakdowns and quotations for proposed repair work.
    (d) Arrange for the replacement of items of plant at the end of their useful lives.
    (e) Check light fittings in common lobbies, corridors, driveways, car park, etc. and replace blown tubes and bulbs.
    (f) Arrange for regular runs of the standby generator to ensure its operational efficiency.

6.   *Recreational facilities*
These facilities usually take the form of swimming pools, gymnasia, tennis and squash courts. The associated activities will be:

(a) Regulate the use of the facilities so that they are used in an orderly manner.
(b) Arrange regular testing of water samples from the swimming pool and maintenance of the filter pumps, to ensure the safety and well-being of users.
(c) Maintain the health fitness equipment in the gymnasium and ensure its proper usage to prevent accidents.

The management corporation can employ the requisite administrative staff direct to undertake the responsibilities that have been listed, but it is common practice to engage professional firms with wide experience to relieve the management corporation of the day to day management problems. The primary duty of managing agents is to provide a complete management service for the property on behalf of the management corporation.

Training for Maintenance

The knowledge and skills required for maintenance could form additional modules to the basic training already received or about to be received in new work. The nature and scope of this training will vary between the professions, management, supervisors, crafts and less skilled operatives.

The professions, managers and supervisors concerned with new construction need to understand the effect of their activities upon the life and use of the work they produce. Operatives should appreciate the effect of substandard work on maintenance costs, while those concerned wholly or mainly with maintenance work should have a detailed understanding of these aspects.

The Construction Industry Training Board

This Board was set up under the Industrial Training Act 1964 and is the Board primarily responsible for training in the building maintenance field. Industrial training boards have three main objectives

(1) to secure an adequate supply of properly trained persons at all levels;
(2) to secure an improvement in the quality and efficiency of industrial training;
.(3) to share the cost of training more evenly between firms.

A Board can provide its own training courses or arrange for other organisations to provide training. It imposes a periodic levy on employers within the scope of the Board to raise funds, from which it makes grants to those employers whose training programmes meet the standards set by the Board.

The Construction Industry Training Board (CITB) recognised the importance of effective training arrangements covering the work of building maintenance and can influence this through the grant system. More directly it provides courses for operatives in various trades related to maintenance and repair work, as well as courses for supervisors and training courses in general management. The trade courses cover craft skills in maintenance work, recording and measurement of work, awareness of the costs of minor works, setting out, recognition of faults and elements of public relations. Unfortunately the professions in the design team are outside the scope of CITB, although they include a large number of potential recruits for maintenance management courses, which are particularly aimed at eliminating some of the avoidable defects in the maintenance system.

Training of Managers

Most of the degree courses in building and related subjects in universities, polytechnics and other colleges devote less than 5 per cent of the timetable to building maintenance. There is however an increasing awareness of the importance of this subject among educational establishments and the number of research projects and post-graduate courses in this area is increasing. There is an identifiable need for an adequate study of management aspects including maintenance control, operational research, management techniques, computer studies, marketing of maintenance services, labour relations and control of human resources and maintenance standards.

The greater use of building surveyors would lead to more effective building maintenance, and this entails attracting well-trained persons of the same level of ability as in the other building industry professions, with identifiable promotion prospects based on qualifications, ability and experience. Apart from adequate salary, the building surveyor engaged on building maintenance should obtain substantial job satisfaction by feeling that he is part of a team of professionals undertaking a constructive and worth while assignment.

The Marks and Spencer organisation require all works team surveyors upon engagement to spend several weeks at a store to become familiar with the method of operation of the company, including its aims and trading principles. Subsequently, they are encouraged to discuss matters of a building nature with store management affording a useful exchange of views and improved understanding and integration.[16]

Student designers ideally should be made fully aware of the effect of design and material failures and of the practical considerations of their work. They should also become familiar with data recovery and analysis procedures and thus know where to obtain the appropriate information within a reasonable time scale.

Training of Supervisors

The generally accepted minimum qualifications for supervisors are BTEC Higher National Diplomas and Certificates in Building Studies. The part-time higher certificate course is more appropriate for experienced men in post, while the higher diploma course caters for the bright, younger man who wishes to advance to a position of responsibility. There will be some building maintenance supervisors of long experience who will be unable or unwilling to attend a higher certificate course and for these persons short residential courses might be more appropriate.

Training of Operatives

There is a shortage of genuine maintenance craft operatives. The absence of the stresses and incentives of new construction tends to attract the older or unskilled worker who is prepared to accept lower remuneration in return for a slower rate of work.

The aim of maintenance contractors and organisations should be to attract men who have a high standard of skill, even higher than that required for men on new work. Many of them have to be capable of covering both old and new work and thus need to be adaptable and discerning, to know what to do and when to seek advice or instructions. They must be convinced that their work is as essential as new construction, take pride in producing a high standard of work, master the use of new materials and techniques and be keen to use tools and plant where practicable.

A large proportion of maintenance work calls for operatives who are skilled in the traditional trades and this situation is bound to continue. This applies particularly to the larger commercial and industrial buildings which require a greater skill or higher quality of finish.[17] At the same time there are many operatives who are skilled in more than one trade and generally concentrate on one particular class of maintenance work. Maintenance costs can be reduced by increasing the number of maintenance operatives skilled in more than one craft such as bricklayer–mason–tiler, carpenter–joiner–glazier, painter–glazier–plasterer and other combinations. This type of operative is particularly valuable on the smaller maintenance jobs. No building owner wants a job held up periodically waiting for various craft operatives to arrive, nor do operatives themselves wish to see their work delayed for a task which they know they are competent to do themselves. Hence the maintenance sector should be viewed in a different way from new work when considering the labour aspect. Indeed it is virtually impossible to organise maintenance work effectively if hard and fast demarcation lines are applied.

Training arrangements for operatives normally follow one of three patterns.

(1) School leavers on government training schemes or pre-apprenticeship courses followed by apprenticeship to a craft or modifications of this arrangement.

(2) Training of young craft operatives to improve skills and prepare for responsibility.

(3) Training of older persons in specialist subjects and use of new materials and techniques.

With pre-apprenticeship schemes, maintenance contractors and organisations receive reports from technical colleges and thus have a good indication of the young person's potential. Apprentices should be placed in the care of good, well-experienced craft operatives, be permitted to attend day release courses and encouraged to sit their craft examinations. Block release and day release courses result in more persons being absent from work, but this is a problem which the contractor or maintenance manager must solve and it also requires acknowledgement by the building owner. Younger craft operatives engaged in maintenance work should also be encouraged to learn from the more experienced men.

Commercial and industrial organisations can also operate 'in house' training schemes, such as the group working practice (GWP) adopted by British Steel to develop the natural abilities of both skilled and non-skilled maintenance operatives by formal training, so that each operative acquires a working knowledge of each other's skills.[18]

Maintenance Incentive Schemes

Objectives, Scope and Requirements of Schemes

The aim of a local authority when introducing an incentive scheme is to obtain maximum value for money from its maintenance or, expressed in another way, to achieve maximum efficiency in maintenance. The contractor's aim is to make a profit while the building owner wants to keep cost down to a minimum, so there may be some conflict in their aims.

An incentive scheme may be defined as a means of relating a worker's remuneration to his performance; the employer will benefit from increased productivity and the employee will gain from higher wages provided that he is prepared to work consistently.

The following principles should be observed when formulating an incentive scheme.

(1) The scheme must be fair to all parties.

(2) The bonus earned should be closely related to the effort expended.

(3) The standard of performance required to secure bonus payment must be realistic and attainable by the average operative.

(4) Bonuses should not be affected by matters outside the operative's control.

(5) Bonus payments should be made as soon as possible after the work has been performed.

(6) The scheme must be relatively permanent—withdrawing a scheme because of temporary adverse economic conditions is damaging to industrial relations.

(7) Operatives should be able to calculate their own bonuses.

(8) The scheme must be approved by trade unions, operatives and supervisors.

(9) There must be a suitable grievance procedure.

A number of criticisms have been levelled at incentive schemes, which are now listed, but most of them can be avoided in practice with well-designed schemes.

(1) The employer is involved in additional work and costs in formulating, introducing and operating the scheme.

(2) Bad feeling may be generated in an indirect work force if the direct labour operatives are constantly receiving a high bonus.

(3) Unsatisfactory schemes cannot be lightly withdrawn because of the initial costs incurred and the bad feeling which would result among operatives.

(4) Disputes may arise which could lead to strikes.

(5) Some operatives may devote their efforts to exploiting loopholes in the scheme.

(6) Not all forms of maintenance work lend themselves to incentive schemes.

(7) The quality of work may suffer unless there is extensive supervision and/or inspection of the work.

Methods of Assessment of Bonus

A number of methods have been used.

(1) *Expenditure* or cost of work done, based on the cost of wages, materials and overheads as compared with estimates. This is not a satisfactory method because of the variations in approach but it is important as a measure of the success of schemes.

(2) *Materials used* as a measure of work done, but this also introduces a number of variables which make for difficulties in assessment.

(3) *Orders preformed*—the least reliable method because of variations in the scope and content of orders.

(4) *Standard times or ratings*—based on work study or records of previous jobs.

Work study provides the most common approach and consists of breaking down a job into elements and computing average times for each element or activity, performed in an efficient manner under proper supervision. A standard time for any job is obtained by totalling the times for the various elements. For example an item of replacing a cistern could be broken down into (a) inspect cistern, (b) disconnect and remove cistern and (c) install and connect new cistern. Allowance will also have to be made for travelling between jobs and for ordering and collecting materials, together with possibly a pre-investigation allowance.

The alternative historical method is simpler but involves some loss of accuracy depending on the care with which the time sheets have been compiled. The standard times approach is generally the most satisfactory method but the standard times must be very carefully computed; high values are to the employer's disadvantage and low values will bring complaints from the workforce. The system must be simple and readily understood, adopting targets that are small, easily identifiable by the operative and agreed before work is commenced. The additional sums payable to operatives are based on a percentage of the time savings against standard times, the percentage credited to the operative is often about 75 per cent with the remaining 25 per cent going to meet the cost of implementing the scheme. The number of operatives in a gang working for bonus should be kept as small as possible, and operatives should have the opportunity to increase wages by 25 to 35 per cent.

Price and Harris[19] have shown how portable personalised microcomputers can be used to collect data on site and to up-date existing output values immediately. It is necessary to use site efficiency factors which isolate the basic times for operations.

Incentive Schemes for Small Builders

It is difficult to recommend a single scheme for general use as schemes need to be tailored to suit the requirements of individual firms. Targeting based on estimator's prices, particularly where analytical estimating is practised, is recommended for all priced work, carefully checked and closely linked with a system of control of labour and profitability. Records should be retained to build up a library of standard data for future use of both estimating and targeting purposes.

A system of immediate profit sharing where profits are calculated and shares paid out on completion of each job is worthy of consideration by firms employing not more than about ten men. This approach is unlikely to be suitable for daywork.

A government report[20] showed that small firms could operate successful incentive schemes giving benefit to management sufficient to allow the whole of the savings in man-hours and profits from increased production to be passed on to operatives. The chief gains to management accrued from savings in oncosts; better control of labour, materials and transport; improved feedback of production information (leading to speedier and more accurate invoices to the client and better estimating for future jobs); and stability of labour through increased earnings without inflationary consequences. Other benefits to management consist of improved accounting procedures introduced in conjunction with the incentive scheme; closer control of cost and better information on cash flow and profitability.

A number of other advantages gained by contractors introducing incentive schemes for building maintenance work have also been identified. The labour force becomes much more efficient, thus reducing wasted time, utilising the full working hours, requesting instructions instead of waiting; planning and organising their own work and thinking about and improving their own methods. The management gains from reduced labour oncosts and site oncosts, increased ratio of materials to labour with increased profit, faster turnover of capital and general overheads spread over greater volume of work. In addition, management can see the productivity level of gangs or individuals and knows with greater certainty the time required for jobs and so is able to plan future jobs with greater confidence. Further useful information on incentive schemes is provided by Lee.[21]

Incentive Schemes in Operation

In one local authority incentive scheme for building maintenance work target times based upon work study were established for the range of work done, and bonus could be earned on all work for which target times had been set. Target times were expressed in standard hours and included allowances for rest and personal needs, minor interruptions and delays. Each operative completed a record sheet showing how each day was spent, and when checked and signed by the foreman the sheets were used to calculate bonus, which was paid one week in arrears. In order to safeguard the health of the men and to maintain the quality of work, a maximum bonus equivalent to a performance of $1.25 \times$ standard times was operated. All times that was spent on unmeasured work or lost was excluded from the bonus calculation and paid at the basic rate of pay.

The first responsibility for quality and safe methods of working lay with the man or team. Bonus was not paid, nor was the operation of a bonus scheme accepted as an excuse for substandard work or for unsafe working methods, for which checks were made.

After one year's operation, the scheme showed an increase in productivity of 86 per cent, increase in operative's earnings of 27 per cent and a reduction in unit labour cost of 37 per cent. Apart from these benefits the authority was able to recruit and retain a better standard of operative, secure more economic use of materials, obtain a better standard of service, and receive a weekly statement of performance presented against a prepared budget.

References

1 DOE. *Practice in Property Maintenance Management*. HMSO (1970)
2 B. A. Speight. Formulating maintenance policy. *Chartered Surveyor* (April 1970)
3 R. D. Milne. *Building Estate Maintenance*. Spon (1985)
4 S. Gillon, M. Dorfman and A. Moye. *The Local Government, Planning and Land Act 1980: A Layman's Guide*. Town and Country Planning Association (1981)
5 J. Edwards. Local authority maintenance operations under the direct labour organisation regulations. *Managing Building Maintenance*. CIOB (1985)
6 B. P. Holloway. Considering planned maintenance. *DOE Fourth National Building Maintenance Conference*. HMSO (1973)
7 B. A. Speight. Maintenance policy, programming and information feedback. *Building Maintenance and Preservation*. Butterworths (1980)

8   H. Graham. Financial control of building maintenance. *Managing Building Maintenance*. CIOB (1985)

9   R. F. Stevens. *BRE Current Paper 55/74: Maintenance Standards and Costs* (1974)

10 K. Thurley. Improving the organisation of maintenance in married quarters estates. *Development in Building Maintenance—1*. Applied Science Publishers (1979)

11 R. Pettitt. Computer aids to housing maintenance management. *Managing Building Maintenance*. CIOB (1985)

12 A. Kelly. *Maintenance Planning and Control*. Butterworths (1984)

13 Chartered Institute of Building. *Programmes in Construction* (1980)

14 S. B. H. Lim. The condominium and its management problems. *Maintenance Man*. Ngee Ann Polytechnic, Building Department, Singapore (1980)

15 T. L. Goh. Management and maintenance of condominium housing development. *UNIBEAM*. National University of Singapore, Building and Estate Management Society (1984)

16 D. J. Cripps. Building maintenance—a client's viewpoint. *Managing Building Maintenance*. CIOB (1985)

17 E. Bampton. Improving productivity in the execution of maintenance work. *Managing Building Maintenance*. CIOB (1985)

18 R. A. Macleod, D. A. Flegg and R. Prout. Minimising the cost of maintenance in a large integrated steelworks. *Terotechnica 2* (1981)

19 A. D. Price and F. C. Harris. *Methods of Measuring Production Times for Construction Work*. CIOB Technical Information Service No. 49 (1985)

20 DOE. *Incentive Schemes for Small Builders*. HMSO (1974)

21 R. Lee. *Building Maintenance Management*. Collins (1987)

14 SUPERVISION OF MAINTENANCE WORK

This final chapter is concerned with the supervision of maintenance work as executed, to ensure that it is of a satisfactory standard and in accordance with the drawings and specification. With the larger contracts it is customary to employ a clerk of works who is constantly in touch with the work, but with smaller schemes periodic supervision only can be obtained often through the medium of architects, surveyors, inspectors or other supervisory staff.

Clerk of Works

General Background

There are three main categories of clerks of works.

(1) Maintenance clerks of works who usually hold permanent posts and are concerned with cathedrals, hospitals, local authority and government buildings.

(2) Estate clerks of works who are responsible for the upkeep of buildings, fences, roads, water services, drainage and other related facilities on large country estates.

(3) Clerks of works who are primarily concerned with new building work and who are likely to move from one project to another as each is completed.

Most clerks of works are building craft operatives and preferably have served as trades and general foremen. They must have an extensive practical knowledge of building materials, principles of construction and the execution of techniques in all trades. They must be able to make basic calculations and take measurements, interpret drawings and other contract documents, write concise letters, prepare accurate and well-presented technical reports and enjoy satisfactory relationships with all the persons with whom they come in contact. In addition, clerks of works must be thoroughly familiar with the conditions of contract and statutory building controls. They need a suitable office from which to work. A most useful handbook for clerks of works was produced by the former Greater London Council.[1]

Duties

The primary duty of a clerk of works is to ensure that all the materials and workmanship are in accordance with the drawings, specification and any other relevant documents. He must refrain from making exorbitant demands or altering details or materials without the approval of the architect, surveyor or maintenance manager to whom he is responsible. The method of carrying out the work is the sole responsibility of the contractor, and the clerk of works should not under any circumstances direct the contractor as to the method to be employed. Where the clerk of works is dissatisfied with materials or workmanship, he should notify the foreman in charge as early as possible, and in the event of no changes being made, the architect or other responsible person should be notified, and he can issue written instructions to the contractor. Under the JCT Standard Form of Building Contract, the clerk of works is defined as "acting solely as an inspector on behalf of the employer under the direction of the architect."

The clerk of works can assist in the smooth operation of the contract by keeping a diary or record of work undertaken by the main contractor and sub-contractors, plant and principal materials delivered to the site, important activities, actions and discussions, weather conditions, any loss of production, issue of architect's instructions, site visitors and other relevant factors. He often takes measurements in conjunction with the contractor's representative, of work such as foundations and drains which will subsequently be covered up, and he may check timesheets and lists of materials used on daywork and possibly setting out. A comprehensive job diary can help the architect or surveyor in assessing any claims for loss or expense submitted by the contractor. The clerk of works should write up the diary as soon as possible after the events recorded and not later than the end of the day.[1]

A typical extract from the diary of a clerk of works follows.

27 June	Weather: mainly dry but two short showers
Workforce	15 bricklayers, 8 carpenters, 4 plumbers, 2 plasterers, 2 scaffolders, 22 labourers. Newtown Flooring Company (4)
Materials	4 t cement 8 m^3 sand 12 m^3 19 mm aggregate 8000 Himley mixed russet bricks 30 metal casements
Visitors	Mr Johnson (architect) Mr Palmer (Newtown Flooring Company)
Drawings received	CE/BH1 and CE/BH2 (boiler house details)
Architect's instructions	Foundations to Block D to be stepped. Eaves overhang on Block C to be increased from 250 to 300 mm

Reports   Weekly report 23 (21–27 June) given to Mr Johnson

Block A—PVC flooring laid in rooms 24, 28, 29 amd 30.
Setting walls in rooms 15, 16, 18, 19, 22 and 23.
Fixing of door linings—some inadequate fixings reported to contractor.
Block B—Roof construction in progress. Service pipe installation commenced in toilets
Block C—Second lift of brickwork to north and east walls. Brickwork up to dpc in annexe
Block D—Excavation for foundations commenced
Drainage Excavation for main drain discontinued owing to groundwater entering trench

Reports

The clerk of works will also prepare reports for consideration by the architect or other supervisory officer and these are of three main types—weekly, periodic and special.

Weekly reports keep the architect fully informed on progress and other relevant matters and may include such aspects as number of men in each trade employed on site, approximate value of work done during the week, weather conditions, amount of rejected material (if any), general remarks on progress and quality of work and whether work by sub-contractors in on target, deliveries of materials and whether these are of satisfactory quality and delivered on schedule, and drawings received and required. Progress of work may be indicated on duplicated plans, possibly by colour with appropriate dimensions added. A bar chart showing the programme and progress of the works is also very useful. On large jobs, photographs provide useful records of progress. The clerk of works should retain a copy of each report on his files.

Periodic reports may be prepared immediately prior to the time when certificates will be issued by the architect authorising interim payments to the contractor. The report contains details of the work carried out since the previous certificate was issued and also lists the materials on site. Where there is no quantity surveyor, the clerk of works may assess the approximate value of the work done and materials supplied, including additional work and daywork.

Special reports may be prepared from time to time to draw the architect's attention to matters requiring his decision. They could for instance cover the substitution of unobtainable materials or components, or deal with delays in the execution of critical items or the failure by the contractor to carry out architect's instructions.

Site Meetings

Site meetings will be held regularly on larger projects and are generally convened by the architect or other person responsible for the supervision of the work. The main objective is to ensure that satisfactory progress is maintained and to provide the opportunity for clearing outstanding points. It is important that all parties directly involved are represented including sub-contractors.

Setting Out

Under the normal terms of contract, the contractor is responsible for setting out the work and is liable for rectifying any mistakes. The clerk of works often checks the setting out although he cannot be held liable for verifying incorrect setting out. It is, for instance, vital that frontage lines shall be determined correctly. They may be prescribed in relation to adjacent or opposite buildings, the road kerb and on large and open sites, the centre of a road, fence or hedge. The frontage line should be adequately pegged and agreement obtained from all interested parties. Once the frontage line is fixed, the clerk of works or other supervisory person should satisfy himself that suitable and accurate methods are being employed to set out the remainder of the work.

The positions of walls are normally established with profiles where there is sufficient space, otherwise marks or incisions on abutting walls or other suitable method will be used. With sloping sites it is important that all measurements are taken horizontally, preferably with steel tapes. The setting out of a steel-framed building requires extreme accuracy as the stanchions and beams are cut to lengths at the fabricator's works, and any error in setting out can involve expensive alterations on site. It is usual to erect continuous profiles around the building and to set out the column spacings along them. Wires stretched between the profiles will give the centres of stanchions around which templets are formed.[2]

When laying drains, painted sight rails should preferably be fixed across the trench, usually at manholes or inspection chambers, at a height equal to the length of the boning rod above the invert level of the drain. A line sighted across the tops of the two adjacent sight rails will represent the gradient of the drain at a fixed height above invert level. At any one time there should desirably be at least three sight rails erected on the length of drain under construction.[2]

Wooden pegs or steel pins are driven into the trench bottom at intervals of at least 900 mm less than the length of the straight-edge in use. The use of a boning rod will enable each peg or pin to be driven until its head represents the pipe invert at that point. The underside of the straight-edge resting on the tops of the pegs or pins will give the levels and gradient of the pipe. The pegs or pins are withdrawn as the pipes are laid. To obtain a true line in a horizontal plane, a side line is strung tightly between steel pins at half pipe level, with the pipe sockets just free of the side line. Pins will normally be located at each manhole or inspection chamber, but intermediate pins will also be needed on very long lengths.[2]

The correct determination of levels for foundation concrete, floors and other features is as important as securing the correct lines and positions. A datum of some kind must be established from which all heights can be determined. It may be an arbitrary level such as a manhole cover or a specific, suitably marked, point on a kerb, wall or other feature. In the case of an open site, the datum point may consist of a wooden peg or steel pin set in concrete. Alternatively a datum of known level may be established related to the nearest Ordnance bench mark, when it is often referred to as a temporary bench mark. Levels are transferred from the datum point to the particular location on site by a dumpy, tilting or automatic level and levelling staff, or by the use of a straight-edge, spirit level and pegs.

Supervision of Building Work

Adequate supervision of new construction and of alteration and repair work is needed to ensure that the materials and workmanship comply with the contract particulars and relevant statutory requirements. In the absence of such supervision, inferior materials, poor workmanship and the omission of important details can occur resulting in subsequent trouble and expense to the building owner.[3] The remainder of this chapter is concerned with the supervision of building work and is subdivided into the appropriate works sections.

Demolition work

It is generally necessary to give the local authority notice of intention to demolish buildings and for the demolition work to be under the supervision of a competent person who is experienced in the type of demolition work which is to be undertaken. The local authority may require any of the following works to be carried out.

(1) Shore up adjacent buildings.

(2) Weatherproof any surfaces of an adjacent building which are exposed by the demolition.

(3) Remove material or rubbish resulting from the demolition and clearance of the site.

(4) Disconnect and seal at specified points any sewer, drain or water pipe in or under the building to be demolished.

(5) Remove any such sewer, drain or water pipe, and seal any sewer, drain or water pipe with which the sewer, drain or pipe to be removed is connected.

(6) Make good to the satisfaction of the local authority the surface of the ground disturbed by the preceding operations.

Safety aspects must be fully considered and adequate precautions taken. Before and during demolition work, all electrical services must be disconnected from the main supply, and only apparatus used for the demolition may be electrically charged. A thorough investigation should precede the demolition work to ensure that there can be no risk of fire or explosion through leakage or accumulation of gas or vapour, and no possibility of flooding from water mains, sewers or culverts.

If the building to be demolished adjoins the public highway, adequate close boarded *hoardings* to the approval of the local authority are required. Hoardings must be painted white for ease of observation and be free from sharp corners or projections which could be a nuisance to pedestrians. Each end of a hoarding should desirably be fixed at an angle from the wall to the front face of the projection. Adequate lighting should be provided during the hours of darkness at each end of the hoarding and along its length. Attention should also be given to the special needs of blind persons.

Where space is very restricted, it may be necessary to erect *gantries*, which are also subject to local authority control. A gantry is a temporary elevated working platform at about first floor level over a public footpath, on which materials may be received and from which rubble and surplus materials may be removed. They must be properly designed to support the loads which they will have to carry and to prevent danger or nuisance to the public from falling materials, dust or water. They can be constructed of timber, steel or a combination of them, with a minimum headroom over the footpath of 2.40 m and with the front of the gantry at least 450 mm back from the kerb face. Gantries require lighting at night and vertical members should be painted white for at least 1.50 m above ground.

When demolishing buildings no floor, roof or other part of the building shall be overloaded with debris or other material. Before and during demolition, the building to be demolished and adjoining buildings must be carefully observed and shoring of one or more of the forms described in chapter 2 should be erected at the first sign of weakness. The demolition work must be carried out methodically and special precautions are needed where the cutting of reinforced concrete, steelwork or ironwork forming part of the structure may suddenly twist, spring or collapse on completion of the cutting operation. Reinforced concrete buildings need special consideration with demolition commencing on the upper floors, thereby reducing the loads on columns as work progresses.

Demolition should be carried out so as to cause as little inconvenience as possible to adjoining owners, occupiers, the public or the employer, with all reasonable precautions taken to avoid unnecessary noise and vibration. The work should be carried out in accordance with BS 6187,[4] and the National Federation of Demolition Contractors Ltd have issued a specification for demolition and associated works[5] and an appropriate form of direct contract.[6]

Excavation

All topsoil must be removed from the area to be occupied by the building, and is normally stored ready for re-use. Trenches must not be excavated deeper than necessary to ensure firm trench bottoms for receipt of foundations and pipes, but always being on the lookout for soft spots. The widths of trenches need checking to ensure adequate spread of foundations, and check measurements taken to affirm that the walls will be correctly positioned centrally on the foundations. Trenches must be kept clear of water and backfilling carried out uniformly on both sides of walls in shallow layers, not exceeding 150 mm deep, and adequately consolidated.

The safety of excavations is very important. Where the sides of excavations require temporary support, to prevent the risk of earth or other material falling into the excavation with consequent danger to workmen and others on the site, adequate timbering or other support must be provided by suitably skilled persons. The supports should be inspected once every seven days by a competent person to ensure that they are in good condition and free from movement. These inspections should be duly recorded in a register kept in the site office.

The supporting material must be free of projecting nails which could be dangerous to persons on the site. Materials must not be deposited or stored near the edges of excavations where they could cause collapse of the side of the excavation with resultant dangers. Where the work may affect the stability of adjoining property, adequate precautionary measures must be taken. Any explosives used for excavation work must be under the control of a competent and experienced person. When a charge is fired, it is essential that no-one is exposed to risk of injury. The police must be informed that explosives are stored on the site and of the purpose for which they will be used.

Concrete Work

The *foundations* of a building are generally considered to be the most important part of the construction of a building and every possible care must be taken in the inspection of foundation work and the concrete used in foundations. The difficulty and expense in making good defects caused by settlement are adequate reasons for ensuring that the foundations are in all respects satisfactory. The effect of different ground conditions on foundation design is described in chapter 2 and particular attention must be paid to made-up ground, ground subject to mining subsidence, underlying caves, proximity of trees, sloping clay sites and unequal loading.

Where small quantities of concrete are *mixed by hand*, the cement must be evenly distributed throughout the aggregate. Pulling a long pronged rake through the heap before mixing helps considerably. The wet mixing must be carried out on a platform or other hard, clean surface. The amount of water added is influenced by the temperature and humidity of the atmosphere and the nature and absorptivity of the receiving point. A good consistency is obtained when a handful of concrete, pressed tight, sticks together and does not crumble or flow. Alternatively, the concrete may be machine mixed or ready mixed.

With *machine mixing*, the water content should be measured in the mixer tank, and the ingredients should be mixed to a uniform consistency. The normal mixing time is at least one minute after all the materials, including water, have been placed in the mixer. The mixer driver should monitor visibly the workability, homogeneity and cohesiveness of each mix, as well as the consistency of production.[7] When the drum is discharged, the first barrow load is coarser than the remainder with the heavier particles gravitating to the bottom. This is particularly relevant when the concrete is destined for comparatively thin slabs around steel reinforcement, and some hand mixing after depositing may be needed. After mixing, the concrete should be deposited without delay and not dropped uncontrolled through a height exceeding 1 m.

The cement and aggregates must be correctly *gauged* to give the specified mix. Gauge box sizes must be checked, and boxes kept clean and filled level when in use. The sizes of gauge boxes for aggregates are normally based on multiples which relate to a 50 kg bag of cement, but must not be so large and heavy that operatives avoid using them. Weigh batching gives much more accurate results, but it is necessary to check constantly that the gauging apparatus is being used correctly and that allowance is made for the moisture content of the sand.[2]

When the *laying of concrete* in foundations finishes part way along a trench at the end of a day, the end of the concrete should be left rough and inclined as a key for the next day's work. Before concrete is placed between *formwork*, the formwork should be checked as to dimensions, levels and strength. Formwork to beams should be so secured that it will not move under pressure from the concrete and a slight camber of about 1/360 span is customary. The side forms are usually removed after two or three days but a longer period is required for soffits of beams, depending upon such factors as the loading and the time of year. If a fair face is required to the concrete, the formwork must be treated so that the concrete comes away clean. All formwork must be clean before use.

Concrete in beams must be laid continuously between supports with any breaks over columns or walls. Where new work is joined to old the joint must be hacked, brushed clean, wetted and given a coat of grout. Concrete in walls must be spread in thin layers and be well tamped or vibrated. Concrete in casings to steelwork should be kept reasonably wet and be well vibrated. Expansion joints should be provided in large areas of concrete.

Bar sizes, shapes and positions will normally be shown on bar schedules and the bars in position need checking to ensure that they are in accordance with the schedule, that intersections are securely wired and that the distances of bars from edges of members comply with the drawings. After the bottom layer of concrete is laid and the bar reinforcement nearly covered, the bars should be lifted slightly with a hook and shaken to ensure that a full bed of concrete is obtained under the reinforcement. Temporary fixings keep the bars in the correct positions.

Freshly laid exposed concrete must be adequately *cured* by covering with bubble plastic sheets, quilts of plastic with fibres or other suitable material, to protect it from the sun and drying winds for at least seven days. Concrete should be at least 5°C when placed and should not fall below 2°C until it has hardened. In cold weather special precautions should be taken such as keeping aggregates and mixing plant under cover, covering exposed concrete surfaces with insulating material, using a richer mix of concrete and/or rapid hardening cement, heating water and aggregate, placing concrete quickly, and leaving the formwork in position for longer periods.

It is sometimes necessary to carry out *site tests* on materials to determine their suitability. The following tests relating to cement and sand serve to illustrate the approach.

Cement: (1) Examine to determine whether it is free from lumps and of a flour-like consistency (free from dampness and reasonably fresh).

(2) Place hand in cement and if of blood heat then it is in satisfactory condition.

(3) Settle with water as paste in a closed jar to see whether it will expand or contract.[2]

Sand: (1) Handle the sand; it should not stain hands excessively, ball readily or be deficient in coarse or fine particles.

(2) Use a standard sieve test—if more than 20 per cent is retained on a 1.25 mm sieve, it is unsuitable for use.

(3) Apply a silt or organic test—a jar half filled with sand and made up to the three-quarters mark with water; shake vigorously and leave for three hours; the amount of silt on top of the sand is then measured and this should not exceed six per cent.[2]

Coarse aggregates for concrete are normally required to comply with the grading requirements of BS 882[8] and these require checking to ensure a well-balanced mix. The maximum size of coarse aggregate is determined by the class of work; rarely exceeding 20 mm for reinforced concrete but increasing up to 40 mm for foundations and mass concrete work.

Concrete strengths are influenced by a number of factors

(1) proportion and type of cement;
(2) type, proportions, gradings and quality of aggregates;
(3) water content;
(4) method and adequacy of batching, mixing, transporting, placing, compacting and curing the concrete.[2]

Concrete mixes can be specified by the volume or weight of the constituent materials or by the minimum strength of the concrete; the latter approach being advocated in BS 8110.[9] The water/cement ratio is a most important factor in concrete quality. A common but rather imprecise test for measuring the workability of concrete is the slump test using a 300 mm high open-ended metal cone which is filled with four consolidated layers of concrete, the cone lifted and the slump or drop of the concrete measured (25 mm for vibrated mass concrete to 150 mm for heavily reinforced non-vibrated concrete). For greater accuracy the compacting factor test should be used.[2]

Brickwork

The *bricks* should conform to the sample deposited with and approved by the architect or other responsible person. Rough checks for suitability on the site include striking two bricks together and the resultant sound on impact should be hard and clear, and certainly not dull. Good bricks should withstand transport to the site without too many breakages. Arrises should be true and dimensions within the generally accepted tolerances, otherwise the contractor may claim an extra for sorting and gauging bricks. Lightness of colouring or a pink tinge indicates underburning and is a serious fault, as the discoloured bricks are likely to disintegrate fairly rapidly if used externally and subjected to severe weather conditions. Overburnt bricks may detract from the appearance of facework as they are likely to be misshapen.

In general the production and use of *mortar* on site is inadequately controlled. Using modern workability aids or masonry cements, very weak mortars can easily be produced and used. For example, some garden walls collapsed on one town development site resulting from a mortar gauging of 1:16. If sulphates are present in the bricks or soil they dissolve in water and attack Portland cement, forming calcium sulphoaluminate, resulting in expansion of mortar and eventual disintegration. The remedy in this situation is to use sulphate resisting cement or stronger mortar—1:3 or 1:4—and to avoid bricks with a high sulphate content. The composition of mortar should be carefully considered and should ideally have a density as close as possible to the density of the bricks. A good general-purpose mortar is cement:lime:sand (1:1:5–6), masonry-cement:sand ($1:4\frac{1}{2}$) or cement:sand with plasticiser (1:5–6). BRE Digest 160[10] provides useful guidelines for the selection of mortars for different situations and using alternative types of brick or block.

There are many matters to observe when supervising bricklaying. Brickwork must be laid to the specified bond and dimensions, and quoins, piers and reveals should be checked for plumb and brick courses laid to gauge. Rankin[11] has advocated guideline tolerances of 3 mm in reveal widths in the full height, 1 mm in 1 m of reveal soffit and 2 mm in 1 m in plumb of reveal. Work showing three times these tolerances would be unacceptable. It is unsatisfactory for brickwork to gain for one scaffold height and then lose to obtain the right level at an upper floor, producing uneven and untidy joints. Bricks with frogs should be laid frog up with good bed and cross joints well flushed up with mortar, and perpends kept perpendicular.

Brickwork to both external and internal walls should be carried up at approximately the same rate, leaving indents for half-brick walls and chases for block partitions. Quoins should be either racked back or part racked and toothed.

Bricks must be well wetted during hot weather and the top of newly constructed work suitably protected during frost or heavy rains. Defacement through scaffold splash must be avoided. Brickwork should be suspended during frosty weather, but where bricks and mortar are free from frost, work can commence at the beginning of the day if the temperature is not below 2°C. Special precautions may be taken to permit bricklaying to proceed during temporary frosts. An old bricklayer's saying is not without relevance—'when mortar hardens on the trowel, sets like beads on the line, and forms a crust on the mortar on the board, then it is time to pack up and go home'.

Cavity walls need particularly close supervision as there is a tendency for mortar droppings to bridge the cavity, lodge on wall ties and stand on dpc trays. Wall ties may slope inwards and be deficient in number or quality. Waterproofing arrangements around door and window openings need checking for soundness.

In the case of *facings* where a variation of colour is required it is usually advisable to use the bricks straight from the stack, provided this does not result in patches of one colour on laying. Where the facework is to be separately pointed, the joints should be raked out for a depth of at least 12 mm and care taken to maintain an even colour of mortar for this work, by ordering sufficient sand for this purpose. All perpends should be well filled and properly pointed with mortar of uniform colour and finish, and putlog holes solidly and neatly filled. Faced brickwork must be free from mortar stains.

For *underpinning*, bricks, must be hard and well soaked during hot weather, and laid with solidly filled joints. Toothings shall be well formed for linking with the adjoining sections of brickwork and must be kept clean. The pinning up process deserves special attention and is normally executed by ramming in a 25 mm layer of fairly dry cement mortar with a piece of board.

Damp-proof courses must be provided at the appropriate positions and be continuous. Damp courses in rolls should be adequately lapped at joints, be well bedded and kept back about 20 mm from the face of the wall to allow for pointing.

Most brickwork is laid from scaffolding which requires particular attention from the safety aspect. The old wooden poles and putlogs have been largely replaced by the more efficient steel or alloy tubes with patent scaffold fittings. They must be maintained in sound condition and be inspected regularly.

Masonry

All stone should be free from cracks, vents and discoloration. For instance some Portland stone shows a brownish tinge, limestones may contain dark spots and sandstones may exhibit laminations and discoloured streaks. Sound stone normally gives a clear ring when tapped with a hammer.

All dowels, cramps and plugs should closely fit the sinkings in the stones and where of iron should be galvanised or dipped in bitumen. Some stones, particularly sandstones, need protecting after laying from wet weather. Masonry needs checking to ensure that all stone members such as lintels have adequate bearings and that they are worked to the required details. As a general rule stone needs to be laid on its natural bed and be properly cleaned down on completion. Some stones need treatment on their back faces to prevent surface staining by cement. When using an unfamiliar stone, the supervisor is advised to visit existing buildings where the same type of stone has been used in similar conditions.

Roofing

Slates and *tiles* should be checked against approved samples to ensure that they are of the correct type and quality. Plain clay tiles should for instance be free from cracks and twists, be hard baked, have well-formed nibs and nail holes and have the appropriate cambers. For random slating the slates are sorted using the largest and thickest at the bottom. Slates and tiles to be nailed should be fixed with two nails to each slate or tile, and the nails should have a high resistance to deterioration such as copper or aluminium. Plain tiles are normally nailed at every fourth or fifth course and at the top course and at eaves. One major problem with extensions is the matching of materials. In the case of roofing tiles it may be advisable to remove old tiles from a rear roof to replace defective tiles on front elevations.

The battens must be checked to ensure that the correct gauge is being maintained with appropriate staggering of joints, and that the required constructional details are being obtained at eaves, ridges and verges, with particular attention being paid to watertightness. Roofing underfelt should be checked to ensure freedom from damage, adequate laps and overhang at eaves. A check should be made to ensure that soakers are fixed around chimney stacks and in other appropriate locations.

Asphalt to flat roofs should be laid in two thicknesses with 150 mm laps to a finished thickness of not less than 20 mm. Care must be taken to ensure that the asphalt is of the correct type and quality and is satisfactorily and evenly laid at the appropriate temperature to suitable falls. Special care is needed at the junctions of flat roofs with parapet walls to secure an effective and watertight joint.

When using *built-up bitumen* felt roofing, it is essential to use three layers bonded in hot bitumen for all except temporary buildings. Upstands and skirtings are best formed by turning up the second and top layers for a minimum height of 150 mm over an angle fillet and they should preferably be masked by a metal or semi-rigid asbestos/bitumen sheet (SRABS) flashing. A check should be made to ensure that each layer of felt complies with the specification, that each is smoothly and evenly laid to the required falls, and is free from cracks, holes or other defects.

With *copper, lead* and *zinc flat roofs* it is important to check that the correct quality and thickness of metal sheet is being used and that rolls and drips are properly formed with adequate laps. A check should be made to ensure that the sheeting is laid to even and adequate falls and that sheets are free to move on two edges.

Gutters and *downpipes* must be checked to ensure that they are of the correct sizes and properly jointed and fixed. It is not uncommon to see 2m lengths of gutter supported by one bracket per length instead of two. Cast iron downpipes need fixing sufficiently clear of wall faces to permit painting of the backs of pipes. Balloon gratings to gutter outlets are sometimes omitted despite being clearly specified.

Thermal insulation to roofs needs checking to ensure that it is of the correct thickness and properly laid to eliminate gaps and particularly to seal off vulnerable eaves. Any cold water apparatus above the insulating layer must be adequately protected.

Carpentry

Carpentry timber should be carefully examined to ensure that it is of satisfactory quality, conforms to the specification and is of the required dimensions. Structural timber should be examined for its general character —straightness of grain, size and type of knots, existence of waney edges, any discoloration, shakes and other defects. The timber should be adequately seasoned with a moisture content roughly equivalent to the humidity of the atmosphere in which it will be placed. Timber which shows signs of decay or contains considerable sap, bad knots or shakes should be rejected. Guide notes issued by the Building Research Establishment (Princes Risborough Laboratory) will assist in the identification of different species of timber. Most carpentry timbers are supplied to nominal sizes and suitable allowance needs to be made for any planed faces when checking dimensions. All timber on the site should be adequately protected from the weather.

Wood is readily attacked by *fungi* which flourish where timbers become wet and where there is no ventilation. These conditions can for instance arise in exterior timber framed panels, where small amounts of water can enter without showing on the interior surface but resulting in rotting at the bottom where water tends to collect. To overcome this problem panels should be double sealed with well-designed joints and finish to openings. In the more vulnerable locations described in chapter 4 timber should be suitably treated with preservative.

Floor joists must be laid truly level and be adequately fixed to wall plates where provided, with adequate bearing. Any joists that have a curve in the direction of their length should be laid with the convex edge uppermost to counteract the tendency to sag. Joists must be checked for size and spacing. Trimmers are normally 25 mm thicker than other joists and the jointing of trimming members should be carefully checked. Any herringbone strutting must be carefully formed and be taken across the floor from wall to wall. Checks must be made to ensure that timbers in position are not weakened excessively by plumbers and electricians cutting large notches for pipes and cables.

Wall plates must be well bedded in an appropriate mortar with the ends half lapped and nailed. *Rafters* must be properly birdsmouthed over wall plates with the depth of the birdsmouth not exceeding one-third of the depth of the rafter. Rafters need checking to ensure that they are laid in a true plane to the appropriate pitch and spacing and are of the specified sizes. All other roof members and joints need to be checked together with any trimming around openings, with rafters properly scribed against ridge boards and hip and valley rafters.

The more common weaknesses which occur with *trussed rafters* include the use of unsuitable (ungraded) timber, careless placing of fasteners, use of faulty fasteners, excessively tight designs resulting in considerable deflection and lack of bracing. The latter two defects can cause opening up of joints to roof tiles and entry of water.

Joinery

All *joinery timbers* shall be checked for quality and the moisture content should not exceed the prescribed limits to restrict subsequent shrinkage.

Grounds for skirtings, architraves and like features must be securely fixed without excessive packing. Where fixing bricks are used they should be inspected to ensure that they will take nails. Floorboards should not exceed the specified width and should be well cramped up and adequately nailed in the prescribed manner. Joints to skirtings must be well formed.

Doors should be fitted to give 3 mm edge joints for painted work and 1.5 mm for polished work. Check to ensure that they are hung to swing in the correct direction and that all hinges bear equally, have the correct length of screw with heads countersunk flush. Adequate allowance must be made for floor finishes. Keyholes must be in true alignment with locks for ease of insertion and withdrawal of keys. Doors need checking to ensure that they are of the required type, construction and dimensions. Hardwood jambs should ideally be screwed and pelleted. The deviation from squareness of door frames should not exceed 1 mm from a 500 mm square edge, and the twist or bow in a door not exceed 5 mm.[11]

Casements must be out of winding and must not stick. Any sash bars must be straight and in alignment, while sliding sashes must have suitable weights and correct lengths of cord or have the prescribed spring devices. Fasteners to sliding sashes should be sunk flush, sashes must slide freely and the bottom sash must fit closely to the sill. Deviation from level of window sills should not exceed 1 mm in 1 m, and plumb of window frame 2 mm in 1 m.[11]

All reasonable steps should be taken to prevent shrinkage of finished work, with joiner's work ideally framed together about two months before fixing. The building should be dried out as quickly as possible by using the central heating system or fires and opening windows on drying days.

Particular attention should be paid to throatings, grooves and similar labours to ensure that they conform to the prescribed details and are properly set. Ironmongery must conform to the specification or to submitted and approved samples.

Plasterwork

Plasterboard ceilings must be securely fixed with suitable galvanised nails to give a true plane surface. Each board or lath should be nailed with not less than four nails to each support equally spaced across the width and driven no closer than 13 mm from its edges. End joints should be staggered in alternate courses with cut ends located over supports. Plasterboard ceilings are normally finished with one or two coats of plaster and angles at junctions of wall and ceiling need checking to ensure that they are reinforced with a strip of jute scrim.

Most *internal wall surfaces* are finished with two coats of plaster and the first coat should be ruled to an even surface and lightly scratched to form a key for the finishing coat. The thickness of all coats must be watched. Straight edges and accurate screeds and grounds are needed to produce a good finishing or floating coat. The finishing coat must be applied with an even amount of material and pressure. Irregular or wavy patches should be replastered. Defective work should be identified at an early stage; irregularities can be felt even if not clearly visible. Good plasterwork requires skilled craft operatives, good materials and adequate time to do the work satisfactorily. There must be a good key on all surfaces to be plastered and there is no substitute for this. Galvanised metal angle beads should ideally be used at all plastered external angles. Rankin[11] advocated that the level of a ceiling should not exceed 10 mm from a 3 m straight edge and no deficiencies in levels should be visible when viewed in normal daylight from a distance of 1 to 1.5 m.

Plaster should be stored in a dry place and be separated from concrete floors by wood battens. The finished plasterwork should be truly vertical, free from cracks, blisters and other imperfections. Different plasters must not be mixed under any circumstances and the manufacturer's instructions should be closely followed. Plaster on the site should be checked to ensure that it is of the type specified. The supervisor must anticipate problems if the plasterer is using dirty water, dirty tools and an already opened sack of gypsum from the previous job.

Paintwork

All *steel* and *ironwork* should be cleared of mill scale, oil, grease, dirt and most rust before painting. The preparation of these surfaces is probably more important than the type of paint to be applied. To remove mill scale and rust, methods such as wire brushing and chipping, hand or mechanical acid pickling, blast cleaning or flame cleaning may be used. When repainting previously painted metal surfaces, care should be taken to remove most traces of rust, following which rust-inhibiting priming paint should be applied to the cleaned surface taking care to cover all parts. The backs of all metal gutters should be suitably metal primed before the fixing and the inside surfaces after fixing.

All *knots* in timber must be sealed with shellac knotting before the timber is primed. Priming paint should not be applied too thickly otherwise it will not soak into the wood so readily. All cracks and holes must be suitably stopped prior to the application of paint. These preliminary operations can be omitted when repainting previously painted woodwork provided the original paint film is sound.

All *coats of paint* must be properly applied in accordance with the specification and the manufacturer's instructions. All surfaces must be clean, dry and free from dust or grease before paint is applied. Check to ensure that the correct number of coats of paint are applied in the prescribed sequence. Brush marks applied parallel with the grain improve the appearance of the work. Surfaces should be rubbed down with fine glass paper before each succeeding coat is applied. Check to ensure that all joinery is properly primed, where required, all knots sealed and nails or screws driven below the surface. Special attention should be given to the priming of the backs of frames, linings, skirtings, fitments and the like which will come into contract with the structure.

All edges of doors, except bottoms, should be painted and where doors are painted different colours each side, check that the striking edge is the same colour as the inside face of the door and the hanging edge as the outside.

The commonest cause of *blistering* of paint is resin exuding from the wood, although painting on a frosty or damp surface can have a similar effect. An excess of oil in the paint may also cause blisters while too little oil may turn the paint into powder on drying. Each coat should be allowed to dry thoroughly before the following coat is applied. External surfaces are preferably painted during a dry period when at least two dry days have preceded the day when painting is undertaken.

New plaster surfaces may require drying out for several months before they are ready to take oil paints. Occupants are not generally prepared to wait this long and there are two alternative approaches—to apply an alkali-resisting primer followed by two coats of oil or paint or to use two or three coats of emulsion paint.

Paperhanging needs checking to ensure that the paper is well secured but not pasted to architraves, skirtings and the like. Check that the pattern is matched and the paper hung plumb.

Glazing

Check to ensure that the glass used is of the correct type and thickness and that the putty is of suitable quality. Ideally the back putty to glass panes should not be cut out until a week after front puttying. Ensure that there is clearance between the edges of the glass and the enclosing wood or metal casement. Large panes must be well sprigged before front puttying. Front putties should be neatly and evenly formed to the appropriate lines. The edges of plate glass in shop fronts and showcases should have their edges blackened.

Plumbing

All materials and components need checking to ensure that they comply with the specification. Sanitary appliances deserve close examination to ensure that they are not mis-shapen, cracked, crazed or pitted and that fittings such as water waste preventers are free from mechanical defects.

A constant check should be maintained while plumbing work is in progress to ensure that pipes are of the correct dimensions and laid in the correct positions with the prescribed provision of valves. All pipes should be properly jointed and graded where appropriate. Hot water pipes must be adequately fixed but at the same time permit movement for expansion and contraction. Care must be taken to ensure that floor joists are not weakened excessively through notches being cut to receive pipes. Pipes should be located so as to be as inconspicuous as possible but nevertheless be readily accessible for repairs, and be positioned on internal walls where practicable.

Special precautions need to be taken with a *single stack plumbing* system to ensure satisfactory operation. For example, the stack must be of adequate size with airtight joints, and branch pipes must fall gradually and continuously in the direction of flow and be free from abrupt changes of direction. The vertical distance between the lowest branch connection and the invert of the drain should not be less than 750 mm or 450 mm for three-storey houses with a 100 mm stack and two-storey houses with a 75 mm stack. Waste pipes from wash basins, baths and sinks should be laid to the minimum prescribed slope, have a maximum length of 1.70 m and be provided with suitable traps. WCs should be sited as near to the stack as possible and the WC branch should have a generous sweep and enter the stack an an angle of 95° from the vertical. Bath and wash basin connections opposite a WC connection should be located at least 200 mm above the WC connection.[12]

Great care is needed in fitting *plastic wastes* to sanitary fitments. Most plastic connections to stack pipes have a rubber bush which on tightening has a tendency to twist on the inside of the stack. This results from the plumber's failure to make a proper connection but is not always seen and the first indication often is a blockage in the system.

Drainage

Pipes, junctions, bends and other fittings should be checked for soundness and British Standard kite marks. Gullies, inspection chamber covers and other special fittings should also be checked for compliance with the specification.

The drainage work needs to be checked against the *drainage plan* which will show the pipe runs and sizes, and the location and possibly also the invert levels of inspection chambers, manholes and other access points. The pipes must be laid in straight lines between inspection chambers or manholes and to even and self-cleansing gradients. Drains which are very shallow, excessively deep or close to building foundations will need surrounding with concrete. The alignment of drains can be checked with a line and gradients with a straight-edge and spirit level, or alternatively a mirror may be placed at one end of the drain and a lamp at the other.

Drain pipes should be kept clean and must be properly jointed to provide watertight joints. Many flexible jointed pipes are now used but where cement and sand joints are used with standard clay pipes, the mix should not be richer than 1:2 to avoid excessive shrinkage on drying. Drain pipes must always be laid on a firm bed to prevent or substantially reduce subsequent settlement with consequent strain on joints. Flexible pipes should generally be laid on a granular bed not less than 100 mm thick, with the granular material extending upwards to the top surface of the pipe.[2]

The Building Regulations[13] require that all drains after laying, and backfilling of trenches, shall withstand a suitable test for watertightness. The most effective test is the *water test* in which a suitable plug is inserted in the lower end of the length of drain which is then filled with water. For house drains, a knuckle bend and length of vertical pipe may be jointed temporarily at the upper end to provide the requisite test head. A drop in the level of water in the vertical pipe may be due to one or more of the following causes

(1) absorption by pipes or joints;
(2) sweating of pipes or joints;
(3) leakage from defective pipes, joints or plugs;
(4) trapped air.

Hence it is advisable to fill the pipes with water for two hours before testing, top them up and then to measure the loss of water over a 30 minute period, normally applying a test pressure of 1.5 m head of water at the upper end and not more than 4 m at the lower end. It may thus be necessary to test steeply graded drains in stages to avoid exceeding the maximum head. The Building Regulations[13] recommend that the leakage of water over 30 minutes should not exceed 0.05 litres for each metre of drain for a 100 mm drain and 0.08 litres for a 150 mm drain. Where there is a trap at the upper end of a branch drain, a rubber or plastics tube should be inserted through the trap seal to draw off the confined air as the pipes are filled with water. Alternative tests are the air test and the smoke test which are both detailed in *Building Technology*.[2]

Inspection chambers and *manholes* (designed to permit the entry of a man) provide access to drains for maintenance. Their distance apart on straight lengths of drain should not exceed 45 m for inspection chambers and 90 m for manholes,[13] and they should also be provided at changes of direction and gradient, at drain junctions where cleaning is not otherwise possible and heads of drains. They must be of adequate size to permit ready access for inspection, cleansing and rodding; have a removable, durable, non-ventilating cover; have step irons or a ladder to provide access where the depth requires this; and have suitable smooth impervious benching when there are open channels.

The dimensions of inspection chambers and manholes will be largely determined by the size and angle of the main drain, the position and number of branch drains and the depth to invert. Brick manholes should always be built in unrendered engineering brickwork finished fair on the inside face, to withstand the humid, corrosive conditions and to prevent cracked rendering falling into the drains. The bricks should be laid in cement mortar (1:3). Half-brick walls are permissible for depths not exceeding 900 mm in granular soils above the water table.[14]

The main channel should be formed of half-round channel pipes and branches are best in the form of three-quarter section standard branch bends discharging in the direction of flow in the main channel. Check the benching to ensure that it is smooth and hard and sloped at about 1 in 12 where men may be required to stand on it. Frames and covers must be well bedded at a suitable level.

Records

A person supervising building work should keep records of site visits, noting dates, weather conditions, labour force, materials delivered to site, work in progress and any other important aspects. He will also record his own observations on the quality of the work and any action that he has taken. He will record details of any old drains, service pipes or other feature opened up during the execution of the works. It is good practice to amend the drawings and specification in red to incorporate any changes that may have been made during the course of the works.

References

1 Greater London Council. *Handbook for Clerks of Works*. Architectural Press (1983)
2 I. H. Seeley. *Building Technology*. Macmillan (1986)
3 *BRE Digest 176*. Failure patterns and implications (1975)
4 British Standards Institution. *BS 6187: 1982 Code of practice for demolition*
5 National Federation of Demolition Contractors Ltd. *Specification for Demolition and Associated Works in the Clearance of Existing Buildings and Structures* (1984)
6 National Federation of Demolition Contractors Ltd. *Form of Direct Contract* (1982)
7 G. Taylor. *Concrete Site work*. Telford (1984)
8 British Standards Institution. *BS 882: 1983 Specification for aggregates from natural sources for concrete*
9 British Standards Institution. *BS 8110: Structural use of concrete. Part I: 1985 Code of practice for design and construction*
10 *BRE Digest 160*. Mortars for bricklaying (1973)
11 I. Rankin. *Quality Control and Tolerances for Internal Finishes in Building*. CIOB Technical Information Service No. 2 (1982)
12 *BRE Digest 249*. Sanitary pipework: Part 2: Design of pipework (1981)
13 *The Building Regulations 1985: Approved Document H1*. HMSO (1985)
14 British Standards Institution. *BS 8301: 1985 Code of practice for building drainage*

APPENDIX 1: METRIC CONVERSION TABLE

Length   1 inch = 25.44 mm [approximately 25 mm], then (mm/
100) × 4 = inch
1 ft = 304.8 mm (approximately 300 mm)
1 yd = 0.914 m (approximately 910 mm)
1 mile = 1.609 km (approximately $1\frac{3}{5}$ km)
1 m = 3.281 ft = 1.094 yd (approximately 1.1 yd)
(10 m = 11 yd approximately)
1 km = 0.621 mile ($\frac{5}{8}$ mile approximately)

Area $1 \text{ ft}^2 = 0.093 \text{ m}^2$
$1 \text{ yd}^2 = 0.836 \text{ m}^2$
1 acre = 0.405 ha [1 ha (hectare) = 10 000 m^2]
$1 \text{ mile}^2 = 2.590 \text{ km}^2$
$1 \text{ m}^2 = 10.764 \text{ ft}^2 = 1.196 \text{ yd}^2$ (approximately 1.2 yd^2)
1 ha = 2.471 acres (approximately $2\frac{1}{2}$ acres)
$1 \text{ km}^2 = 0.386 \text{ mile}^2$

Volume $1 \text{ ft}^3 = 0.028 \text{ m}^3$
$1 \text{ yd}^3 = 0.765 \text{ m}^3$
$1 \text{ m}^3 = 35.315 \text{ ft}^3 = 1.308 \text{ yd}^3$ (approximately 1.3 yd^3)
$1 \text{ ft}^3 = 28.32$ litres (1000 litres = 1 m^3)
1 gal = 4.546 litres
1 litre = 0.220 gal (approximately $4\frac{1}{2}$ litres to the gallon)

Mass   1 lb = 0.454 kg (kilogram)
1 cwt = 50.80 kg (approximately 50 kg)
1 ton = 1.016 t [1 tonne = 1000 kg = 0.984 ton]
1 kg = 2.205 lb (approximately $2\frac{1}{5}$ lb)

Density $1 \text{ lb/ft}^3 = 16.019 \text{ kg/m}^3$
$1 \text{ kg/m}^3 = 0.062 \text{ lb/ft}^3$

Velocity
$$1 \text{ ft/s} = 0.305 \text{ m/s}$$
$$1 \text{ mile/h} = 1.609 \text{ km/h}$$

Energy
$$1 \text{ therm} = 105.506 \text{ MJ (megajoules)}$$
$$1 \text{ Btu} = 1.055 \text{ kJ (kilojoules)}$$

Thermal conductivity $1\text{Btu/ft}^2 \text{ h}°\text{F} = 5.678 \text{ W/m}^2°\text{C}$ (where W = watt)

Temperature
$$x°\text{F} = \tfrac{5}{9}(x - 32)°\text{C}$$
$$x°\text{ C} = \tfrac{9}{5}x + 32°\text{F}$$
$$0°\text{C} = 32°\text{F (freezing)}$$
$$5°\text{C} = 41°\text{F}$$
$$10°\text{C} = 50°\text{F (rather cold)}$$
$$15°\text{C} = 59°\text{F}$$
$$20°\text{C} = 68°\text{F (quite warm)}$$
$$25°\text{C} = 77°\text{F}$$
$$30°\text{C} = 86°\text{F (very hot)}$$

Pressure
$$1 \text{ lbf/inch}^2 = 0.007 \text{ N/mm}^2 = 6894.8 \text{ N/m}^2$$
$$(1 \text{ MN/m}^2 = 1 \text{ N/mm}^2)$$
$$1 \text{ lbf/ft}^2 = 47.88 \text{ N/m}^2 \text{ (newtons/square metre)}$$
$$1 \text{ tonf/inch}^2 = 15.44 \text{ MN/m}^2 \text{ (meganewtons/square metre)}$$
$$1 \text{ tonf/ft}^2 = 107.3 \text{ kN/m}^2 \text{ (kilonewtons/square metre)}$$

For speedy but approximate conversions:
$$1 \text{ lbf/ft}^2 = \frac{\text{kN/m}^2}{20} \text{ hence } 40 \text{ lbf/ft}^2 = 2 \text{ kN/m}^2$$

and $\text{tonf/ft}^2 = \text{kN/m}^2 \times 10$, hence $2 \text{ tonf/ft}^2 = 20 \text{ kN/m}^2$

Floor loadings   office floors — general usage: $50 \text{ lbf/ft}^2 = 2.50 \text{ kN/m}^2$
office floors — data-processing equipment: $70 \text{ lbf/ft}^2 = 3.50 \text{ kN/m}^2$
factory floors: $100 \text{ lbf/ft}^2 = 5.00 \text{ kN/m}^2$

Safe bearing capacity of soil
$$1 \text{ tonf/ft}^2 = 107.25 \text{ kN/m}^2$$
$$2 \text{ tonf/ft}^2 = 214.50 \text{ kN/m}^2$$
$$4 \text{ tonf/ft}^2 = 429.00 \text{ kN/m}^2$$

Stresses in concrete
$$100 \text{ lbf/in.}^2 = 0.70 \text{ MN/m}^2$$
$$1000 \text{ lbf/in.}^2 = 7.00 \text{ MN/m}^2$$
$$3000 \text{ lbf/in.}^2 = 21.00 \text{ MN/m}^2$$
$$6000 \text{ lbf/in.}^2 = 41.00 \text{ MN/m}^2$$

Costs
$$£1/\text{m}^2 = £0.092/\text{ft}^2$$
$$1 \text{ shilling (5p)/ft}^2 = £0.538/\text{m}^2$$
$$£1/\text{ft}^2 = £10.764/\text{m}^2 \text{ (approximately } (£11/\text{m}^2))$$
$$£5/\text{ft}^2 = £54 \text{ m}^2$$

$$£10/ft^2 = £108/m^2$$
$$£15/ft^2 = £161/m^2$$
$$£20/ft^2 = £215/m^2$$
$$£25/ft^2 = £269/m^2$$
$$£30/ft^2 = £333/m^2$$
$$£40/ft^2 = £430/m^2$$

APPENDIX 2: OCCUPANCY COST ANALYSIS—LABORATORY

LABORATORY

Building function: Medical research laboratories	Owner/Occupier: Charity organisation
Location: Central London	Date of erection: 1961—extended 1972

UPPER MANAGEMENT CRITERIA AND BUDGET PROCEDURE

Maintain correct room temperature and very clean working conditions; ensure continuity of all services and facilities to prevent any interruption of laboratory activities. Preserve a capital asset; maintain the building in as good a condition as possible.

An analysis is made of the budget during the previous and current budget years to check on accuracy and to examine expenditure trends. Annual maintenance budget is included with other departmental estimates and submitted for approval. Expenditure reviewed monthly—checks for likely over-expenditure.

MAINTENANCE MANAGEMENT AND OPERATION

Superintendent engineer, deputy engineer, records clerk and 2 secretaries responsible for maintenance. Annual inspections by superintendent engineer. Painting frequency: normally every 5 years—more frequently if required. Costs recorded under 30 budget heads—individual maintenance job cards not costed.

Work done by DEL 50% and contracted out 50%. Directly-employed labour establishment is 5 supervisors, 18 craft operatives and 14 labourers.

Selected contractors for minor works use the organisation's own contract conditions. Competitive tenders for large works employ JCT or IMechE contracts.

Contract supervision: daily inspection on minor works; progress reports and interim payments on large works.

BUILDING FUNCTION AND PARAMETERS

Medical research laboratory—laboratories 36%, circulation areas 29%, plant rooms 16%, staff facilities 14%, offices 5%. 400 occupants (average).

Design criterion: sound construction for minimum maintenance to avoid interruption of research work.

Gross floor area: 14 831 m^2
Area of pitched roofs (on plan):—
Area of flat roofs (on plan): 988^2
Area of external glazing: 1583 m^2

Storeys above (and including) ground floor: 9 No.
Floors below ground floor: 3 No.
Floor to ceiling height: 2.70 m generally to underside of suspended ceiling.

FORM OF CONSTRUCTION

Structure: Reinforced concrete frame. Cavity brick external walls. Timber double-glazed windows. Reinforced concrete roof with screed and 2-coat asphalt covering. Plastered brick internal partitions on ground floor, elsewhere generally special demountable timber partitions. Solid concrete floors.

Finishings and fittings: Granite flooring in entrance hall, all rooms and corridors linoleum covered, tiling in plant room and cloakrooms. Teak laboratory bench tops, cupboards and shelving.

Decoration: Plastered surfaces emulsion painted; teak oil on windows and frames externally.

Services: LPHW radiators for corridors and staircases; high velocity dual duct system with full air conditioning for all rooms except plant rooms and cloakrooms. 3 No. 10-person/1500 lb passenger lifts 30 ft/min serving 11 floors; 3 No. 3000 lb and 2 No. 1500 lb goods lifts 150 ft/min serving 11 floors. Emergency lighting, cold rooms, liquid nitrogen plant, compressed air, CO_2 gas, etc.

FINANCIAL STATEMENT: COST PER 100 m² FLOOR AREA

Gross floor area: 14 831 m²

Element	1976/77	1977/78	1978/79	1979/80	1980/81
0. Improvements & adaptations	£ 294.62	£ 369.07	£ 642.39	£1498.74	£2424.54
1. Decoration					
1.1 External decoration	16.81	18.99	19.55	24.79	31.41
1.2 Internal decoration	97.57	85.65	103.53	133.43	234.39
Sub-total	£ 114.38	£ 104.64	£ 123.08	£ 158.22	£ 265.80
2. Fabric					
2.1 External walls	0.67	7.52	17.90	—	31.33
2.2 Roofs	1.35	6.50	2.02	3.37	—
2.3 Other structural items	45.12	45.05	32.06	39.00	42.94
2.4 Fittings & fixtures	136.18	179.69	116.84	134.10	237.94
2.5 Internal finishes	27.41	38.40	36.53	32.84	59.18
Sub-total	£ 210.73	£ 277.16	£ 205.35	£ 209.31	£ 371.39
3. Services					
3.1 Plumbing & internal drainage	93.72	83.07	114.95	151.68	232.41
3.2 Heating & ventilating	410.42	455.74	500.67	838.16	806.13
3.3 Lifts & escalators	45.08	57.63	53.15	56.16	78.32
3.4 Electric power & lighting	261.99	315.73	305.26	400.52	395.75
3.5 Other M & E services	258.88	279.89	370.90	472.33	462.21
Sub-total	£1070.09	£1192.06	£1344.93	£1918.85	£1974.82
4. Cleaning					
4.1 Windows	12.31	12.58	12.40	18.13	18.29
4.2 External surfaces	—	—	—	—	—
4.3 Internal	285.03	284.30	273.89	361.67	480.65
Sub-total	£ 297.34	£ 296.88	£ 286.29	£ 379.80	£ 498.94

431

Appendix 2 cont'd

5. Utilities

5.1 Gas	11.99	16.46	12.16	18.57	239.24
5.2 Electricity	674.84	764.13	684.28	1195.71	1331.91
5.3 Fuel Oil	501.83	464.73	512.62	929.57	913.57
5.4 Solid fuel	—	—	—	—	—
5.5 Water rates	78.13	198.66	93.90	133.17	137.35
5.6 Effluents & drainage charges	7.02	34.05	27.90	35.34	42.07
Sub-total	£1273.81	£1478.03	£1330.86	£2312.36	£2664.14

6. Administrative costs

6.1 Services attendants	—	—	—	—	—
6.2 Laundry	46.65	56.45	58.30	85.52	105.30
6.3 Porterage	173.92	182.70	208.59	233.40	284.74
6.4 Security	200.54	211.50	246.09	340.56	396.30
6.5 Rubbish disposal	2.95	3.43	7.47	99.14	123.30
6.6 Property management	557.59	636.20	706.44	1060.83	1254.30
Sub-total	£ 981.65	£1090.28	£1226.89	£1819.45	£2163.94

7. Overheads

7.1 Property insurance	87.65	85.63	56.64	56.64	80.51
7.2 Rates*	271.65	456.89	380.72	476.30	573.11
Sub-total	£ 359.30	£ 542.52	£ 437.36	£ 532.94	£ 653.62
TOTAL	£4307.30	£4981.57	£4954.76	£7330.93	£8592.65

*50% rates reduction as a research institution.
Source: Building Maintenance Information Ltd.

APPENDIX 3: OCCUPANCY COST ANALYSIS—HALLS OF RESIDENCE

HALLS OF RESIDENCE

Building function: Student residenial accommodation
Location: North-east England urban area

Owner/Occupier: University
Date of erection: 1974

UPPER MANAGEMENT CRITERIA AND BUDGET PROCEDURE

Maintain building in its state within the limits of the budget allocated by the Finance Committee.

Overall annual budget estimate is prepared for all university buildings and grounds, split into 1—elemental heads, 2a—wages and salaries, 2b—materials and contracting services. The maintenance estimate is considered along with other departmental recurrent estimates and adjusted according to allocations. Budget control is the responsibility of the maintenance officer who reviews expenditure monthly

MAINTENANCE MANAGEMENT AND OPERATION

The maintenance officer is responsible for the maintenance of all buildings and grounds assisted by the supervisory staff (electrical, mechanical, buildings and grounds), office manager, secretary and three clerks.

Total estate comprises 90.2 hectares including several small sites away from the main campus; 134 319 m² floor area teaching and residential accommodation. Routine inspections: regular visits made by supervisors. Maintenance implemented by PPM process and requisitions raised by heads of departments and others. Painting frequencies: 5 year cycle externally; 2 and 4 years internally depending on designated use.

Cost records and feedback: individual jobs are cost coded according to 1—building, 2—element subdivided between a—DEL, b—contract, c—PPM.

Work done by DEL 55% and contracted out 45%. Directly-employed labour establishment is 58 in total including 6 chargehands.

Contracted-out work is on a daywork basis, larger contracts use university form or JCT contract.

433

BUILDING FUNCTION AND PARAMETERS

Residential student accommodation in 8 blocks with study bedrooms and shared conditions and sanitary accommodation. Communal reading room. 95% student accommodation, 5% ancillary buildings. 260 occupants.

Gross floor area: 4036 m²

Area of pitched roofs (on plan): 1714 m²

Area of flat roofs (on plan): 23 m²

Area of external glazing: 468 m²

Storeys above (and including) ground floor: 3 No.

Floors below ground floor: —

Floor to ceiling height: 2.30 m

Height to ridge: 10.00 m

FORM OF CONSTRUCTION

Structure: Traditional construction. Loadbearing brick/block cavity external walls, plastered internally. Aluminium horizontal sliding windows; Velux rooflights. Pitched roof with clay pantiles on battens and felt; flat roof with 3-layer felt on chipboard and timber joists. Small area of plain clay tile to addition. Blockwork internal partitions. Solid *in situ* concrete floor slab to sections and suspended timber joists with tongued and grooved floor boarding to remainder.

Finishings and fittings: Carpet, quarry tiles, non-slip tiles, thermoplastic tiles. Kitchen fittings, shelves.

Decoration: Emulsion paint to walls; stain and varnish to some woodwork, gloss to some, including externally.

Services: Copper water services, PVC wastes, saltglazed drains. Gas-fired boilers, one per house; 1 radiator per room 18–21°C. Kitchen equipment. Laundry.

FINANCIAL STATEMENT: COST PER 100 m² FLOOR AREA

Gross floor area: 4036 m²

Element	1978/79	1979/80	1980/81	1981/82	1982/83
0. Improvements & adaptations	£ 92.81	£ —	£ 14.99	£ —	£ —
1. Decoration					
1.1 External decoration	—	—	1.41	8.28	0.37
1.2 Internal decoration	103.64	17.64	157.56	91.87	136.92
Sub-total	£ 103.64	£ 17.64	£ 158.97	£ 100.15	£ 137.29
2. Fabric					
2.1 External walls	2.38	3.77	3.37	6.37	4.16
2.2 Roofs	9.54	3.44	9.09	9.12	11.65
2.3 Other structural items	22.64	3.69	15.12	22.18	66.40
2.4 Fittings & fixtures	13.95	15.81	12.44	15.54	45.81
2.5 Internal finishes	7.88	4.46	8.15	6.24	29.69
Sub-total	£ 56.39	£ 31.17	£ 48.17	£ 59.45	£ 157.71
3. Services					
3.1 Plumbing & internal drainage	9.42	24.40	35.06	24.38	43.09
3.2 Heating & ventilating	9.61	18.19	31.71	39.56	38.60
3.3 Lifts & escalators	—	—	—	—	—
3.4 Electric power & lighting	9.96	25.05	60.46	65.34	59.94
3.5 Other M & E services	10.21	12.44	26.81	28.77	31.74
Sub-total	£ 39.20	£ 80.08	£ 154.04	£ 158.05	£ 173.37
4. Cleaning					
4.1 Windows	5.03	5.28	8.23	7.51	9.46
4.2 External surfaces	—	—	—	—	—
4.3 Internal	290.34	423.44	422.52	520.46	516.55
Sub-total	£ 295.37	£ 428.72	£ 430.75	£ 527.97	£ 526.01

435

Appendix 3 cont'd

5. Utilities

5.1 Gas	189.12	322.70	400.02	411.17	446.11
5.2 Electricity	59.27	97.40	127.94	143.36	162.36
5.3 Fuel Oil	—	—	—	—	—
5.4 Solid fuel	—	—	—	—	—
5.5 Water rates	25.79	27.92	27.65	27.75	25.05
5.6 Effluents & drainage charges	—	—	—	—	—
Sub-total	£ 274.18	£ 448.02	£ 555.61	£ 582.28	£ 633.52

6. Administrative costs

6.1 Services attendants	—	—	—	—	—
6.2 Laundry	26.63	64.72	82.71	104.86	58.67
6.3 Porterage					
6.4 Security	96.41	122.02	138.53	146.13	174.06
6.5 Rubbish disposal } 6.6 Property management	85.18	93.93	117.14	120.66	130.20
Sub-total	£ 208.22	£ 280.67	£ 338.38	£ 371.65	£ 362.93

7. Overheads

7.1 Property insurance	39.52	44.65	52.50	60.93	58.82
7.2 Rates	256.32	289.39	348.81	369.80	406.42
Sub-total	£ 295.84	£ 334.04	£ 401.31	£ 430.73	£ 465.24
TOTAL	£1272.84	£1620.34	£2087.23	£2230.28	£2456.07

Source: Building Maintenance Information Ltd.

436

APPENDIX 4: ENERGY COST ANALYSIS OF HOSPITAL

CI/SfB 412

ACUTE HOSPITAL

Building function: 521 bed, mainly acute hospital.

Location: Edmonton, London N18.

Description of site: Sheltered location.

Gross floor area: 68 900 m^2

Gross internal cube: 231 900 m^3.

Date of erection: Older buildings originally erected as a workhouse dated 1840–1985 with additions dated 1899 as infirm wards. Some newer buildings 1960, 1972, 1981.

Form of construction: Older prewar construction is brick built with pitched slate roofs; postwar buildings concrete framed.

Mechanical and electrical services: 5 No. Lancashire boilers fitted with economisers and all rated at 10 000 lb/hour. Average electrical maximum demand is 820 kVA per month. Standby generator maximum rating is 1200 kW.

Hours of occupation: Generally wards and residential blocks 24 hours per day, 7 days; remainder 8 hours per day, 5 days per week.

Hours of heating: In accordance with occupancy.

Hours of lighting: In accordance with occupancy and natural light.

Measures undertaken to reduce energy consumption: Installation of automatic doors; wall and roof insulation; strict upkeep of pipe and ventilation system lagging and installation of various heat recovery systems. Imminent installation of a building automation system.

Appendix 4 cont'd

ENERGY USAGE

Source	1979/80		1980/81		1981/82	
	Quantity	Cost	Quantity	Cost	Quantity	Cost
Gas	12 593 GJ	£ 26 032	13 256 GJ	£ 35 601	12 936 GJ	£ 32 348
Electricity	18 288 GJ	£120 194	18 240 GJ	£135 118	17 986 GJ	£188 267
Fuel oil (3500 sec.)	183 006 GJ	£296 776	178 788 GJ	£412 288	180 390 GJ	£400 676
Total therms equivalent	2 027 221 therms		1 993 072 therms		2 002 814 therms	
ENERGY CONSUMPTION Therms per 100 m² per annum	2942.27		2892.70		2906.84	

ENERGY USAGE

Source	1982/83		1983/84		1984/85	
	Quantity	Cost	Quantity	Cost	Quantity	Cost
Gas	12 268 GJ	£ 37 491	5 692 GJ	£ 37 751	8 412 GJ	£ 37 515
Electricity	22 760 GJ	£225 280	24 345 GJ	£229 872	22 845 GJ	£231 814
Fuel oil (3500 sec.)	159 149 GJ	£436 465	157 602 GJ	£525 319	132 409 GJ	£517 750
Total therms equivalent	1 840 409 therms		1 778 443 therms		1 551 226 therms	
ENERGY CONSUMPTION Therms per 100 m² per annum	2671.13		2581.19		2251.42	

Source: BMI Ltd, *Study of Energy in Buildings* (1986).

APPENDIX 5: CLASSIFICATION OF MAINTENANCE OPERATIONS AND REPAIRS

Primary code	Secondary code
1. *External decoration*	
2. *Internal decoration*	
3. *Main structure*	31. Foundations and basements
	32. Frame
	33. External walls, chimneys and flues (excluding boiler flues)
	34. External windows and doors including glazing
	35. Roof structure
	36. Roof coverings
	37. Roof lights and glazing
	38. Gutters and rainwater pipes
4. *Internal construction*	41. Ground floors (where not part of foundation slab)
	42. Upper floors
	43. Staircases and steps
	44. Internal walls and partitions
	45. Doors and screens including glazing
5. *Finishes and fittings*	51. Ceiling finishes
	52. Wall finishes
	53. Floor finishes
	54. Shelves, built-in furniture and miscellaneous joinery
	55. Ironmongery
	56. Cloakroom and similar miscellaneous fittings
6. *Plumbing and sanitary services*	61. Cold water service pipes, storage tanks, cisterns and valves
	62. Hot water service pipes, storage tanks and cylinders, domestic boilers, valves and insulation
	63. Sanitary fittings including taps and traps
	64. Waste, soil and vent pipes
7. *Mechanical services including heating and ventilation and gas installations*	71. Boilers, firing, instrumentation and automatic controls, flues
	72. Steam and hot water distribution including heat exchangers and heating appliances

Primary code	*Secondary code*
	73. Workshop equipment, lifting appliances and special industrial equipment
	74. External water supply, treatment and storage plant
	75. Lifts
	76. Air conditioning, ventilation and refrigeration
	77. General utilities
	78. Gas installations and equipment (except kitchen equipment)
8. *Electrical services and kitchen equipment (all types)*	81. Electrical generation and prime movers
	82. Electrical transmission and distribution
	83. Electrical installations (wiring switch and control gear)
	84. Electrical appliances and fittings (except kitchen equipment)
	85. External lighting and airfield lighting
	86. Lighting protection, ELV systems and equipment
	87. Kitchen equipment
9. *External and civil engineering works*	91. Roads, car parks, parade grounds, hardstandings and runways
	92. Paths, playgrounds and general paved areas
	93. Fences, gates and boundary walls
	94. Drains and ditches
	95. Sewage disposal
	96. Water storage tanks and reservoirs
	97. Railway tracks, platform docks, wharves and jetties, sea defence walls
	98. Horticultural and arboricultural works
	99. Miscellaneous external works
0. *Miscellaneous and ancillary works*	01. Adaptations and minor new works
	02. Replacement of fire damage
	03. Routine cleaning

Source: DOE. *Maintenance Manuals for Buildings*. HMSO (1970)

INDEX